International Conference on Climate Nexus Perspectives: Water, Food and Biodiversity (I2CNP 2020)

E3S Web of Conferences Volume 183 (2020)

Khenifra, Morocco
4 June 2020

Editors:

Lahcen El Youssfi
Sidi Imad Cherkaoui
Adil Aghzar
Mohamed Farissi
Alex Godoy Faundez
Diego Rivera Salazar
Ousmane Seidou
Temitope Olawunmi Sogbanmu
Wesley Doorsamy
Yassine Ouaggajou

ISBN: 978-1-7138-1413-9

Printed from e-media with permission by:

Curran Associates, Inc.
57 Morehouse Lane
Red Hook, NY 12571

Some format issues inherent in the e-media version may also appear in this print version.

For additional information, please contact EDP Sciences – Web of Conferences at the address below.

EDP Sciences – Web of Conferences
17, Avenue du Hoggar
Parc d'Activité de Courtabœuf
BP 112
F-91944 Les Ulis Cedex A
France

Phone: +33 (0) 1 69 18 75 75
Fax: +33 (0) 1 69 28 84 91

contact-edps@webofconferences.org

Additional copies of this publication are available from:

Curran Associates, Inc.
57 Morehouse Lane
Red Hook, NY 12571 USA
Phone: 845-758-0400
Fax: 845-758-2633
Email: curran@proceedings.com
Web: www.proceedings.com

TABLE OF CONTENTS

Conference Title: International Conference on Climate Nexus Perspectives: Water, Food and Biodiversity » I2CNP 2020

Preface

It is our pleasure to present and share the Proceedings of the Conference titled: International Conference on Climate Nexus Perspectives: Water, Food and Biodiversity» I2CNP 2020, organized by the research team of environment and natural resources management at the Higher School of Technology of Khenifra, Sultan Moulay Slimane University in Morocco and its national and international partners.

The event attracted an enormous interest from participants which included doctoral students, researchers, professors, practitioners, and professionals translating into a considerable number of participations despite the restrictive conditions of confinement and lockdown. International researchers that took part in the conference were from diverse countries such as: Nigeria, South Africa, Cameroon, Benin, Egypt, Algeria, Oman, Iran, India, Canada, Chile, Romania, Croatia, Spain, Belgium, France and Morocco.

The focus of the conference was on the prevalence of threats to the climate on the one hand and the crises related to biodiversity, water and food security on the other. Not only was the emphasis made on these key factors, but it was also made on the rising global environmental issues/consequences of the degradation of ecosystems. The latter requires a response integrating all the interactions of the nexus. Therefore, the organization of this conference aimed to address intricate questions linked to integrated natural resource management, proposing sustainable solutions that allow better adaptation and resilience to climate change.

We would like to address our sincere thanks to the members of the organizing and scientific committee as well as all participants.

Dr. Lahcen El Youssfi
On behalf of the Editorial Team
Khenifra, 4th June 2020.

Statement of Peer review

In submitting conference proceedings to *Web of Conferences*, the editors of the proceedings certify to the Publisher that

1. They adhere to its **Policy on Publishing Integrity** in order to safeguard good scientific practice in publishing.
2. All articles have been subjected to peer review administered by the proceedings editors.
3. Reviews have been conducted by expert referees, who have been requested to provide unbiased and constructive comments aimed, whenever possible, at improving the work.
4. Proceedings editors have taken all reasonable steps to ensure the quality of the materials they publish and their decision to accept or reject a paper for publication has been based only on the merits of the work and the relevance to the journal.

Title, date and place of the conference

International Conference Climate Nexus Perspectives :Water, Food and Biodiversity » I2CNP 2020

4 June 2020 (Online)

Proceedings editor(s):

Lahcen El Youssfi
Sidi Imad Cherkaoui
Adil Aghzar
Mohamed Farissi
Alex Godoy Faundez
Diego Rivera Salazar
Ousmane Seidou
Temitope Olawunmi Sogbanmu
Wesley Doorsamy
Yassine Ouaggajou

Date and editor's signature

14 July 2020

New data on the ecological distribution of *Euphorbia resinifera* O.Berg in the Beni Mellal-Khenifra region

Abderrahim Ettaqy[1,2], *Abdelhakim* Taha[3], *Aziz* ElGhiouane[3], *Anas* ElKhou[3], *Abdelaali* Boulli[2], *and Younes* Abbas[3*]

[1]Regional Direction of water and forest, Forest and Water Department, Beni Mellal, Morocco
[2]Ecology and sustainable development team, Faculty of Sciences and Technics, SMSU Beni Mellal, Morocco
[3]Polyvalent Laboratory R&D, Polydisciplinary Faculty, SMSU, Beni Mellal Morocco

Abstract. *Euphorbia resinifera* O. Berg, also called Euphorbusresin, is a species belonging to the Euphorbiaceae family. It is endemic to the Beni Mellal Khenifra region and extends discontinuously between El Ksiba and Demnate. Currently, this plant is known as a meliferous species and its product is considered as a terroir product with a Protected Geographical Indication (PGI). However, few studies have conducted on its ecology and particularly its mapping in relation to the environmental conditionswhere it thrives. That's why in this study, we tried to establish, for the first time in the region of Beni Mellal, a distribution mapping of E. *resinifera* according to the ecological needs of the plant, the exploration of high-resolution satellite imagery and on the field by observing a representative sample of the study area. Results showed that it spread from Demnate to Elksiba in the Atlas piedmont area but appears as scattered individuals from Demnate to AitOurir due to a complexity of factors mainly related to the soil characteristics, paleogeography and the climate exacerbated by the negative results of anthropogenic impacts. It extends also into the central high Atlas valley as small isolated units which reach 1900 m of altitude. This could explain why other ecological requirements should be considered in terms of the distribution of this endemic species in the context of climate change.

Introduction

Climate change is becoming the main driver of environmental change in the Mediterranean region [1]. In this region, multiple paleo-geographic events and contrasting climatic cycles have led to the emergence of an unusually high biodiversity [2, 3]. The Mediterranean climate of Morocco has become a hotbed of climatic hazards which can lead to successive periods of drought and the aridity of a large part of its territory, which can create favorable conditions for the development and domination of different plant species and climax. Among these species, *Euphorbia resinifera*,O. Berg, also called *Tithymalus resiniferus* (O. Berg H. Karst) or Euphorbus resin which belongs to the Euphorbiaceae family, is an endemic species of Morocco and considered as a plant well adapted to drought. In the Atlas region of Beni Mellal, it extends from Elksiba to Demnate. The oldest illustration of a succulent Euphorbia dates back almost 1500 years, it was identified and proposed by Lawant and Winthagen [4] as *Euphorbia resinifera* Berg from north-west Morocco. Currently, this plant is best known for its use in the melliferous field, since the euphorbic honey derived from this plant is considered a local product with Protected

* Corresponding author: a.younes@usms.ma

Geographical Indication (PGI) in the Tadla-Azilal region[5]. Despite the botanical and morphological differences, this species has long been confused with *E. Officinarum* L. ssp*echinus* present in the sub-region south of Morocco. Recently, molecular tools have revealed that they are two different species of euphorbia [6].

In Morocco, *E. resinifera* colonize all the sunny slopes in areas where it is encountered [7]. It grows mainly in the rocky and arid rocks of the low limestone mountains between 600 to 1 500 m [8]. In 1937[9], Gattefossé and Vindt have estimated its area to about 130 000 hectares. But, in 1966 the Moroccan Ministry of Agriculture and Agricultural Reform limited the geographical area of this plant to the following zones: Haouz, Middle Atlas, Umm-er-Rbia, Central High Atlas (M'Goun) and JbelLkest. The area in the province of Beni-Mellal is estimated to 8 215 ha [5]. Until now, the figures put forwards about *E. resinifera*'s area question us about the actual extent of the species in its geographical range.

From a phytocoenological point of view, *E. Resinifera* is considered as a thermo-mediterranean species, included in the *Acacia gummifera* series and the *E. resinifera* sub-series corresponding to *Rubio longifoliae-Euphorbictumresiniferae* [10]. But when it is considered as a secondary essence, it is found in the *Tetraclino-Euphorbietumresiniferae* association characterized by a temperate semi-arid climate growing on the dolomites and limestone where it finds on this substrate the solid support, the heat and the dryness which it needs [11]. It is located in the regions of Bezou, Beni Mellal and Demnate between 900 and 1 100 m [10]. According to Benabid[12], the species is part of the Macaronesian floristic procession that characterizes the southwestern coast of Morocco and the Canary Islands. The species grows in the marginal northeast area of these Macaronesian plantdistributions on the rocky foothills of the Atlas of Beni Mellal and Demnat. Bioclimatic atmospheres can be semi-arid and temperate sub-humid, in the sub-Mediterranean and thermo-Mediterranean vegetation, between sea level and 800 m. The association of plants described by Fennane [11] is part of the vegetation subset of *E. resinifera* which is included in the *Acacia gummifera* vegetation series.

Currently, field observations show that *E. resinifera* is mainly observed in the Beni Mellal-Khenifra region and there is a lack of data related to its redistribution, particularly its spatiotemporal evolution. It is in this perspective that this work was approached and focused on its distribution area with the aim to:

- Determine the range of *E. resinifera* in its native region in the Atlas of Beni Mellal and the Central High Atlas in Morocco through the use of remote sensing and fieldobservation;
- Identify some ecological traits and characteristics that drivethe distribution of this species;
- Define a reference state GIS based approach for cartography and spatiotemporal analysis in the scope of *E. resinifera*stands preservation actions.

Material and methods

To address the objectives of this study, the designed methodology is divided into three major phases:

1. Delimitation of the potential area of *E. resinifera*

For this purpose, the visual interpretation of satellite imageries with very high spatial resolution was used. These are single-spectral satellite scene covering the whole study area. All the images were acquired from the free Google Earth platform and are displayed in real colors and visually interpreted in a GISenvironment.Several satellite images were used, with suitable dates and geographical areas, to cover the entire study area and benefit from maximum precision.

Indeed, the digitalization of the *E. resinifera* range was based on the shape, texture and structure of the euphorbia patches which are, in most cases, clearly visible to the interpreter's eye.

2. Validation by field observations

After finalizing the extent of *E. resinifera* distribution, it would have been appropriate to validate this work by careful field verification. As a result, several control points were randomly selected to verify both the presence of the species and to inquire about any relevant data regarding the descriptors of the euphorbia developmental environment. The table 1 and figure 1 below summarize this phase:

Table 1: Coordinates and ecological characteristics of the studied sites

Check point n°	Locality	Latitude(dd)*	Longitude(dd)	Altitude(m)	Density of *Euphorbus* resin
1	Cemetery of Beni Mellal	32,3460211	-6,31589088	604	Sparce density
2	-	32,5341135	-6,14929764	613	Clear density
3	-	32,5435261	-6,13884432	660	Dense
4	Cemetery of IghremN'Laalam	32,564955	-6,129555	627	Absent
5	-	32,5860822	-6,032355	873	Sparce density
6	Tighboula	32,5860822	-6,032355	785	Absent
7	-	32,5827137	-5,99817105	1068	Absent
8	El Ksiba	32,5718496	-6,03810582	1041	Sparce density
9	El Ksiba	32,5660948	-6,066856	1041	Sparce density
10	-	32,5511	-6,0856	944	Sparce density
11	Ikhourba	32,50054	-6,153433	662	Dense
12	Tagzirat	32,4195655	-6,2203465	690	Clear density
13	Foumelanser	32,36939	-6,25465792	746	Medium density
14	Sidi brahim	31,6442	-7,4599	700	Absent
15	Tazart	31,64961	-7,44381	732	Absent
16	Aitkaaboch	31,726	-7,0541	1042	Sparce density
17	5 KmSouth of Demnate	31,75535	-6,99303	1067	Clear density
18	Demnate	31,75962	-7,0067	937	Dense
19	Marrakech-Azilal Road	31,8819	-6,90124	900	Clear density
20	Ouzoud	31,9449	-6,7641	1041	Sparce density
21	KsarManar	32,021319	-6,470691	1450	Medium density
22	Oued Ahanssal	32,021793	-6,456787	1430	Dense
23	AitMhamed	31,87962	-6,456148	1700	Absent
24	AitMazigh	32,082366	-6,335851	1260	Absent
25	-	32,090846	-6,304826	1600	Clear density
26	-	32,06846	-6,266642	1700	Absent
27	Tabarouchte	32,12474	-6,197198	1360	Clear density
28	Tillouguitecentre	32,033247	-6,205586	1256	Medium density
29	TillouguiteZt-Ahansal road	31,979669	-6,09110244	1850	Sparce density
30	TillouguiteZtAhansal road	31,979471	-6,08375702	1920	Sparce density

*Geographical coordinates expressed inWGS84 system.

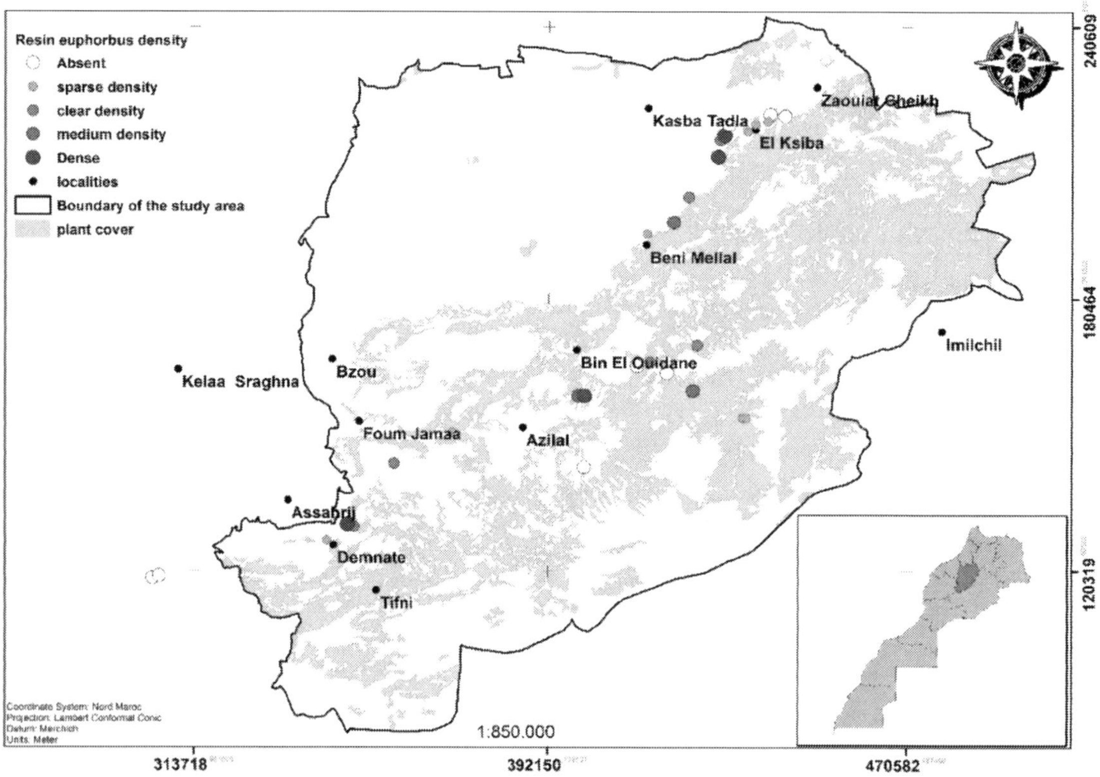

Fig. 1. Field positions and status of sampled control points of *E. resinifera*.

3. Spatial analysis by superposition

Overlapping this limit to the range of edaphic and morpho-climatic factors involved in the distribution area of the studied species. In this case the spatial analysis was based on the bioclimatic range (regarding Emberger's map for Morocco in 1955) [13], elevation,slope exposures and substratum (geology).

Results

1. Geographic range

The results of the image-interpretation validated with an accuracy of 73.33% and readjusted to refine the limit of the geographical area of the Euphorbus resinare presented in the following distribution map (Figure 2):

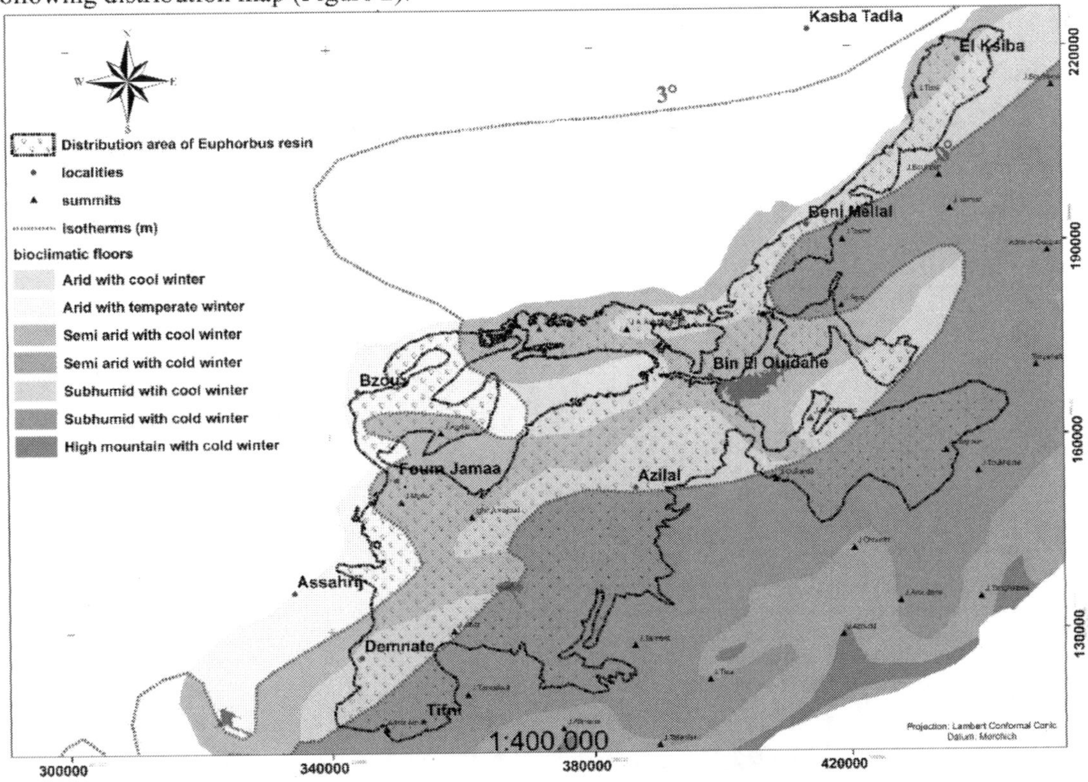

Fig. 2. Distribution of Euphorbus resin inthe Beni Mellal-Khenifra region.

According to the map, the geographic area of the euphorbia extends on an area of 316 500 ha, from Demnate (central high Atlas) in the southwest to the Elksiba zone (middle Atlas) in the north-east, with infiltrations in the center of the high Atlas following the continental gradient where the euphorbia grows under specific microclimatic conditions: rocky soils, hot exposures and a clear to patchy density of the tree cover. Most of this area is located at the level of the north Atlassian piedmont. From a bioclimatic point of view, the area of euphorbia is found in arid to sub-humid bioclimates but with temperate, cool and exceptionally cold variants.

2. Hypsometryrange

If several research studies have shown that Euphorbus resin plant associations elevation spread from 680m to 1 100m.It's interesting to note, from this study, that the upper part of its area reaches 1 200 m ranging thus between 600-1 200 (figure 3). Also, this study newly reports the presence of this species was observed at altitudes above 1 900m. This area extension indicates a possible migration to high altitudes as a result of climate change leading to an increase in temperatures (shiftingof the 0°C isotherm to the east).

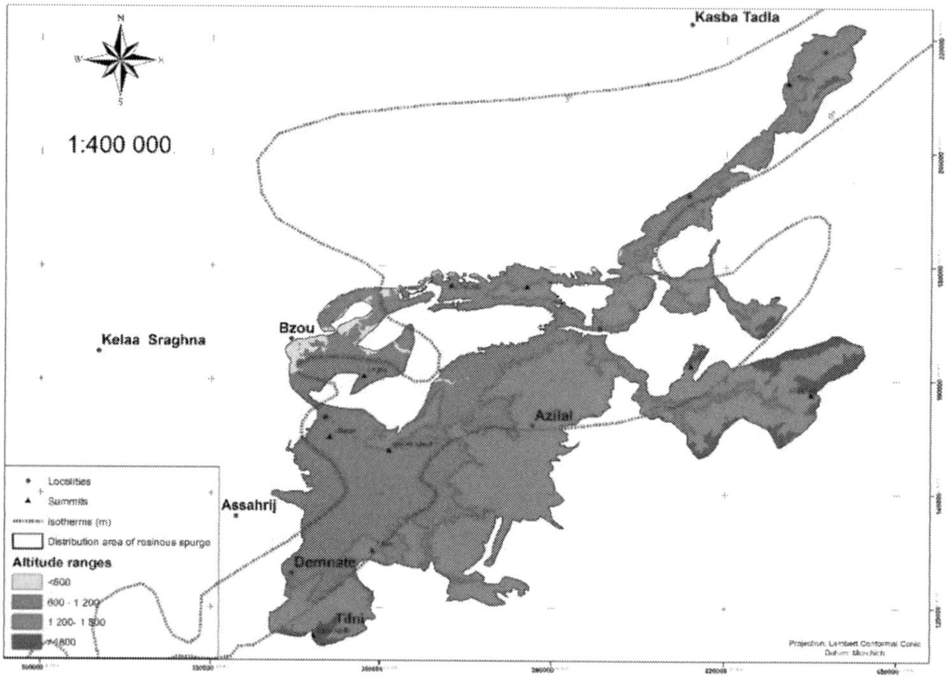

Fig. 3. Hypsometry and isotherms for the Euphorbus resin distribution

3. Range geology

The geological map below (figure 4) illustrates a clear location showing that the majority of the resin Spurge geographical area is dominated by Lias geological formations characterized by limestone, dolomitic and dolomite-limestone substratum. This finding corroborate that the spurge formations develop on calcareous substrates, dolomite, rocky soils and cracked slabs, which shows the rupicolous character of the species.

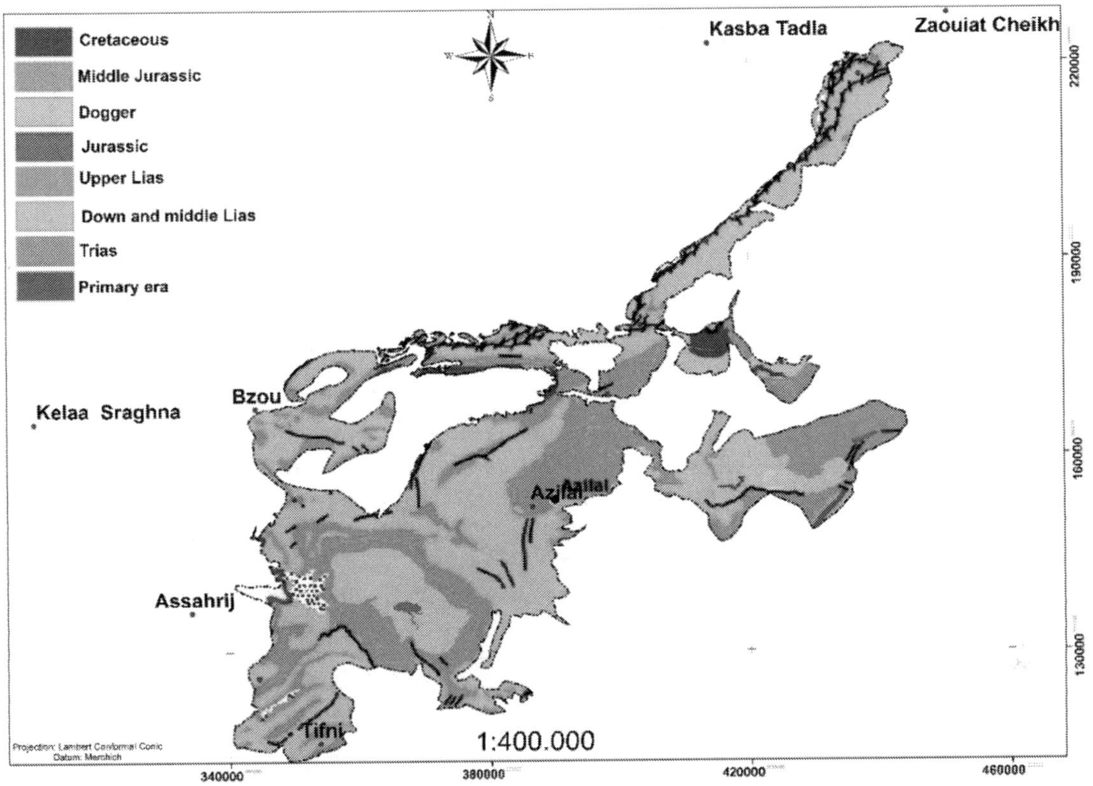

Fig. 4. Geological range of Euphorbus resin in the Beni Mellal-Khenifra region.

Discussion

This study aims to complete the gaps in the geographical distribution of an endemic and economically important plant in Morocco, particularly in the region of Beni Mellal-khenifra. Field surveys revealed that *E. resinifera* is mainly concentrated, from the northeast of El Ksiba (Middle Atlas) to the southwest of Demnate(High Central Atlas) with infiltrations in the center of the High Atlas following the continental gradient, where the spurge thrives on rocky soils with warm exposures and a clear to sparse density of the canopy cover. This result is in agreement with that of Benabid and Fennane[10], where they noticed that the bulk of this area is located at the level of the North Atlasian piedmont. From a bioclimatic point of view, it seems that this species prefers arid, semi-arid and locallysubhumid atmospheres with temperate, cool and exceptionally cold variants. This justifies its presence on the altitudinal range from 600 to 1 200m as indicated by Benabid[12] and designated as a part of the large Mediterranean-Saharan regional transition zone [14].

However, it's surprising that the species was also found at altitudes higher than 1 900m, suggesting that this species existed in this altitudinal range before. Furthermore, previous studies have never mentioned its presence at similar altitudes [9, 10]. Thus, future research on the ecophysiologicaltraits of the species is necessary to better documentthis unusual behavior. Another explanation can be attributed to an eventual migration of this species to high altitudes under the effect of climate change leading to an increase in temperatures (decline of the 0°C isotherm to the east). As it was described by White [14],working on Africa's vegetation with a global analysis, the three succulent shrubby species of Euphorbia are confined to the arid 'étage' and the drier, low-lying parts of the semi-arid 'étage', with, in the case of two species (*E. Officinarum*and *E. echinus*), a considerable southwards extension into the Atlantic Coastal Saharainfluenced by marine currents. They reach their maximum development at the western end of the Mediterranean-Sahara transition zone.*E. resinifera*, meanwhile, occurs in the north-east and south-east exhibitions which are sunny with less oceanic influences.The presence of *E. resinifera* at the altitude up to 1 900m suggests the possibility of a new phytoecological association of *E. resinifera* other than what has been proposed by several researchers [10, 11, 15]. However, caution should be exercised in the analysis, especially in plant and ecological terms, as Mediterranean vegetation is dominated by two main characteristics, namely heterogeneity and vulnerability [16].

The overlay of the euphorbia delimitation map with the geological map of the area showed that it is a rupicolous species. Indeed, most of the geographical area of this resiniferous species is dominated by geological formations of the Lias characterized by a calcareous, dolomitic and dolomite-calcareous substratum. This chain that dates from the Lias is probably different from those of the High Atlas and the Middle Atlas, maybe it is indigenous, while the other chains are allochthonous. In these formations, *E. resinifera* finds on this substrate the solid support, cracked slabs, heat and drought that are indispensable to its growth [11]. It is now clear that the three species of Euphorbia, *E. resinífera*, *E. beaumierana* and *E. echinus* even they are cactoid shrubs with thick, succulent, ridged, spinous stems forming dense clumps which normally grow 0.6-1 m tall, but in range they replace each other geographically [14]. Indeed, there was likely an adaptive strategy in Euphorbia that enabled the occupation of increased arid niche space accompanying Miocene expansion of arid ecosystems. These opportunities evidently facilitated recent, replicated bursts of diversification in Euphorbia [17].

In this study, the question was raisedregarding the scatter of this plant in AitOurir and Tizin'Testregion, where its presence was reported by Gattefossé and vindt[9] and Boulli [18]. The probable causes of this dispersion is a result of interaction between different factors, notably: the biological constraints generated by an inability to regenerate by natural seedlings; climate changes that have significantly altered the original climate conditions (original paleoclimate) for the regeneration and prosperity of the species and, finally, because of the human activities that, through its impacts and management choices, determine the dynamics of the ecosystems and the magnitude of the phenomena of rarefaction or extinction of the species [19]. In our case, the stands of *E. resinifera* are exposed to the exploitation of sand quarries and rock blocks found mainly in the limestone zone, as well as the conversion of the land into cereals and vegetable crops. All these observations challenge us to rethink the future of this plant considered as plants of multiple uses. Even to stabilize its natural balance, especially its annual floral cycle prerequisite for the survival of bee communities in the region.

Conclusion

This study allowed us to establish for the first time in Morocco a map of the regional distribution of *E.resinifera*, an important species with a high socio-economic and environmental interest. Field surveys show that it concentrates mainly, from the El Ksiba area in the north-east (Middle Atlas) to the south-west of Demnate(High Central Atlas) with infiltrations in the center of the High Atlas following the continental gradient, where it thrives on rocky soils with sunny and warm exposures and a clear to sparse density of canopy cover. Contrary to what has been observed by several researchers, this species was encountered at altitudes exceeding 1 900m. This suggests a possible migration to high altitudes under the effect of climate change leading to an increase in temperatures and also, the possibility of a new phytoecological association of *E. resinifera* other than what has been proposed by several authors. Nevertheless, its dispersionin some areas is a complex phenomenon subject to soil, paleogeogeography and climate factors exacerbated by intense anthropogenic action. Efforts must be made and undertaken to reflect on the future of this emblematic species.

References

1. FAO and Plan Bleu, Food and Agriculture Organization of the United Nations, Rome and Plan Bleu, Marseille. 308p (2018).
2. P. Quézel, In C. Gomez-Campo (éd.). Plant conservation in the Mediterranean area, W. Junk, Dordrecht, Pays-Bas. Geobotany **7**, p. 9-24. (1985).
3. F. Médail and P. Quézel, Annals of the Missouri Botanical Garden, n° **84**, p. 112-127 (1997).
4. P. Lawant and D. Winthagen, Bradleya, number **19**: 3-14 (2001).
5. MAPM (Ministère de l'agriculture et de la pêche maritime), p. 34 (2011).
6. B. Dorsey, T. Haevermans, X. Aubriot, J. Morawetz, R. Riina, V. Steinmann and P. Berry, TAXON **62** (2): 291–315 (2013).
7. J. Gavinet, PFE Ingénieur Forestier, Agro Paris Tech, ENGREF (France) :101p (2007).
8. MARA (Ministère de l'Agriculture et de la Réforme Agricole), livret-guide TOME I, le Milieu Marocain, 42 p (1966).
9. J. Gattefossé and J. vindt, Terre et Vie: 539-543 (1931).
10. A. Benabid and M. Fennane, Lazaroua, **41**: 21-97 (1994).
11. M. Fennane, Bul. Inst. Sc. Rabat, N° **12**: 99-148 (1988).
12. A. Benabid, Ibis Press, Paris, 357 p (2000).
13. L.Emberger, Atlas du Maroc n°6b, (1955) ;
14. F. Whirte, Mémoire accompagnant la carte de végétation de l'Afrique, 391p (1983).
15. M. Barbéro, P.Quézel, S. Rivas-Martínez, Phytocoenologia Band 9 Heft **3** : p. 311 – 412 (1981).
16. P. Quézel, U.N.E.S.C.O. Note tech. du M.A.B: 9-33 (1976).
17. J.W. Horn, Z. Xi, R. Riina, J.A.Peirson, Y. Yang, B.L. Dorsey, P.E. Berry, C.C. Davis and K.J.Wurdack, Evolution **68-12**: 3485–3504 (2014).
18. A. Boulli, Thèse de Doctorat de 3ème Cycle, UCA, Fac. Semlalia, Marrakech.
19. F. Médailand K.Diadema, Ann. Géo., n° **651**: 618-640.

E3S Web of Conferences **183**, 01002 (2020) https://doi.org/10.1051/e3sconf/202018301002
I2CNP 2020

Economic valuation of ecosystem services in Africa

Aya Jamouli [1*], *Khali* Allali [2]

[1]*Hassan II Institute of Agronomy and Veterinary Sciences, Rabat, Morocco*
[2]*National School of Agriculture, Meknes, Morocco*

Abstract. Ecosystems provide vital services that improve and support livelihoods and human well-being. Consequently, scientific research on ecosystem services (ES) has increased, over the past two decades globally, as well as in Africa. This study provides an overview of existing economic ES valuation methods in Africa using the Web of Science databases. The results highlighted that South Africa recorded the high number of ES valuation publications. The most evaluated ES category was provisioning then regulating services. In terms of economic valuation methods, the market price was the most popular, followed by the contingent valuation and the choice experiment methods. Recommendations are provided for future research in this filed.

1 Introduction

Ecosystems provide humans a diversity of services known as "Ecosystem services" [1]. They are generally classified into provisioning services such as food and raw materials, regulating services such as water and climate regulation, supporting services such as soil formation, and cultural services such as recreation and tourism. These categories are linked directly to human well-being [1,2]. The benefits provided by ES are crucial in maintaining human survival and contribute to poverty alleviation [2,3]. Therefore, The loss and disruption of these services, particularly in rural areas of developing countries [4], can negatively affect human well-being [5].

Africa hosts a high number of Least Developed Countries [6], and more than 50 percent of its rural population depends on ecosystem services [7]. However, the ecosystems of this continent are facing various anthropogenic pressures [8]. Common drivers are rapid population growth, increasing demand for food, water and energy, climate change, and overexploitation [9]. This shows the importance of assessment and valuation of such services in safeguarding and managing the loss of ecosystems [9,10].

Several approaches and methods have been developed [11]. According to [12,13], three main approaches can be distinguished: economic methods for estimating monetary values, biophysical methods for mapping and modeling ES, and socio-cultural methods for understanding social values of ES.

Many studies were conducted to assess and review the valuation of ES at different scales, such as: McDonough et al. [14] at the global scale, Rendon et al. [15] for Europe, Schuhmann and Mahon [16] for the Caribbean, Jiang [17] for China, Pittock et al. and Alamgir [18,19] for Australia, Van den Belt and Blake [20] for New Zealand, and Perez-Verdin et al. [21] for Mexico.

In Africa, several reviews were carried out by Vihervaara et al. [22], Seppelt et al. [23], Martinez-Harms and Balvanera [24], Egoh et al. [25], Crossman et al. [26], and Wangai et al. [27]. However, no study has attempted to review the economic valuation of ES in Africa.

Expressing the value of ES in monetary terms in Africa can help to face the challenge of biodiversity loss through the following two ways [9], [28–31]:

*Corresponding author: jamouliaya@gmail.com

-Raise awareness about the relative importance of ES;
-Improve decision making on the use and management of ES.

In this context, we conduct a literature review on the economic valuation of ES in Africa. It summarizes the existing studies of ES research and provides key issues for future research. This study focuses on the publication trend between 2005 and 2019, the methods that are highly used and the ES category that is most frequently performed and with which methods.

2 Methods

Based on [32], the search process for our literature review consists of four phases: data collection, selection criteria, content analysis, and statistics analysis (Fig. 1).

2.1 Data collection

The ISI Web of Knowledge database (http://www.isiknowledge.com) was used for our data collection search using numerous combining terms: "Ecosystem service*" AND "Africa*" AND (valu* OR "Economic analysis" OR "Monetary assessment") in the titles, the abstracts, and as keywords. We gathered all English academic publications from 2005 to 2019. The starting date was the publication year of The Millennium Ecosystem Assessment (MEA) report [33]. Since then, a significant number of ecosystem services research have been published [12].

This ISI website search allowed us to identify 389 studies over the past 15 years. We suppose that our review covers a majority part of the literature, given that the ISI Web of Science indexes 11,877 major journals from 81 different countries across the world [34].

2.2 Selection criteria

The abstracts, titles, and methodology sections of the original database were reviewed to decide the final papers to be considered in our economic valuation of ES in Africa. In this study, we retained only the papers that:
-Were conducted in an African country;
-Provided an ES that can be categorized according to the MEA framework [33];
-Used an economic valuation method.

This set of criteria resulted in 28 scientific articles that were summarized and reviewed in detail (Appendix A lists all case studies with their characteristics).

2.3 Content analysis

The content of the 28 ES studies selected was classified according to the following aspects:
-Publication year;
-Geographic location;
-Ecosystem type;
-Ecosystem service category;
-Economic valuation method.

2.4 Statistics analysis

Descriptive statistics was used in the statistical analysis of the information collected. The results are, generally, organized and presented as a number of case studies and publications for each feature mentioned above. All analysis and graphs were conducted using Excel software. Geographic distributions of economic valuation studies were mapped using ESRI ArcGIS 10.7.

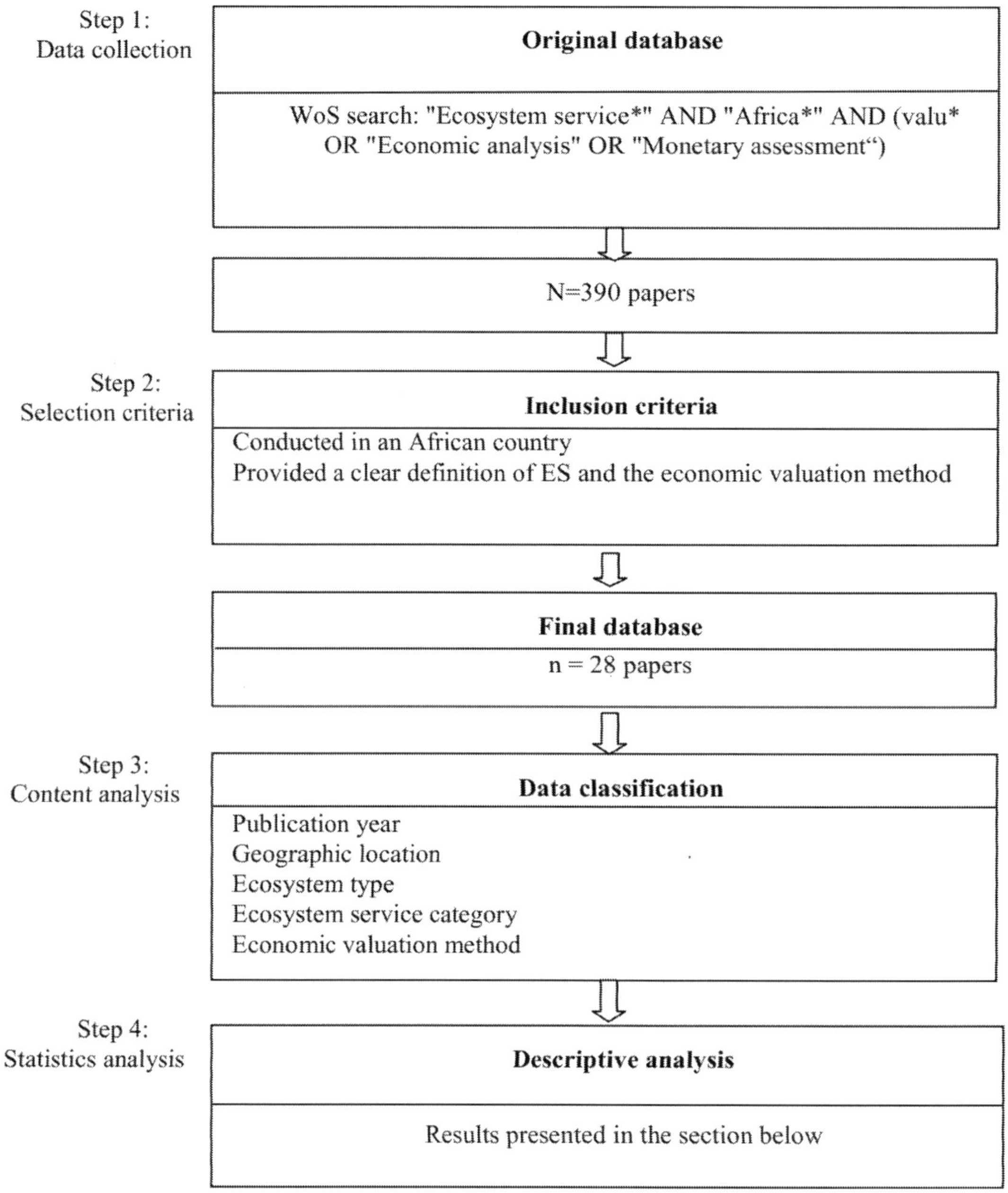

Fig. 1. Steps of the literature review

3 Results and discussion

3.1 Publication trends

The results show an increase trend in the number of publications since 2012 (Fig. 2). An average of one publication per year was noticed from 2005 to 2012, and it increases to an average of three publications per year after 2012. No publication has been selected for 2005, 2006, and 2009. This global progress can be explained by the release of several initiatives and projects around ES and biodiversity, including the Millennium Ecosystem Assessment (MEA), The Economics of Ecosystems and Biodiversity (TEEB), Intergovernmental Platform on Biodiversity and Ecosystem Services (IPBES), and the Ecosystem Services Partnership (ESP). They provide crucial information on the ES concept and attract significant attention of the researches from all over the world [12], [35], including Africa.

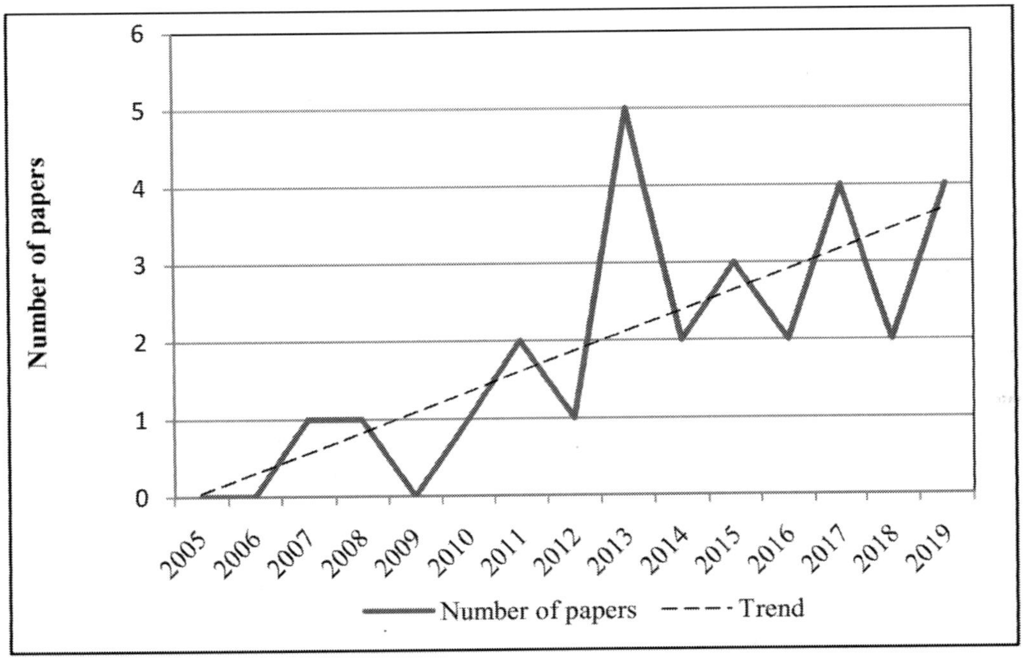

Fig. 2. Number of studies per year between 2005 and 2019

3.2 Geographical distribution

As illustrated in Fig. 3, the spatial distribution of economic ES studies shows that more than 50% of the studies have been conducted in South Africa and Kenya alone (13 and 6, respectively). The remaining papers are distributed between Mozambique, with a total of four studies; Tanzania, with three publications and a single study per country for Senegal, Ghana, Madagascar, and Benin. The first African study related to ES issues was conducted in South Africa in 2005, using a spatial mapping of provisioning and supporting ES [27]. Since then, it has become the leading African country in terms of ES articles published between 1987 and 2017 [36].

Moreover, Southern and Eastern Africa gather almost all case studies, while there were no papers from the northern part of the continent, even in the review provided by [27]. Although it is difficult to explain the lack of research in these countries, it might be due to the limited human and financial resources or the exclusion of non-English studies [37].

https://doi.org/10.1051/e3sconf/202018301002

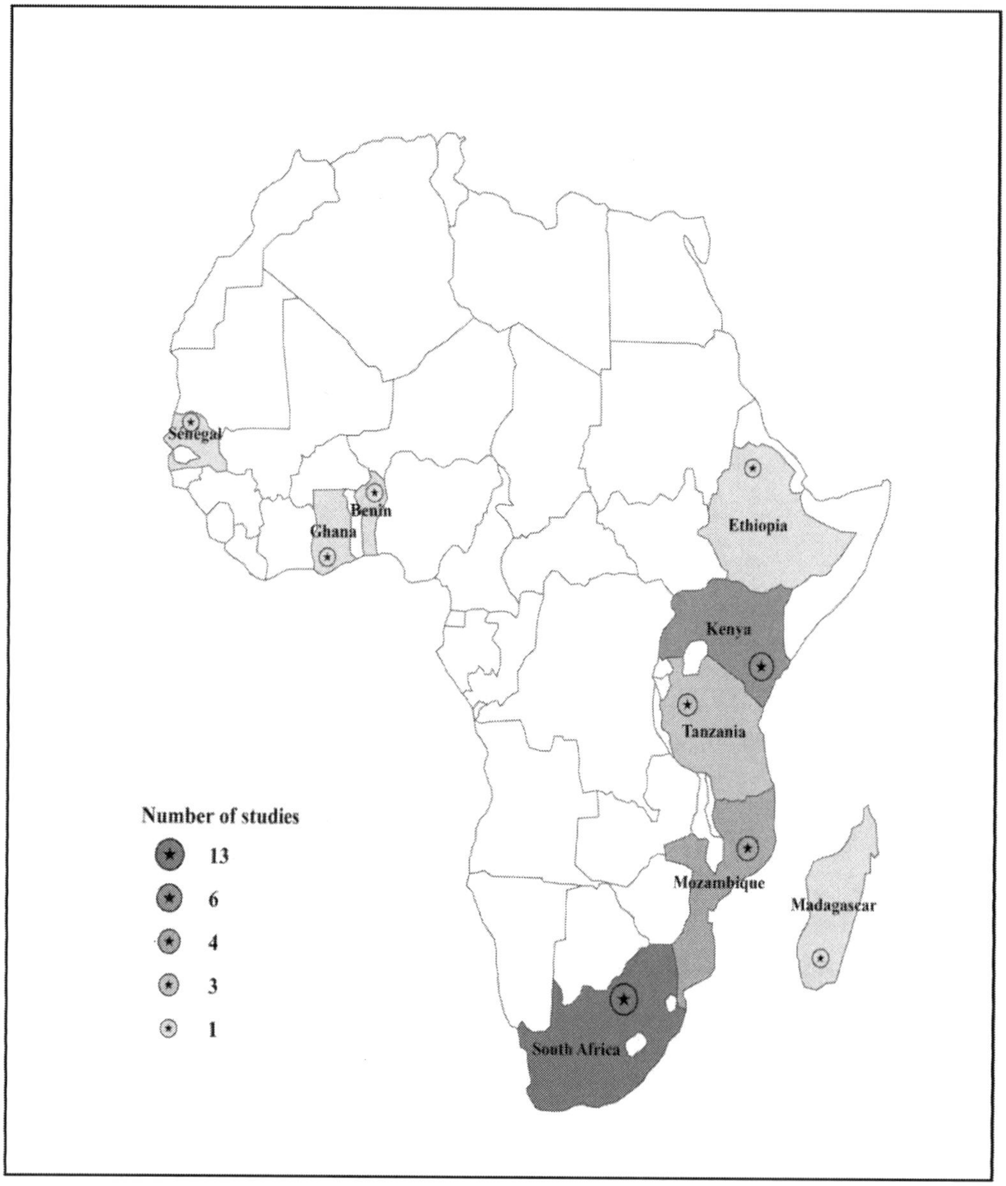

Fig. 3. Geographic distribution of ES studies in Africa (there can be more than one country per study)

3.3 Ecosystem types

Within the 28 publications, eight ecosystem types were investigated (Fig. 4). The most dominant types were wetland and forest ecosystems (5 studies for each), followed by marine, agro-ecosystem, and urban (3 studies for each). Freshwater and dryland are also considered (2 studies for each). The rest of the studies cover more than one ecosystem (e.g., Grassland, woodland, wetland). However, one study [38] did not clearly define the type of ecosystem and it uses the geographical name of a region (The Orange River Basin in the Lesotho Highlands).

We should note that some ecosystems are difficult to classify. For example, the urban forest is not clear whether to be categorized into a forest or an urban ecosystem. Consequently, future studies should adopt an international framework for a consistent definition of ecosystems (e.g., MEA or TEEB) [17].

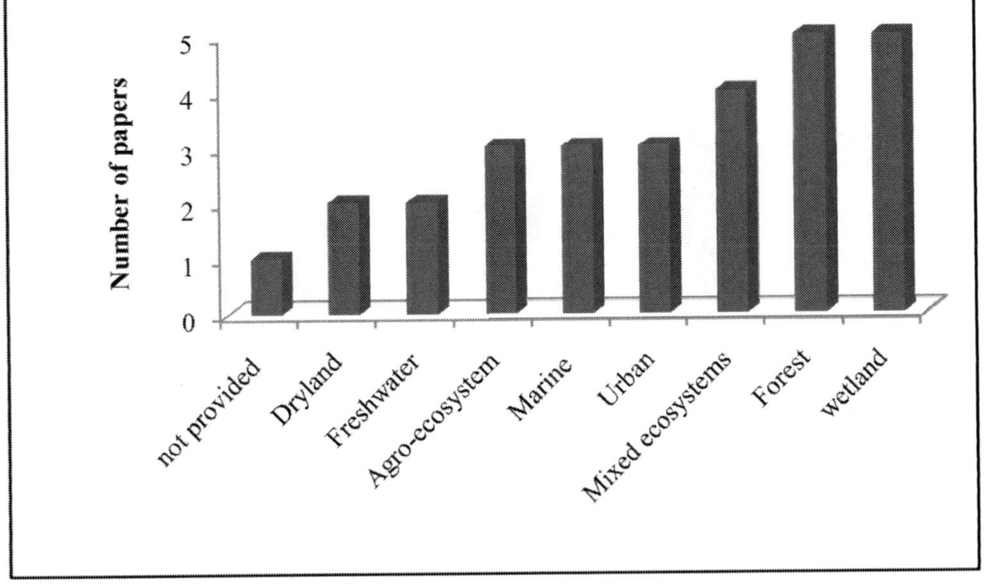

Fig. 4. Number of studies per ecosystem types

3.4 Ecosystem services categories

In terms of ecosystem services studied (Fig. 5), provisioning services were the most assessed (13), followed by regulating (12) and cultural services (11). However, supporting services received the least attention from researches (2).

Overall, 19 of 28 studies assessed only a single ES. The remaining nine studies examined multiple services, of which the combination of provisioning and regulating services was the most common. [39] and [40] applied trade-offs and synergies between different types of ES. However, this type of analysis was lacking in those studies based on one category.

Regardless of the number of ES, more than half of the 82 ES categories are provisioning services (food, water, raw materials, etc.), confirming the findings by [27], [36] who reported that provisioning ES were the most services studied. This particular attention can be explained by the fact that many African countries depend directly on this category of services to meet their basic needs [25], [41–43].

 https://doi.org/10.1051/e3sconf/202018301002

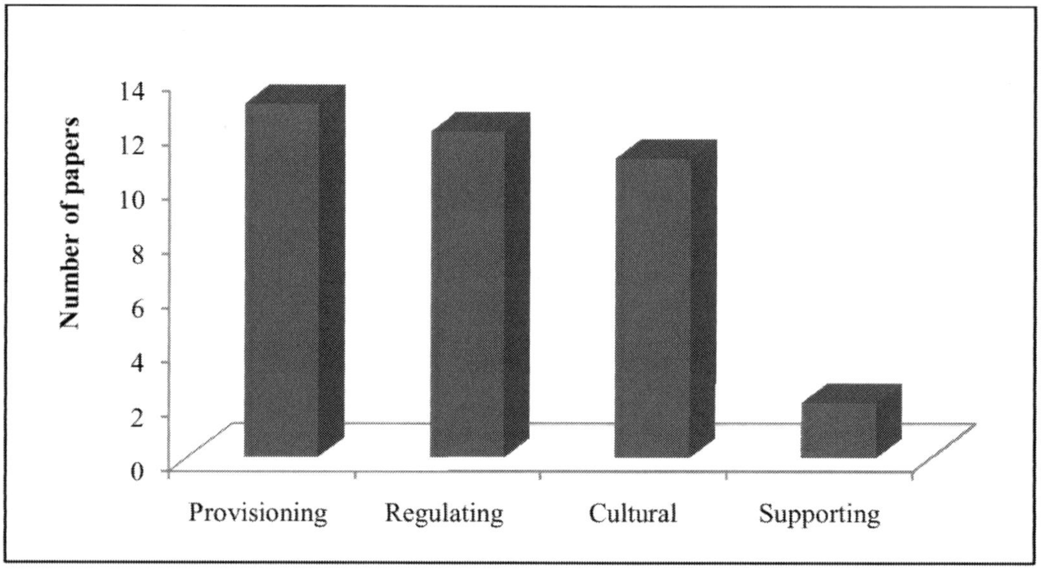

Fig. 5. Number of studies per ES category

3.5 Economic valuation methods

This study identified nine main valuation methods, which include revealed and stated preferences (Table 1). The number of papers using each method is shown in Fig. 6. In general, the market price is the most used, followed by the contingent valuation and the choice experiment (used at a similar frequency). The number of ES categories by each economic method is shown in Fig.7. Choice experiment and benefits transfer were applied to assess large categories of services compared to the other methods. In contrast to [44], [45], the market price method has been frequently used to evaluate the provisioning services.
The next sections focus on presenting two of the most used methods per approach.

Table 1. Summary of the reviewed economic valuation methods

Approach	Method	Description from reviewed studies
Revealed preference	Market price	Derives economic values of ES using prices in markets.
	Production function	Relates the marketed output to the inputs used in production. Ecosystem service is considered as an additional input.
	Replacement cost	Uses costs as a proxy for the value of ES, including methods based on the costs to substitute services, the costs of avoiding damages, or the travel and time costs for access to ecosystems.
	Avoided cost	
	Travel cost	
	Hedonic pricing	Estimates economic values of ES using the prices of similar properties, usually a house.
	Benefits transfer	Transfers ES values from existing benefits from the literature to another context.

Stated preference	Contingent valuation	Asks people their willingness to pay (WTP) for specific ES.
	Choice experiment	Asks people to choose between a set of attributes.

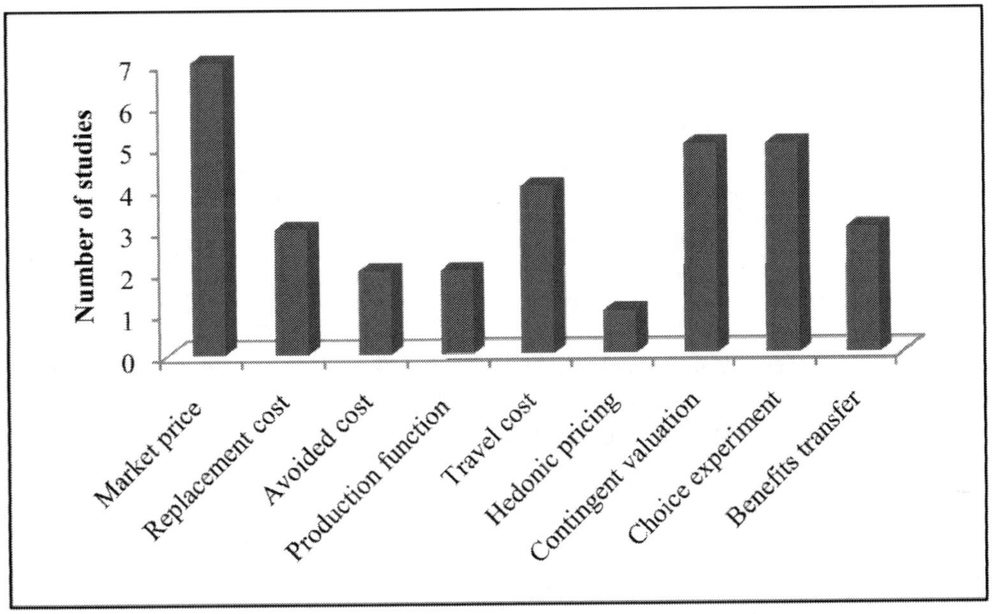

Fig. 6. Number of studies per economic valuation method

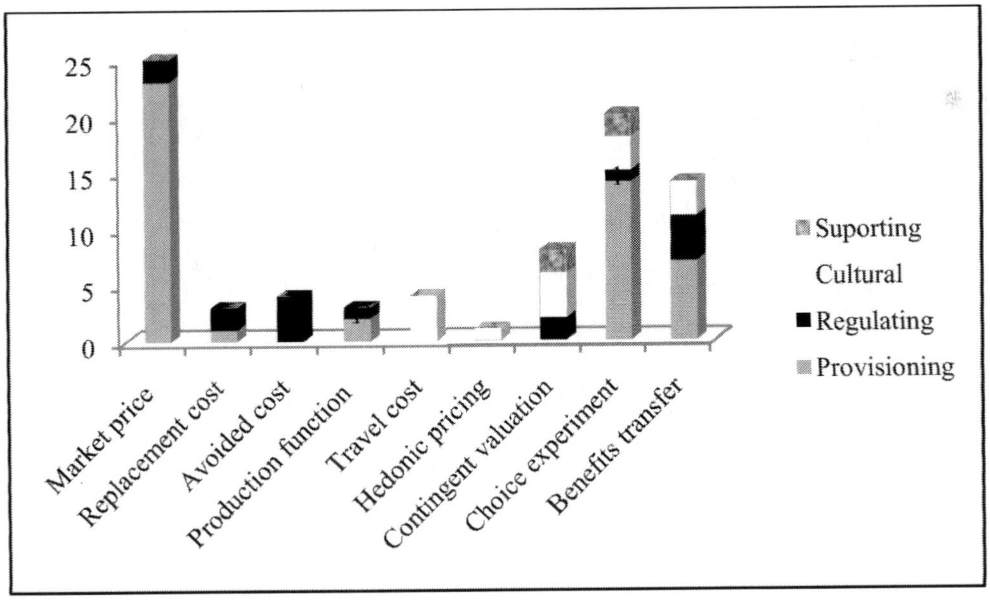

Fig. 7. Number of studies per methods and ESs categories

3.5.1 Revealed preference methods

Seven revealed preference methods, found in seven studies, were identified, of which the market price dominates the database (Table 1, Fig. 6). It directly provides the value of services from markets [46], [47]. For example, [48] calculated the value of grazing service based on the commercial value of the grass. In addition, four papers adopted the travel cost method to provide the value of recreational activities. As an example, to estimate the recreational value of a freshwater system, [49] calculated the time it takes to get to the site and the time spent there.

3.5.2 Stated preference methods

Contingent valuation and choice experiment are the two stated preference methods founded (Table 1, Fig. 6). The contingent valuation measures people's willingness to pay (or to accept) [50]. For instance, this method was used to estimate the economic value of non-market benefits of an urban forest in Ghana [51]. Furthermore, choice experiment asks respondents to choose their preferred option between different attributes [9]. [52] investigated the Mozambique population' preferences for water, firewood, and land, applying the choice experiment method. Therefore, stated preference methods had received more attention than the revealed preference methods. This result is consistent with [45], [53].

3.5.3 Combination of methods

Four case studies used a combination of economic methods. For example, the ES valuation of [39] was based on market pricing and benefit transfer methods. We also found that three studies used economic and mapping methods. For instance, [54] combined economic valuation (market price and benefit transfer) and mapping assessment to provide the values of key northern Mozambique's marine ecosystem. Integration of economic and biophysical models needs also to be considered as in [55].

3.6 Limits of the review

This study was limited to English peer-reviewed studies and it is based exclusively on the databases of the Web of Science. Thus, all conclusions presented here should be interpreted with caution.

4 Conclusion

The main objectives of this study were to present an overview of economic valuation of ES in Africa and to offer some recommendations for future research. Twenty eight studies conducted in different parts of the continent were analyzed. Most the studies was conducted in South Africa and focused on services provided by wetland and forest ecosystems. Provisioning services was the most assessed category and the market price was the frequently cited method.
Recommendations for future research on ES in Africa should:
-Improve the knowledge of ES and their incorporation into decision-making and the management of natural resources;
-Combine multidisciplinary approaches, especially economic and mapping methods;
-Explore trade-offs and synergy between different ES;
-Adopt an international framework for a solid classification of ES;

-Assign equal attention to different ES category, especially those that are neglected (e.g., cultural diversity, knowledge);
-Conduct further economic valuation studies in Africa, especially in the northern region of the continent using English language.

Acknowledgements

This research is part of a PhD work funded by CNRST (National Center of Scientific Research, Morocco), Pr.Laghfiri Malaynine is acknowledged for his helpful suggestions.

References

1. M. A. Board, *Millennium ecosystem assessment*, Washington, DC New Isl, **13** (2005)
2. H. Suich, C. Howe, G. Mace, *Ecosystem services and poverty alleviation: a review of the empirical links*, Ecosyst. Serv, **12**, 137–147 (2015)
3. J.A. Fisher, G. Patenaude, K. Gir, K. Lewis, P. Meir, P. Pinho, M.D. Rounsevell, M. Williams, *Understanding the relationships between ecosystem services and poverty alleviation: a conceptual framework*, Ecosyst. Serv, **7**, 34–45 (2014)
4. H. Tallis, P. Kareiva, M. Marvier, A. Chang, *An ecosystem services framework to support both practical conservation and economic development*, Proc. Natl. Acad. Sci, **105**, 9457–9464 (2008)
5. M. Erhard, A. Teller , J. Maes , A. Meiner, P.M. Berry, A. Smith, R. Eales, L. Papadopoulou, A. Bastrup-Birk, E. Ivits, *Mapping and assessment of ecosystems and their services: mapping and assessing the condition of Europe's ecosystems: progress and challenges*, (2016)
6. UNDP, *Human development report 2019: beyond income, beyond averages, beyond today*, United Nations Dev. Progr, 352 (2019)
7. E. Archer, L. Dziba, K.J. Mulongoy, M.A. Maoela, M. Walters, R.O. Biggs, M.C.C. Salem, F. DeClerck, M.C. Diaw, A.E. Dunham, O. Failler, *The regional assessment report on biodiversity and ecosystem services for Africa: summary for policymakers*, IPBES (2018)
8. J. Hugé, A.J. Rochette, S. de Béthune, C.P. Paitan, K. Vanderhaegen, T. Vandervelden, S. Van Passel, M.P.M. Vanhove, B. Verbist, D. Verheyen, T. Waas, *Ecosystem services assessment tools for African Biosphere Reserves: A review and user-informed classification*, Ecosyst. Serv, **42**, 101079 (2020)
9. M. Masiero, D. Pettenella, M. Boscolo, S. . Barua, I. Animon, J.R. Matta, *Valuing forest ecosystem services A training manual for planners and project developers.* Forestry Working Paper, Licence: CC BY-NC-SA 3.0 IGO (2019)
10. F. Eigenbrod, B.J. Anderson, P.R. Armsworth, A. Heinemeyer, S.F. Jackson, M. Parnell, C.D. Thomas, K.J. Gaston, *Ecosystem service benefits of contrasting conservation strategies in a human-dominated region*, Proc. R. Soc. B Biol. Sci, **276**, 2903–2911 (2009)
11. K.J. Bagstad, D.J. Semmens, S. Waage, R. Winthrop, *A comparative assessment of decision-support tools for ecosystem services quantification and valuation*, Ecosyst. Serv, **5**, 27–39 (2013)
12. R. Costanza, R. De Groot, L. Braat, I. Kubiszewski, L. Fioramonti, P. Sutton, S. Farber, M. Grasso, *Twenty years of ecosystem services: How far have we come and how far do we still need to go?*, Ecosyst. Serv, **28**, 1–16 (2017)
13. P.A. Harrison, R. Dunford, D.N. Barton, E. Kelemen, B. Martín-López, L. Norton, M. Termansen, H. Saarikoski, K. Hendriks, E. Gómez-Baggethun, B. Czúcz,

Selecting methods for ecosystem service assessment: A decision tree approach, Ecosyst. Serv, **29**, 481–498 (2018)

14. K. McDonough, S. Hutchinson, T. Moore, J. M.S. Hutchinson, *Analysis of publication trends in ecosystem services research*, Ecosyst. Serv, **25**, 82–88 (2017)
15. P. Rendon, M. Erhard, J. Maes, B. Burkhard, *Analysis of trends in mapping and assessment of ecosystem condition in Europe*, Ecosyst. People, **15**, 156–172 (2019)
16. P.W. Schuhmann, R. Mahon, *The valuation of marine ecosystem goods and services in the Caribbean: A literature review and framework for future valuation efforts*, Ecosyst. Serv, **11**, 56–66 (2015)
17. W. Jiang, *Ecosystem services research in China: A critical review*, Ecosyst. Serv, **26**, 10–16 (2017)
18. J. Pittock, S. Cork, S. Maynard, T*he state of the application of ecosystems services in Australia*, Ecosyst. Serv, **1**, 111–120 (2012)
19. M. Alamgir, P.L. Pert, S.M. Turton, *A review of ecosystem services research in Australia reveals a gap in integrating climate change and impacts on ecosystem services*, Int. J. Biodivers. Sci. Ecosyst. Serv. Manag, **10**, 112–127 (2014)
20. M. Van Den Belt, D. Blake, *Ecosystem services in new Zealand agro-ecosystems: A literature review*, Ecosyst. Serv, **9**, 115–132 (2014)
21. G. Perez-Verdin, E. Sanjurjo-Rivera, L. Galicia, J.C. Hernandez-Diaz, V. Hernandez-Trejo, M.A. Marquez-Linares, *Economic valuation of ecosystem services in Mexico: Current status and trends*, Ecosyst. Serv, **21**, 6–19 (2016)
22. P. Vihervaara, M. Rönkä, M. Walls, *Trends in ecosystem service research: early steps and current drivers*, Ambio, **39**, 314–324 (2010)
23. R. Seppelt, C.F. Dormann, F.V Eppink, S. Lautenbach, S. Schmidt, *A quantitative review of ecosystem service studies: approaches, shortcomings and the road ahead*, *J. Appl.* Ecol, **48**, 630–636 (2011)
24. M.J. Martínez-Harms, P. Balvanera, *Methods for mapping ecosystem service supply: a review*, Int. J. Biodivers. Sci. Ecosyst. Serv. Manag, **8**, 17–25 (2012)
25. B.N. Egoh, P.J. O'Farrell, A. Charef, L.J. Gurney, T. Koellner, H.N. Abi, M. Egoh, L. Willeme, *An African account of ecosystem service provision: Use, threats and policy options for sustainable livelihoods*, Ecosyst. Serv, **2**, 71–81 (2012)
26. N.D. Crossman, B. Burkhard, S. Nedkov, L. Willemen, K. Petz, I. Palomo, E.G. Drakou, B. Martín-Lopez, T. McPhearson, K. Boyanova, R. Alkemade, *A blueprint for mapping and modelling ecosystem services*, Ecosyst. Serv, **4**, 4–14 (2013)
27. P.W. Wangai, B. Burkhard, F. Müller, *A review of studies on ecosystem services in Africa*, Int. J. Sustain. Built Environ, **5**, 225–245 (2016)
28. M.T. Ahmed, *The Grand Ethiopian Renaissance Dam and the sustainability of water resources: would monetary compensation warrant some more restrictions on dams?*, Sustain. Water Resour. Manag, **5**, 1327–1334 (2019)
29. LUC, *Calculating the economic values of the ecosystem services provided by Natura 2000 features and sites in Wales Prepare*, Ecosyst. Serv. *Contract*, March, 21 (2015)
30. DEFRA, *An introductory guide to valuing ecosystem services* (2007)
31. P. Kumar, *The economics of ecosystems and biodiversity: ecological and economic foundations*. UNEP/Earthprint (2010)
32. D.P. Andrew Booth, Anthea Sutton, *Systematic approaches to a successful literature review*. Los Angeles, London, New Delhi, Singapore, Washington D.C (2012)
33. M. E. Assessment, *A report of the Millennium Ecosystem Assessment*, Ecosyst. Hum. Well-Being Desertif. Synth, 26 (2005)
34. Clarivate Analytics. Journal Citation Reports (2019)

35. K. McDonough, S. Hutchinson, T. Moore, J. M.S. Hutchinson, *Analysis of publication trends in ecosystem services research*, Ecosyst. Serv, **25**, 82–88 (2017)
36. X. Zhang, R.C. Estoque, H. Xie, Y. Murayama, M. Ranagalage, *Bibliometric analysis of highly cited articles on ecosystem services*, PLoS One, **14**, 1–16 (2019)
37. L. Neimann Rasmussen, P. Montgomery, *The prevalence of and factors associated with inclusion of non-English language studies in Campbell systematic reviews: A survey and meta-epidemiological study*, Syst. Rev, **7**, 1–12 (2018)
38. G.M. Lange, E. Mungatana, R. Hassan, *Water accounting for the Orange River Basin: An economic perspective on managing a transboundary resource*, Ecol. Econ., **61**, 660–670 (2007)
39. C.A.K. Baba, J. Hack, *Economic valuation of ecosystem services for the sustainable management of agropastoral dams. A case study of the Sakabansi dam, northern Benin*, Ecol. Indic, **107**, 105648 (2019)
40. S. Silvestri, L. Zaibet, M.Y. Said, S.C. Kifugo, *Valuing ecosystem services for conservation and development purposes: A case study from Kenya*, Environ. Sci. Policy, **31**, 23–33 (2013)
41. M. Niamir, *Community forestry: herders' decision-making in natural resources management in arid and semi-arid Africa*, Community For. herders' Decis. Nat. Resour. Manag. Arid semi-arid Africa (1990)
42. G. Davies, *Bushmeat and international development*, Conserv. Biol, **16**, 587–589 (2002)
43. M. Karki, S.S. Sellamuttu, S. Okayasu, W. Suzuki, L.A. Acosta, Y. Alhafedh, J.A. Anticamara, A.G Ausseil, K. Davies, A. Gasparatos, H. Gundimeda, *The regional assessment report on biodiversity and ecosystem services for Asia and the Pacific: summary for policymakers* (2018)
44. V.C.S. Hackbart, G.T.N. P. de Lima, R.F. dos Santos, *Theory and practice of water ecosystem services valuation: Where are we going?*, Ecosyst. Serv, **23**, 218–227 (2017)
45. M. Browne, G. Fraser, J. Snowball, *Economic evaluation of wetland restoration: a systematic review of the literature*, Restor. Ecol, **26**, 1120–1126 (2018)
46. De Groot, *Function-analysis and valuation as a tool to assess land use conflicts in planning for sustainable, multi-functional landscapes*, Landsc. Urban Plan, **75**, 175–186 (2006)
47. M. Chaikumbung, H. Doucouliagos, H. Scarborough, *The economic value of wetlands in developing countries: A meta-regression analysis*, Ecol. Econ, 124, 164–174 (2016)
48. S. Mudavanhu, J. Blignaut, N. Stegmann, G. Barnes, W. Prinsloo, A. Tuckett, *The economic value of ecosystem goods and services: The case of Mogale's Gate Biodiversity Centre, South Africa*, Ecosyst. Serv, **26**, 127–136 (2017)
49. T. Vundla, J. Blignaut, D. Crookes, *Aquatic weeds : to control or not to control. the case of the Midmar Dam, KwaZulu-Natal, South Africa*, **12**, 412–429 (2017)
50. V. Ginsburgh, *Contingent valuation, willingness to pay, and willingness to accept*, in Economic ideas you should forget, Springer, 65–66 (2017)
53. W.K. Dumenu, *What are we missing? Economic value of an urban forest in Ghana*, Ecosyst. Serv, **5**, 137–142 (2013)
52. A.K. Kosenius, M. Kniivilä, M. Pitiot, P. Horne, *Location of forest plantations in Mozambique: Gains and losses in water, firewood and land availability*, Land use policy, **88**, 104175 (2019)
53. X. Cheng, S. Van Damme, L. Li, P. Uyttenhove, *Evaluation of cultural ecosystem services: A review of methods*, Ecosyst. Serv, **37**, 100925 (2019)
54. A. Ghermandi, D. Obura, C. Knudsen, P.A.L.D. Nunes, *Marine ecosystem services*

in the Northern Mozambique Channel: A geospatial and socio-economic analysis for policy support, Ecosyst. Serv, **35**, 1–12 (2019)
55. B.L. Keeler, S. Polasky, K.A Brauman, K.A. Johnson, J.C. Finlay, A. O'Neill, K. Kovacs, B. Dalzell, *Linking water quality and well-being for improved assessment and valuation of ecosystem services*, Proc. Natl. Acad. Sci, **109**, 18619–18624 (2012)
56. B. Tibesigwa, J. Siikamäki, R. Lokina, J. Alvsilver, *Naturally available wild pollination services have economic value for nature dependent smallholder crop farms in Tanzania*, Sci. Rep, **9**, 1–10 (2019)
57. K. Bekele, J. Haji, B. Legesse, U. Schaffner, *Economic impacts of Prosopis spp. invasions on dryland ecosystem services in Ethiopia and Kenya: Evidence from choice experimental data*, J. Arid Environ, **158**, 9–18 (2018)
58. M. Jacobson, Y.H. Shr, F. Dalemans, C. Magaju, R. Ciannella, *Using a choice experiment approach to assess production tradeoffs for developing the croton value chain in Kenya*, For. Policy Econ, **86**, 76–85 (2018)
59. J. Dikgang, E. Muchapondwa, *Local communities' valuation of environmental amenities around the Kgalagadi Transfrontier Park in Southern Africa,* J. Environ. Econ. Policy, **6**, 168–182 (2017)
60. S. Niquisse, P. Cabral, *Assessment of changes in ecosystem service monetary values in Mozambique*, Environ. Dev, **25**, 12–22 (2018)
61. A. Seck, A *dichotomous-choice contingent valuation of the Parc Zoologique de Hann in Dakar*, African J. Agric. Resour. Econ, **11**, 226–238 (2016)
62. M.C. Mutoko, L. Hein, C. A. Shisanya, *Tropical forest conservation versus conversion trade-offs: Insights from analysis of ecosystem services provided by Kakamega rainforest in Kenya*," Ecosyst. Serv, **14**, 1–11 (2015)
63. C.A.K. Daly, G. Fraser, J. D. Snowball, *Willingness to pay for marine-based tourism in the Ponta do Ouro Partial Marine Reserve, Mozambique*, African J. Mar. Sci, **37**, 33–40 (2015)
64. J. Cilliers, S. Cilliers, *From green to gold : a South African example of valuing urban green spaces in some residential areas in Potchefstroom*, T. Reg. Plan, **2015**, 1–12 (2015)
65. J. Bayliss, M. Schaafsma, A. Balmford, N.D. Burgess, J.M. Green, S.S. Madoffe, S. Okayasu, K.S.H. Peh, P.J. Platts, W.Y. Douglas, *The current and future value of nature-based tourism in the Eastern Arc Mountains of Tanzania*, Ecosyst. Serv, **8**, 75–83 (2014)
66. D.W. Mulatu, A. van der Veen, P.R. van Oel, *Farm households' preferences for collective and individual actions to improve water-related ecosystem services: The Lake Naivasha basin, Kenya*, Ecosyst. Serv, **7**, 22–33 (2014)
67. W.J. De Lange, R. Veldtman, M. H. Allsopp, *Valuation of pollinator forage services provided by Eucalyptus cladocalyx*, J. Environ. Manage, **125**, 12–18 (2013)
68. B. Conradie, M. Garcia, *An estimate of the recreational value of the a gulhas plain , south africa , with special reference to the value of plant biodiversity a Data collection*, **16**, 170–182 (2013)
69. A. Schäffler, M. Swilling, *Valuing green infrastructure in an urban environment under pressure - The Johannesburg case*, Ecol. Econ, **86**, 246–257 (2013)
70. N.A. Davenport, C.M. Shackleton, J. Gambiza, *The direct use value of municipal commonage goods and services to urban households in the Eastern Cape, South Africa*, Land use policy, **29**, 548–557 (2012)
[71] S. Simonit, C. Perrings, S*ustainability and the value of the 'regulating' services: Wetlands and water quality in Lake Victoria*, Ecol. Econ, **70**, 1189–1199 (2011)

[72] M. du Preez, S. G. Hosking, *He value of the trout fishery at Rhodes, North Eastern Cape, South Africa: A travel cost analysis using count data models*, J. Environ. Plan. Manag, **54**, 267–282 (2011)

73. M. Du Preez, S. Tessendorf, *Application of thr contigent valuation method to estimate the willingness to pay for restoring indigenous vegetation in Underberg, Kwazulu-Natal, South Africa*, **13** (2010)

74. M.H. Allsopp, W.J. de Lange, R. Veldtman, *Valuing insect pollination services with cost of replacement*, PLoS One, **3** (2008)

Appendix A

The 28 studies on the economic valuation of African ES

Study	Year	Country/Area of study	Type of Ecosystem	MEA classification (No of ES assessed)	Number of ES	Valuation method
[39]	2019	Benin	Mixed ecosystem	Provisioning (7) Regulating (2)	9	Market price Benefit transfer
[56]	2019	Tanzania	Agro-ecosystem	Regulating	1	Production function
[54]	2019	Tanzania Mozambique Madagascar	Marine	Provisioning (1) Regulating (1) Cultural (1)	3	Market price Benefit transfer
[52]	2019	Mozambique	Forest	Provisioning	2	Choice experiment
[57]	2018	Ethiopia Kenya	Dryland	Provisioning (2) Cultural (2) Regulating (1)	5	Choice experiment
[58]	2018	Kenya	Forest	Provisioning	1	Choice experiment
[59]	2017	South Africa	Agro-ecosystem	Regulating (1)	1	Avoided cost
[48]	2017	South Africa	Mixed ecosystems	Provisioning (3) Regulating (1)	4	Production function Market prices
[60]	2017	Mozambique	Mixed ecosystems	Provisioning (7)	9	Benefit transfer

				Regulating (1) Cultural (1)		
[49]	2017	South Africa	Freshwater	Cultural	1	Travel cost
[59]	2016	South Africa	Dryland	Provisioning	7	Choice experiment
[61]	2016	Senegal	Forest	Cultural	1	Contingent valuation
[62]	2015	Kenya	Forest	Provisioning (7) Cultural (2)	9	Market Price Contingent valuation Travel cost
[63]	2015	Mozambique	Marine	Cultural	1	Contingent valuation
[64]	2015	South Africa	Urban	Cultural	1	Hedonic pricing
[65]	2014	Tanzania	Wetland	Cultural	1	Contingent valuation
[66]	2014	Kenya	Wetland	Provisioning (3) Cultural (1) Supporting (2)	6	Choice experiment
[67]	2013	South Africa	Agro-ecosystem	Provisioning	1	Replacement cost
[68]	2013	South Africa	Marine	Recreation	1	Travel cost
[51]	2013	Ghana	Urban	Regulating (1) Supporting (2)	3	Contingent valuation
[69]	2013	South Africa	Urban	Regulating (1) Cultural (2)	3	Market price
[40]	2013	Kenya	Wetland	Provisioning	2	Market price
[70]	2012	South Africa	Urban	Provisioning	3	Market price
[71]	2011	Kenya	Mixed ecosystems	Regulating	3	Avoided cost
[72]	2011	South Africa	Freshwater	Cultural	1	Travel cost

[73]	2010	South Africa	Agro-ecosystem	Regulating	1	Contingent valuation
[74]	2008	South Africa	Not provided	Regulating	1	Replacement cost
[38]	2007	South Africa	Wetland	Regulating	1	Replacement cost

Conservation Amid COVID-19 Pandemic: Ecotourism Collapse Threatens Communities and Wildlife in Morocco

SidiImad Cherkaoui[1]*, *Mohamed* Boukherouk[2], *Tarik* Lakhal[3] *Adil* Aghzar[1], and *Lahcen* El Youssfi[1]

[1]Ecological Engineering and Natural Resource Management Research Group,Higher School of technology of Khenifra, BP 170 Khénifra 54000Morocco, Sultan MlySlimane University, Morocco.
[2]Laboratoire de recherches des sciences appliquées à l'environnement et au développement durable (SAEDD), High School of technology of Essaouira, Cadi AyyadUniversity Marrakech, Km 9, Route d'Agadir, Essaouira Aljadida BP. 383,Morocco
[3]Département des sciences de gestion, Faculté des Sciences Juridiques, Economiques et Sociales Agdal, Université, Mohammed V de Rabat.

Abstract. COVID-19 pandemic has had huge impacts on multiple industries and sectors, not just ecotourism and wildlife protection in Morocco. Ecological health and wildlife are a critical resource for the country's tourism sector recovery. Conservation is considered as one of the industries that are hit hard by the COVID-19 pandemic. Many of the threats facing biodiversity and protected areas have been exacerbated during and following, the outbreak with practices such as poaching, wildlife trafficking, and forest logging activities resumed due to the country lockdown and ecotourism collapse. Conservationists should take emergency actions in the short-term to help rural communities and grassroots organizations and review waysto achieve conservation and sustainability goals in the post-pandemic era.

1 Introduction

Travel and tourism account for 10.3 percent of the country's global GDP [1], which makes the sector larger than agriculture. In 2019 alone, it created one in four new jobs [1]. The economic contribution of wildlife tourism is equally impressive. It came to US$343.6 billion (0.4 percent of global GDP) in 2018[2]. Wildlife tourism supported 21.8 million jobs across the world, or 6.8 percent of total travel and tourism jobs [3]. The percentage is much higher in Africa, at a percentage of 36.3 percent [3].

Globally, ecotourism has been central to thousands of conservation projects that have generated jobs and income, empowering rural communities [4, 5]. It has become a key argument in the "conserve or exploit" debate. The World Travel and Tourism Council (WTTC) estimates that up to 75 million jobs are at immediate risk, and anticipates an economic loss of up to US$2.1 trillion lifted [6]. The recovery period after disease outbreaks has in the past averaged around 19.4 months- everything doesn't return to normal the day after the lockdown is lifted[6].In 2020, the World Economic Forum (WEF)[7] has ranked nature loss as one of the top global risks. COVID-19 has proved that nature loss and wildlife consumption are the root cause of the emergence of zoonotic infectious diseases, such as coronavirus, Ebola and HIV/AIDS [8]. Although they are quite different at first glance, they all originated from animal populations under conditions of severe environmental pressures[9].In addition, they all illustrate that destructive human behavior towards nature is endangering human health – a stark reality that was ignored for decades. Research indicates that most emerging infectious diseases and zoonotic spillover are driven by human activities [10, 11].

The tourism sector in Morocco is one of the important contributors to the economy of the country [12]. According to the Moroccan higher planning commission [13], its contribution to the GDP is 6.8% with an estimate of MAD 72.4 billion in 2017. But with the spread of the Coronavirus in the world the tourism industry got highly impacted as most of the countries are in the lockdown [14, 15]. This has created a major downfall in the tourism sector including ecotourism [6].

In this paper, a general commentary is made on the effect of the Coronavirus on the ecotourism sector in Morocco and its impact on rural dependent communities and conservation particularly in protected areas. In fact, many international as well as domestic flights in Morocco are forced to be canceled, as a result of the measures taken by the Moroccan authorities as a response to COVID-19 outbreak. Consequently, ecotourism and related activities are deemed to be entering into a great crisis also establishing socio-economic and conservation concerns. Incomes generating activities that are tightly linked to ecotourism are most affected as tourists are not able to visit Morocco. Also, conservation is also seen to be facing unprecedented threats as is seen in this context. The aim of this study is to shed light on the impact of the occurrence of corona virus on ecotourism, on dependent communities and on conservation especially in key Morocco's biodiversity areas.

Ecotoursim as driver of incentive conservation

Globally, ecotourism results in significant positive changes in perceptions and behaviors of local communities, both towards protected areas and Biodiversity [16, 17]. In addition, many communities living near protected areas benefit from a share of tourism revenues. As the world emerges from the COVID-19 pandemic and undertakes the unprecedented task of repairing and stimulating national economies, investments in conservation could prove game-changing for people and nature in many poor countries that are nonetheless rich in biodiversity [18].

Incentive-based conservation provides employment, ecological services, alternative incomes, health care, education, tourism, and project development [19]. Such approaches are based on the assumption that benefits derived directly from natural resources will encourage conservation and sustainable use of those resources [20, 21]. Ecotourism has proven to improve conservation, awareness, attitudes, and increase pro-conservation behaviors [22, 23]. Community participation in conservation programs and the receipt of economic and social benefits may lead to improved conservation perceptions and approaches, which may, in turn, lead to pro-conservation attitudes [24, 25]. Reduction in negative behaviors (e.g. poaching) and an increase in positive behaviors (e.g. participation in conservation projects) can lead to positive conservation results (e.g. increased species densities and/or range; habitat improvement [26].

In Morocco, ecotourism is an effective means of enhancing protection by positively changing local perceptions, attitudes and behaviors towards wildlife and their natural habitats [27, 28]. It appears to play an important role in terms of the scale and scope of the conservation impacts [29]. However, although economic returns are important, they are not the main determinant of conservation outcomes and should not be the main metric used to assess conservation success of ecotourism sites[30, 24, 31]. Conservation practitioners are aware that even small scale ecotourism ventures can result in significant benefits for conservation [32, 33, 34].

Impact of COVID-19 on tourism in Morocco

In Morocco, tourism has experienced exceptional performances during the last few years[35]. Ambitious projects have increased the beds capacity; increased flights connections and recreational ground activities. Tourism is an important provider of foreign currency and holds an important place in the Moroccan economy with a contribution of about 7% to GDP, more than 1 million jobs and 8 billion US$ incomes in 2019[36]. Tourism is spreading throughout the faraway lands, benefiting from a surge in demand, the proximity of the markets that generate it and, above all, the impetus of the tourism strategy (Visions 2010 and 2020), which included sustainable tourism and nature tourism among its main guidelines[37].

Unfortunately, after this promising evolution during the two decades of the 2000s, the cascading effect of the COVID-19 pandemic is crippling the Moroccan tourism and hospitality industry. Taking drastic security measures, Morocco closed its borders, suspended flights, and trips before declaring the lockdown on march19th. Over a century tourism has never known such a catastrophic situation. The entire tourist chain is at a standstill and the entire activity is paralyzed. According to the National Tourism Confederation, the sector will have a shortage of 98% and about 3.2 billion Euros of loss [38].Tour operators, guides, transporters and restaurants... have seen their activities going to a sudden and abrupt halt.

In Marrakech, the first which concentrates 30% of hotels and more than 42% of tourism activity, the situation is dramatic. The car parks are crowded with vehicles, the hotels emptied, employees released, and services postponed. Elsewhere, in the hinterland and far from the traditional concentration touristy areas, the hiking trails are deserted as well as the lodges and guest houses. All the mountain guides, trek logistics providers lose their jobs and incomes. The considerable development of green tourism in Morocco is completely stopped. The efforts to save nature and raise awareness, which Morocco has deployed over the past decades, are abandoned for human security measures.

The decline in international tourism should continue until the end of the year. A situation that will depend on health and hygienic conditions in the world and the end of lockdown measures taken in the issuing countries, particularly in relation to the resumption of travel abroad. The situation in Morocco will have to suffer during a long period since the policies as well as the general panic will push tourists to focus on domestic travel and destinations nearby. Meanwhile, the debate and the adopted decisions are strictly economically and socially oriented. The main concerns of the government and the stakeholders are how to absorb the losses of months of paralysis and how to deal with the shortage. Efforts have thus focused on urgent social services to compensate for the losses of household incomes on the one hand and to allow tourism operators to deal with last-minute cancellations and their engagements during the lockdown. Although the confinement argues in favor of questioning the consumer society reality and its negative effects on the environment, reality shows a concentration on the economic impacts of COVID19 and less interest to sustainability, nature, and resources safeguarding considerations.

Impact of COVID-19 on wildlife and conservation in Morocco

To better analyze the situation, we looked for information in court minutes, crime statements written by the Department of Water and Forestry and credible reports illustrated with photos that conservation NGOs shared on their websites or on different social media networks.

Indeed, from the Atlantic Moroccan Sahara to the Atlas Mountains, the COVID-19 outbreak has leveled a devastating blow to efforts to protect endangered wildlife in Morocco. For instance, in the Argan *Arganaspinosa* woodland steppes of Amzi in Tiznit area as well as in the Blue Cedar *Cedrusatlantica* forests of Middle Atlas, conservation work to protect some of Morocco's most important ecosystems and flagship species is facing a crisis following a collapse in ecotourism during the pandemic lockdown. The conservation authorities and Wildlife protection NGO have recently expressed concerns about the increasing death toll and trafficking of endangered species in Morocco. Some poachers and traffickers posted pictures of themselves holding poached animals, which they caught in several regions, on their social media account. These acts are widely condemned across the country by the Forest Department, NGO and public opinion.

Amid the pandemic, scientists have repeatedly urged humanity to reset its relationship with nature or suffer worse outbreaks [39, 40, 41]. But the economic consequences of the COVID-19 lockdown have raised fears of a surge in poaching, illegal fishing, and deforestation in life-sustaining ecosystems, with thousands of jobs in the ecotourism sector and related activities at risk across the country. The national focus now is on protecting human lives in this devastating pandemic. However, in many biodiversity hotspots and protected areas, socioeconomic impact is generally observed, particularly in areas where communities rely heavily on ecotourism for their livelihoods. According to conservancies in Morocco, grass roots organizations that depend on ecotourism to fund projects for endangered species and key habitats could be forced to close, after border closures and worldwide travel restrictions abruptly halted millions of dollars of incomes.

While poaching a particular animal may not escalate owing to the current suppression of travels and negative sentiments against animal products, illegal wildlife trafficking and poaching will go up if there is nobody to monitor activities within the protected areas and outside. In Anti-atlas, the Vulnerable Cuvier's Gazelle *Gazellacuvieri*, an endemic antelope of the Maghreb ecoregion[42], washeavy poached between March and Mayfollowing the collapse of the local tourism sector, according to the local populations. The Endangered Barbary Macaques *Macacca Sylvanus*, another endemic species of Morocco and Algeria[43], poaching has recovered it pace in High Atlas, in Eastern Rif region and possibly elsewhere. In Moroccan forested areas, thousands of song passerines were illegally caught while these species are breeding or migrating. Throughout the country European Bee-eaters (*Meropsapiaster)* are being heavily persecuted and killed by beekeepers who took advantage of the country lockdown to enhance persecution practices. This is also true for some apex and Morocco's highly threatened predators such as the African Golden Wolf *Canisanthis*, Honey Badger *Mellovoracapensis* and Stripped Hyena *Hyenahyena*[44]and raptors (Accipitridae, Falconidae) whose persecution and poisoning exacerbate as depredation the reaction of farmers. Moreover, shepherds are taking advantage of the country lockdown to bring their cattle in the protected areas where vegetation cover is generally better than in unprotected areas due to overgrazing and drought[45].

With regard to threatened Antelopes such as Cuvier's Gazelles, Mhorr gazelle *Nangerdamamhor*, Addax *Addaxnasomaculatus*, and Barbary sheep *Ammotraguslevia*, it could be years before these species can fully recover, increasing the risk that people come to rely on other activities to make a living, putting unsustainable pressure on natural resources[46, 47]. Additionally, it is harder to monitor land grabbing and illegal poaching. While the poaching of several species has continued during the lockdown (Table. 1), a recent report found that the illegal wildlife trade had been severely disrupted by movement and travel restrictions in Morocco (Cherkaoui *obs. pers*). Nevertheless, conservationists fear an explosion of illegal hunting if organizations are forced to lay off wildlife wardens and suspend surveillance programs.

In fact, poaching and wildlife trafficking already existed on a national scale even before the coronavirus outbreak [48, 49, 50,51]. Nevertheless, with more people out of work, the wild animals will be more appealing than animals sold by licensed animal shops. Then the right question here is the following: if the wardens/guides have no salaries, how will they effectively monitor human activities in and out of the protected areas?"Members of these communities may lose faith in wildlife conservation if there is no money forthcoming. In addition, people who live around Morocco's protected areas and looked forward to offering guidance to tourists may resort to other income-generating activities such as farming and fueling the never-ending human-wildlife conflicts as animals invade and destroy their livestock or new farms.

At several places across the country, we have recorded a spike in poaching activities, with animals killed between March and April 2020. More importantly, several grassroots and conservation organizations have experienced delays in funding during the pandemic.While wardens are forced to stay at home, illegal land grabbing and intentional wildfires increased in many places. For instance, there's rampant deforestation in Taza, Middle-Atlas, High Atlas and Rif Mountains that rose sharply when the Moroccan government decreed health containment in early March of this year. The pandemic has not helped because there are apparently fewer patrols and rangers in the field. In the contrary, illegal loggers as well as poachers, who obviously do not care about coronavirus, saw their activity increase widely under the conjugated effect of protected areas surveillance drop.

Discussion

Sustainable and wildlife tourism is an increasing contributor to economic activity around the world.The global benefits of conservation from ecotourism have been well documented [52]. Community-based conservation areas in particular provide income support for families through a share in tourism-derived income. Also, the mere existence of species-rich landscapes provides the public opportunities to visit the national parks and preserves [53]. A number of ecosystem services are provided by global biodiversity [54, 55]. Indeed, vast quantities of carbon stored in trees and soils act as a natural hedge against climate change [56]. Existing and potential use of natural products for medicinal purposes is huge [57]. Further, wild species, from the largest to the microscopic, interact in ways that keep up delicate-running natural cycles-whether the provision of fresh water or fertile soil - that make Planet Earth habitable for humans [58]. Biodiversity provides transboundary benefits, and that not all countries can meet the costs of its conservation, it is in the interests of the world to share in the costs [58, 60].

Success would depend on investments that simultaneously reinforce the impact of conservation and mitigate or reduce wildlife trafficking and poaching activities; providing capital preservation and/or returns on investments and generating cash flows through sustainable use of nature by local communities [60]. In addition tothis, both to protect wild animals and to reduce the possibility of a new pandemic like COVID-19, scientists should emphasize the importance of stopping illegal wildlife trafficking, closing wildlife markets, and protecting natural habitats.

In Morocco, protected areas are generally underfunded, resulting in insufficient funds to support communities living closest to key ecosystems and nature [62]. As a result, communities often pursue land-use practices that provide for their immediate needs, but which may undermine wildlife conservation [63]. However, when local communities have enough support, they look after the land and its biodiversity [64]. This is apparent in Morocco's national parks, reserves, and other country's biodiversity hotspot areas[62, 63]. Moreover, the ecotourism activities in Morocco dropped significantly during the Covid-19 lockdown which affected the generating income activities, seasonal employments and green jobs provided by this sector as the tourism companies associated with outbound tourism faces severe loss. Part of rural economy inside and nearby protected areas is largely reliant on the ecotourism sector and every year a significant amount of global tourists come to visit the important biodiversity destinations in Morocco [65].

In business, COVID-19 has prompted increased discussion of the importance of diversified incomes and resilient supply chains. Therefore, a fundamental lesson from the current outbreak situation is that conservancies should not overly reliant on one source of income. Development of inclusive finance models such as small loans for rural communities can be a further option to promote diversified sustainable livelihood opportunities in local economies, minimizing the risk of relying on a specific sector or resource. Besides, wildlife protection in Morocco should be from now onward everybody's best interest for sustainable future and resilient rural communities and landscapes. Currently, there are even more pressing reasons to invest in communities that protect nature through wildlife tourism and conservation. COVID-19 recovery packages must include this investment in public policy. Nature underpins people's survival, wellbeing and, sustainable development.

Moreover, ecotourism will undergo great changes as it is the case for all aspects of human life. As we can already see, hygienic concerns are surfacing and should help to reshape the face of the trip by taking into consideration parameters such as social distancing, cleanliness, respect for other tourists and destinations. The conservative approach will undoubtedly be enriched with new measures and should make it possible to better protect species and resources more. General awareness should be established and the concerns about the well-being of the population and the species will gain more and more.The challenge is to be able to manage this period of emptiness between confinement and the start of travel, which will last a long time. The different ecotourism and conservationist actors must mobilize to manage several facets of the crisis.

Conclusion and perspectives

In the light of COVID-19 outbreak and its impact on Ecotourism sector and biodiversity conservation in Morocco, it is crucial to anticipate on the post-crisis situation. While many revival plans are now being considered by the various tourism organizations, conservation authorities in the country as well as NGOs and communities working in ecotourism must be more proactive and think about a real plan for the transformation of Moroccan tourism which has to take into account the ecological and environmental stakes. The challenge is no longer to seek only the growth of tourism only, but to seek the development of ecotourism with positive impacts on the environment and conservation. It will be a question of finding a balance between the preservation of ecosystems and economic realities.

The COVID-19 outbreak is an opportunity to review priorities and rethink tourism policies for the development of a new tourism model by mainstreaming ecotourism component for sustainability education and conservation. We suggest the following measures to be taken as urgently as possible:

1) To reflect on a new plan for the transformation of tourism by capitalizing on the unprecedented experiences of eco-tourism as a solution the for climatic and health crises;
2) To support rural tourism as a strategy that helps to anchor rural communities in their territory and thus promote the socio-economic development of the hinterland;
3) To promote nature-based solutions;

4) To raise more fund and support to conservation projects and grassroots communities that might favor combating poaching, overgrazing and habitat destruction through a win-win situation;
5) To accelerate the processes of classification of new protected areas or the extension of the surface of the existing ones;
6) To create a new participatory approach, integrating the local population protected areas and natural resources co-management;
7) To promote environmental-friendly agricultural practices such as permaculture, agro-tourism, and organic farming;
8) To Set up a financial contribution (Green tax) on tourism products (transport, catering, accommodation, etc.) to finance ecotourism activities and conservation;
9) To support tourism professionals and stakeholders to develop sustainable tourism packages;
10) To put in place adapted and easily accessible financing tools to enable the resilience of sustainable tourism professionals (accommodation providers, guides, producers, etc.) in the event of a crisis and outbreaks,
11) To maintain or, ideally, increase the, increasing ranger numbers will be critical to ensure that protected areas and valuable species are adequately protected;
12) To promote environmental education in order to inculcate in future generations the economic, social, cultural and environmental importance of nature conservation; and
13) To make an evaluation of public policies on the basis of species and nature conservation indicators.

This being said, stopping natural habitat destruction, poaching, illegal possession and illegal trade of wildlife has become, more than ever, a global priority to prevent zoonosis, such as the virus causing COVID19 or other diseases spillover between wild animals and humans. Moreover, it is crucial to ensure the balance of ecosystems, halt the loss of biodiversity and increase the capacity of humankind to cope with and adapt to climate change.

References

1. World travel tourism council (WTTC), Travel and tourism economic impact, March, (WTTC report, 2019a)
2. Word Bank Group (WBG), Supporting Sustainable Livelihoods Public Disclosure Authorized through Wildlife Tourism. Knowledge Series, Tourism for Development, (Washington, DC 20433, USA, 2018)
3. World travel tourism council (WTTC), The economic impact of global wildlife tourism, (WTTC report, 2019b)
4. S. Farber, R. Costanza, D. Childers, J. Erickson, K.Gross, M. Grove, et al,Bio. Sci.,**56** (2006)
5. A. Stronza, A. C. Hunt, A. C. and Fitzgerald, L. A Ann. Rev. Environ. Res. **44** (2019)
6. World travel tourism council (WTTC), Only the G20 can drive forward a coordinated recovery response to the COVID-19 crisis says WTTC (2020)
7. World Economic Forum (WEF), Nature Risk Rising: Why the Crisis Engulfing Nature Matters for Business and the Economy. Geneva Switzerland(2020)
8. K.E. Jones, N.G.Patel, M.A.Levy, A.Storeygard, D.Balk, J.L.Gittleman, P.Daszak,.Nature**451** (2008)
9. J.A. Patz, P. Daszak, G.M. Tabor, A.A. Aguirre, M. Pearl, J. Epstein, , E. Disease, Env.Heal.Persp., **112** (2004)
10. R.K. Plowright, C.Parrish, H.McCallum, P.J.Hudson, A.Ko, A.Graham, J.Loyd-Smith, Nat. Rev.Microb., **15**(8): 502–510(2017)
11. C.L.Faust, , H.I.McCallum, L.S.P. Bloomfield, N.Gottdenker, T.R.Gillespie, C.J.Torney, A.P.Dobson, R.K. Plowright, Ecol. Lett., **21** (2018)
12. K. Harrou, Tableau de bord sectorial de l'économie marocaine, Ministère de l'Economie et des Finance. Rabat, Morocco (2019)
13. Haut Commissariat au Plan (HCP) www.hcp.ma Site institutionnel du Haut-Commissariat au Plan. Site de l'Observatoire du Tourisme Baromètres de l'activité touristique (2017)
14. M. Chinazzi, J. T. Davis, M. Ajelli, C. Gioannini, M. Litvinova, S. Merler, C. Viboud, Science. **368** (2020)
15. Daily Ft. Tourist arrivals fall 17.7% in Feb. Retrieved from aily FT. 2020. Tourism industry meets to assess COVID-19 impact, way forward (2020)
16. N.M. Ardoin, M.Wheaton, A.W.Bowers, C.A.Hunt, W.H.Durham, J. Sust. Tour.**23** (2015)
17. R.Buckley, Ecological indicators of tourist impacts in parks. J. of Ecot., **2** (2003b)
18. UNTWTO (2020)
19. A.Spiteri, S.K.Nepal, Envi.Manag.**37**, 1-14 (2006).
20. B. B. Bynum, G. T.Green, Ecotourism and natural resource conservation: the 'potential' for a sustainable symbiotic relationship, J. of Ecot. (2015)
21. A. Stronza,The economic promise of ecotourism for conservation. J. Ecot.6(3):210–30 (2007)
22. R.C. Buckley, J.G. Castley, deVasconcellos, F. Pegas, A.C. Mossaz, R. Steven, Plos One **7** (2012)
23. J.Coria, E.Calfucura. Ecol. Econ.**73**:47–55 (2012)
24. C.J. Stem, J.P. Lassoie, D.R.Lee, D.D.Deshler, J.W.Schelhas, Soc. Nat. Resour. **16** (2003)
25. F.V. Pegas, A.Coghlan, A. Stronza, V. Rocha, J..Ecot.**12**(2013)

26. C.M.Holmes, Oryx,**37**:305-315 (2003)
27. F.ARIB, Téoros,**24** (1), 37-41.(2005)
28. Y.Lamnadi, Signs of hope or wrong way? The implementation of ecotourism as a touristic product within Morocco destination. In: Blàzquez, M., Mir-Gual, M., Murray, I. y Pons, G.X. (eds.). Turismo y crisis, turismocolaborativo y ecoturismo. XV Coloquio de GeografíadelTurismo, el Ocio y la Recreación de la AGE. Mon. Soc. Hist. Nat. Balears, 23: 357-367. SHNB-UIBAGE. ISBN 978-84-617-5115-0 (2016)
29. L.Sbai, Evaluation du cadre juridique et institutionnel relatif à l'écotourisme et aux aires protégées au Maroc. Rapport UICN (2012)
30. A.Blamford, A. Bruner, P. Cooper, R. Costanza, S. Farber, R. Green, et al,.Science,**297** (2002)
31. T.P. Romero-Brito, R.C. Buckley, J. Byrne, Plos One**11** (2016)
32. K. Lindberg, Economic aspects of ecotourism. In K. Lindberg, M.E. Wood, and D. Engeldrum (eds). Ecotourism: a guide for planners and managers. Volume 2. Washington, D.C. The Ecotourism Society. 87-117. (1998)
33. H. K.Cordell, M.Bowker,The global economic contribution of protected natural lands and wilderness through tourism. (The Wild Planet Project, May 2007, 2007)
34. C.A.Hunt, W.H.Durham, L.Driscoll, M.Honey, J. Sustain. Tour. **23(3)**:339–57 (2015)
35. I. Aberrhaze, *Tourisme et artisanat du 03 au 07 Décembre 2018*. Portail du Centre National de Documentation du Maroc (2019), Consulté le 22 octobre 2019.
36. Ministère du tourisme, Tourisme en chiffres (2020) www.tourism.gov.ma
37. M. Boukherouk, A.Tlidi, Tourisme, in Noureddine El Aoufi et Bernard Billaudot (Dir), Made in Maroc Made in Monde, Profils sectoriels et émergence industrielle, Economie critique, 2019, Rabat Volume **3p** 343-380 (2019)
38. B. Berrisoule, Coronavirus - Tourisme : Les professionnels comptent leurs pertes, l'Economiste, Edition N°:5721 Le 18/03/2020 (2020) |
39. World Health Organzation (WHO) Managing Epidemics: Key facts about major deadly diseases," World Health Organization (2018). P. M. White, others, spending at least 120minutes a week in nature is associated with good health and wellbeing," Scientific Reports 9 (2019):
40. C. K. Johnson, others, Proc. R. Soc. B. **287** (2020)
41. IUCN SSC Antelope Specialist Group. *Gazellacuvieri*. The IUCN Red List of Threatened Species 2016: e.T8967A50186003 (2016)
42. T.M. Butynski, J. Cortes, S. Waters, J. Fa, M.E. Hobbelink, E. van Lavieren, F. Belbachir, F. Cuzin, K. de Smet, M. Mouna, H. de Iongh, N. Menard, A.Camperio-Ciani, *Macacasylvanus*. The IUCN Red List of Threatened Species 2008: e.T12561A3359140 (2008)
43. S. Aulagnier, F. Cuzin, M. Thévenot (Eds.). 2017 *Mammifères sauvages du Maroc. Peuplement, répartition, écologie*, Société française pour létude et protection des mammifères, Paris (2017)
44. S.I. Cherkaoui, S. Selmi, Z. Amhaouch S. Hanane, J. of Env. Ass. **190** (2018)
45. F.Cuzin, E. Cuzin, A. Sehhar, T. Wacher, *Etude pour l'élaboration de lignes directrices et d'un plan d'action stratégique pour la conservation des ongulés au Maroc. Vol. 1, rapport principal*. Haut-Commissariat aux Eaux et Forêts et à la Lutte Contre la Désertification (HCEFLCD), Projet de Gestions des Aires Protégées (PGAP) et Banque Mondiale, Global Environment Facility (GEF) (2007)
46. T. Abaigar, E.Rodríguez-Caballero,C. Martinez,Z Amhaouch,L. M. Samlali,F. Aparicio,T. EL Bella, A. Essalhi,J. Fernandez,J. Garcia,J. Haya,A. M'Bareck,H. M'Bareck,M. L. Gonzales, F. de Larrínoac. Glob. Ecol. and Cons. **19**(2019)
47. A. Shipp, Traff. Bull.**19**: 65 (2002)
48. E. van Lavieren, .Traff. Bull. **21** (2008)
49. E. Martin,C. Perry-Martin, *Swara* (Jul-Sep): 16-29 (2012)
50. D. Bergin, V.Nijman, Ongoing, *Traff. Bull.***26** (2014)
51. C.Wardle, R.Buckley, A. Shakeela, and J.G.Castley, Journ. of Ecot. ((2018)
52. J.J.VASKE, *Survey Research and Analysis: Applications in Parks, Recreation and Human Dimensions*. Venture Publishing, State College, USA (2008)
53. B.J.Anderson, P.R.Armsworth, F.Eigenbrod,*J. Appl. of Ecol.***46** (2009)
54. S.Quijas, P.Balvanera,Biodiversity and Ecosystem Service. In: Levin S.A. (ed.) Encyclopedia of Biodiversity, second edition, Volume **1**, pp. 341-356. Waltham, MA: Academic Press.52013°
55. F.de Bello, S. Lavorel, S. Diaz, et al,Biod. and Conses.**19** (2010)
56. B.J.Cardinale, K.L.Matulich, D.U.Hooper,et al..Amer. J. Bot.**98** (2011)
57. P.Balvanera, A.B Pfisterer,N.Buchmann,Ecol. Lett.**9**: 1146–1156 (2006)
58. R. Costanza, R. d'Arge, R. De Groot, S.Farber, M. Grasso, B. Hannon, et al, Natur*e*, **387**, 253-260 (1997)
59. D.Li Yong, A.Jain, , Y.Liu, , M.Iqbal, C.Y.Choi, N.J.Crockford, S.Millington, J.Provencher, Cons. Biol., **32** (2018)
60. C.G. Sandbrook, Conserv. Lett. **3** (2010)

61. M. Ribi, International Workshop on sustainable tourism in Moroccan Protected Areas, Waters and Forests Department. pp. 56 (2007)
62. Global Environment Facility (GEF), Morocco Protected Areas -Management Project. The World Bank 19917. Washington, USA. (2010)
63. M. El-BekkaY, A. Moukrim, F. Benchakroun, AJEST, Vol. 7(6), pp. 441-447 (2013)
64. N. Roudiad,Vision 2020 for tourism in Morocco Focus on Sustainability and Ecotourism. Expert Group Meeting on Ecotourism, Poverty Reduction & Environmental Protection. Rabat, Morocco (2013)

Table. 1. List of wildlife suffering from poaching and trafficking toll rise as reported from conservation groups in Morocco.Data collected between 01 March and 31 May 2020

Species or group of species	**IUCN Status**	**Protected**	**Region in Morocco**	**Numberreported**
Macacca Sylvanus Barbary Macaques	**EN**	Yes	N.E, H.A	3
*Hyaenahyaena*Stripped hyena	**NT**	Yes	N.E	2
Canisanthus African golden wolf	LC	No	A.S; S.E	13
Genettagenetta Common Genet	LC	No	All	18
Mellivoracapensis Honey Badger	LC	Yes	S.A	1
Herpestes ichneumon Egyptian Mongoose	LC	No	All	22
*Mustelanivalis*Least weasel	LC	No	All	10
Vulpeszerda Fennec Fox	LC	Yes	S.E	3
*Vulpesvulpes*Red Fox	LC	No	All	> 50
*Felislybica*Lybian wild Cat	LC	No	S.A	3
Gazellacuvieri Cuvier's Gazelle	**VU**	Yes	A.A	11
dorcas gazelle Gazelladorcas	**VU**	Yes	A.S	2
Ammotraguslevia Barbary Aoudad	**VU**	Yes	S.E	3
*Lepus mediterraneus*Maghreb Hare	LC	No	All	> 500
*Atelerixalgirus*Algerian hedgehog	LC	Yes	All	> 500
Chiroptera		Yes	All	> 500
*Atlantoxerusgetilus*Atlas ground squirrel	LC	No	C.A	> 1000
*Testudo graeca*Common Tortoise	**VU**	Yes	All	> 1000
*Uromastyxnigriventris/Uromastyxacanthinura*Moroccan spiny-tailed lizard	NA	Yes	S.E	> 500
*Chamaeleochamaeleon*Common Chameleon	LC	Yes	All	> 500
Najahaje Egyptian cobra	NA	Yes	A.S	6
*Bitisarietans*Puff adder	LC	Yes	C.A, A.S	8
Varanusgriseus Desert Monitor	LC	Yes	A.S, S.E	4
*Bubo ascalaphus*Pharaoh Eagle-Owl	LC	Yes	N.E, S.E, A.S	12
Buteocirtensis Maghreb Buzzard	LC	Yes	All	7
*Falco biarnicus*Lanner Falcon	LC	Yes	S.A	2
Other raptors		Yes	All	> 50
Alectoris Barbara Barbary Partridge	LC	No	All	> 100
StreptopeliaturturTurtle dove	**VU**	Yes	All	> 500
*Upupaepopos*Common Hoopoe	LC	Yes	All	> 500
*Meropsapiaster*European Bee-eater	LC	No	All	> 1000
Golden finch*Cardualiscardualis*	LC	Yes	All	> 1000
Other song birds			All	> 1000

A.A= Anti-tlas; AS= Atlantic Moroccan Sahara; C.A= Central Atlantic; EN= Endangered, HA= High Atlas; IUCN= International Union for Conservation of Nature; LC= Least Concern; M.A= Middle Atlas; NA= Not Assessed; N.E= North-Eastern; NT=Near Threatened; SE= South Eastern; VU= Vulnerable,

Impact of climate change on the phytoplankton community of the Aglou-Sidi Ifni coastline (Moroccan Atlantic coast)

El mahjoub Gallouli[1]*, *Hassan* Oulad Ali[1], and *Ahmed* Aamiri[1]

[1] Biodiversity and Ecosystem Functioning Team, Ibn Zohr University, Faculty of Sciences, Department of Biology, Agadir, Morocco.

Abstract. The objective of this study is to understand the evolution of the typologies (physicochemical and phytoplankton) in Aglou-Sidi Ifni coastline according to hydrometeorological and climatic hazards. To achieve this objective, monitoring of the phytoplankton community and water physicochemical parameters was conducted at three stations along this coastline and during the winter and fall seasons of two different annual cycles (2014-2016). The first cycle experienced heavy flooding while the second cycle experienced severe drought. This study revealed spatial and temporal variability in phytoplankton composition and the relative abundance of its major groups. Temporal variability may be due to changes in the physicochemical characteristics of water masses as a function of continental inputs and wave intensity, which differ from cycle to cycle and season to another. Whereas the spatial variability would be due to the modification of these physicochemical characteristics from one station to another under the influence of the morphology of the coast in this zone, which curves westwards and carries Cape Aglou at its northern end, thus making the station of Sidi Ifni more exposed to currents than the other two stations which are somewhat sheltered.

1 Introduction

Phytoplankton is a major component of aquatic ecosystems in general and oceanic ecosystems in particular [1]. It forms the first link in most pelagic chains and thus conditions the richness of the waters and the resource. Phytoplankton production by photosynthesis is under the control of limiting physicochemical factors such as light and nutrients (nitrogen, phosphorus, silica, etc.), whose availability is largely linked to the state of the water column (water mass mixing, exchanges with the continental environment, rainfall, winds, turbidity, temperature) [2]. At the regional scale, these systems are sensitive to climatic forcing through the concomitant action of pressure fields, wind intensity and direction, temperature and precipitation on the variability of physicochemical elements in the water column [3]. The sensitivity of these organisms to changes in environmental factors is high and small fluctuations in these parameters can have pronounced effects on

* Corresponding author: melgallouli@yahoo.fr

the structure of marine phytoplankton communities [4], and consequently on the other links in the food webs of marine ecosystems. The Aglou-Sidi Ifni coastline of the Moroccan Atlantic coast is characterized by significant fishery richness and a natural deposit of mussels and is home to one of the three first pilot marine protected areas in Morocco. It is the Massa Marine Protected Area (AMP-Massa). This coastline of the Guelmim-Oued Noun region belongs to the arid bioclimatic stage [5]. The climate of this region is predominantly Mediterranean and arid, with a rare rainfall (100-200 mm per year), concentrated mainly in winter and autumn. However, over the last 20 years, the Guelmim-Oued Noun region has experienced a disruption of seasonal rainfall and a clear warming trend of around 0.7 °C [6]. The rainfall recorded during the 2014-2015 hydrological year generated significant floods and runoffs, both in terms of volume and maximum flow. On the other hand, the hydrological year 2015-2016 was dry with a temperate winter and without continental input for the Aglou-Sidi Ifni coastline. This work was carried out during the autumn and winter seasons of these two cycles and focuses on the study of the impact of these changes on the phytoplankton community of this coastline.

2 Materials and methods

2.1 Sampling

The water samples used for the quantitative study of phytoplankton and for the determination of nutrient salts were taken, using a two-litre Niskin bottle, one metre from the surface. While the phytoplankton samples for the qualitative phytoplankton study was collected using plankton net. These samplings were carried out during the autumn and winter periods of two annual cycles (April 2014-March 2016). Three stations of the Aglou-Sidi Ifni coastline (Fig. 1) were surveyed regularly during this period with a bi-monthly step: Aglou (29°48'36"N - 09°54'34"W), Sidi Boulfdail (29°42'38"N - 09°59'58"W) and Sidi Ifni (29°23'19"N - 10°11'53"W) , the positions of these stations were validated by Global Position System (GPS).

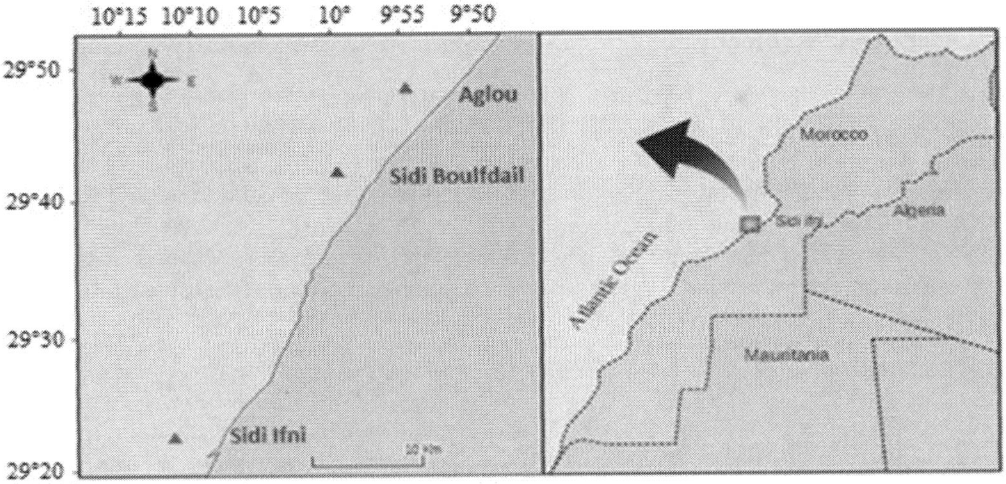

Fig. 1. Map of sampling stations.

2.2 Physicochemical parameters

2.2.1 In situ measurements

Water temperature, salinity and dissolved oxygen were measured in situ at the surface using a multi-parameter instrument (WTW Multiline P4). Transparency was measured in metres using an immersed Secchi disc at each station.

2.2.2 Analyses carried out in laboratory

The elements analysed in laboratory are: nitrites ($N\text{-}NO_2^-$), nitrates ($N\text{-}NO_3^-$), ammoniacal nitrogen ($N\text{-}NH_4^+$) and orthophosphate ions ($P\text{-}PO_4^{3-}$). These analyses were performed using an absorption spectrophotometer according to the nutrient analysis protocol described in the Manual of Marine Chemical Analysis [7]. Chlorophyll "a" was determined by measuring the absorbance of the extract with a spectrophotometer at wavelengths of 665 nm and 750 nm before and after acidification by the monochromatic method [8].

2.3 Phytoplankton

Biological samples are immediately fixed with Lugol and preserved with 5 % formaldehyde. In the laboratory, the determination and enumeration of phytoplankton is carried out according to the Utermöhl method [9]. Individuals of each species are counted in 10 ml sedimentation tanks, under an inverted microscope of the ULWCD 0.30 OLYMPUS type (G: 100; 400), after the sample has been decanted for at least 6 h [10], according to a method inspired by Hasel and described by Sournia [11]. The specific identification of phytoplankton is based on visual observation of the general characteristics and remarkable morphological attributes of the cells and using appropriate identification keys [12-17]. The final result of the count in the sedimentation chamber is converted to a concentration in number of cells per litre (Cell/l).

2.4 Statistical treatment

In order to determine the impact of floods and droughts as well as the morphology of the coast on the spatial and temporal variability of phytoplankton, a PCA (Principal Component Analysis) was carried out, using the R software [18], from the data matrix of this study. The variables considered in this statistical study are the above-mentioned physicochemical parameters, the concentration of chlorophyll "a" and cell densities of total phytoplankton and its two main groups (diatoms and dinoflagellates) in samples collected at the three stations during the winter and fall seasons of the two annual cycles of this study.

3 Results

3.1 Physicochemical parameters

The evolution of the water temperature is almost the same in the three stations with high values during the autumn periods and low values during the winter periods. The minimum temperature values were recorded on February 2015 with 16.4 °C at Aglou, 16.1 °C at Sidi Boulfdail and 16 °C at Sidi Ifni. The maximum values were recorded on October 2015 with 21.8 °C, 21.5 °C and 21 °C respectively. The Sidi Ifni station is cooler than the other two stations (Fig. 2). The values recorded during the drought period (A15-W16) are higher than those recorded during the flood period (A14-W15).

The analysis of variance revealed a very highly significant difference ($p < 0.001$) between the seasonal averages of water temperature at each of the three stations along the Aglou-Sidi Ifni coastline. In addition, Tukey's test of multiple comparisons of these averages disclosed a spatiotemporal heterogeneity in the temperature of the coastal waters of Aglou-Sidi Ifni coastline.

The low salinity values of the waters of the Aglou Sidi Ifni coast were recorded on December 2015 (35.45 g/l in Aglou, 35.47 g/l in Sidi Boulfdail and 35.44 g/l in Sidi Ifni) and the maximum values were recorded in November 2015 in the Sidi Boulfdail station (35.65 g/l) and on October 2015 in the Aglou (35.65 g/l) and Sidi Ifni (35.62 g/l) stations. The salinities recorded during the drought period seasons are higher than those recorded during the same seasons of the flood period (Fig. 2).

The analysis of variance revealed a very highly significant difference ($p < 0.001$) between the seasonal averages of salinity recorded in the three stations of the Aglou-Sidi Ifni coast.

The evolution of dissolved oxygen concentration is similar at the three stations of the Aglou-Sidi Ifni coast with concentrations greater than or equal to 8 mg/l in winter 2015 and concentrations less than 8 mg/l and greater than 6.5 mg/l during the other periods of this study. The dissolved oxygen concentrations recorded during the seasons of flood period are higher than those recorded during the same seasons of the drought period (Fig. 2).

Analysis of the variance of dissolved oxygen in the waters of the Aglou-Sidi Ifni coastline divulged a very highly significant seasonal variability ($p < 0.001$) of this parameter at the three stations of this coastline.

The maximum transparency (14.5 m) was recorded at the Aglou and Sidi Boulfdail stations in winter 2016 and the minimum transparency (6 m) was recorded at the Sidi Ifni station in autumn 2014.

The highest seasonal averages of water transparency were recorded in winter 2016 and the lowest in autumn 2014 (Fig. 2). The analysis of variance divulged a highly significant difference between the seasonal averages of transparency in the Sidi Ifni station ($p < 0.01$). However, this difference is non-significant in the Aglou and Sidi Boulfdail stations ($p > 0.05$).

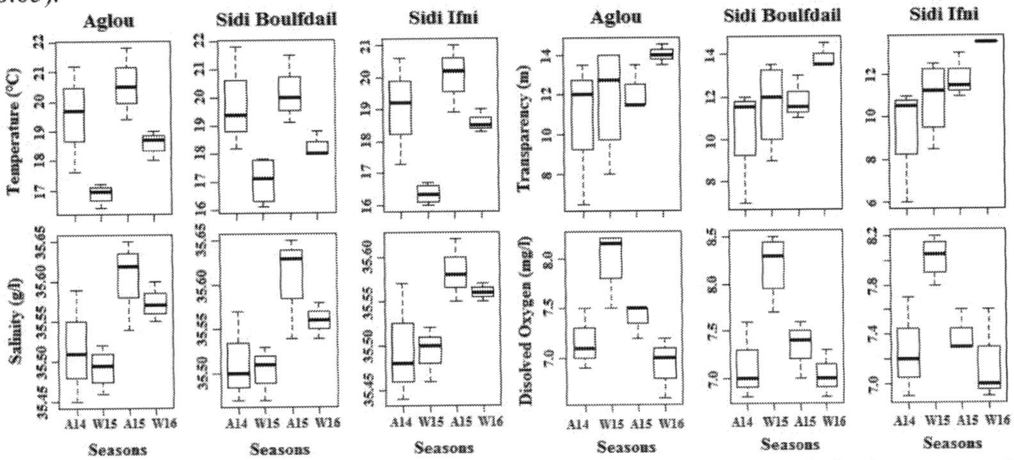

Fig. 2. Seasonal and inter-annual variability of the temperature, salinity, disolved oxygen and transparency in the three stations of the Aglou-Sidi Ifni coastline. The seasons are represented by their initials (A: autumn, W: winter) followed by two digits that indicat the year.

High nutrient concentrations were recorded in winter 2015 with 391 µg/l $N\text{-}NO_3^-$, 36.34 µg/l $N\text{-}NO_2^-$, 10.26 µg/l $N\text{-}NH_4^+$ and 152 µg/l $P\text{-}PO_4^{3-}$.While the lowest concentrations were recorded in winter 2016 with 186 µg/l, 19.32 µg/l, 5.4 µg/l and 35.15 µg/l respectively (Fig. 3) .

Analysis of variance revealed a very highly significant difference ($p < 0.001$) between the seasonal averages of $N\text{-}NO_2^-$ and $N\text{-}NH_4^+$ concentrations recorded in the three stations. A highly significant difference ($p < 0.01$) was found between the seasonal means of $N\text{-}NO_3^-$ concentrations in each station. While the difference between the seasonal averages for $P\text{-}PO_4^{3-}$ was significant ($p < 0.05$) in Aglou station and non-significant ($p > 0.05$) at the other two stations.

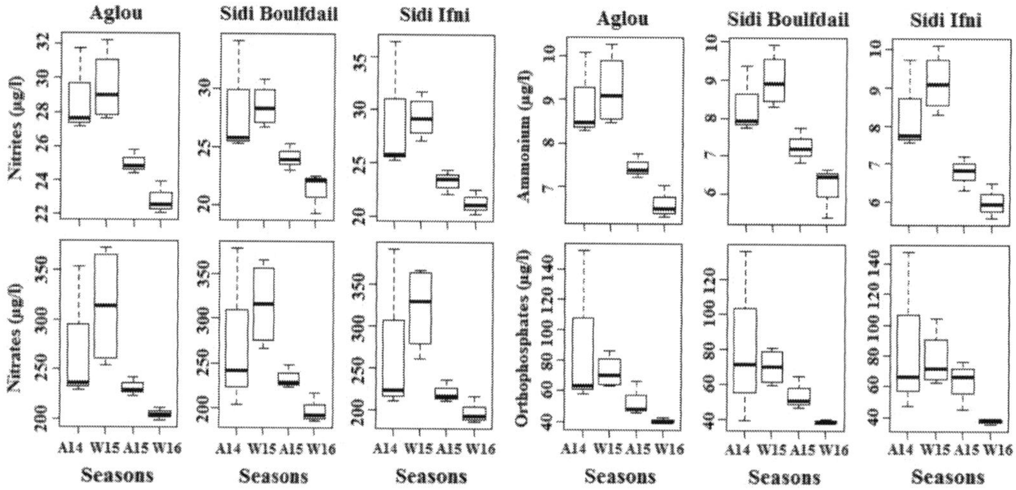

Fig. 3. Seasonal and inter-annual variability of the nitrates, nitrites, ammonium and orthophospates in the three stations of the Aglou-Sidi Ifni coastline. The seasons are represented by their initials (A: autumn, W: winter) followed by two digits that indicat the year.

3.2 Phytoplankton

3.2.1 Specific Richness (SR) and Chlorophyll "a"

This study disclosed significant phytoplankton diversity (233 taxa) within the Aglou-Sidi Ifni coastline. The number of taxa recorded in the Aglou-Sidi Ifni coastline is higher than that recorded in the majority of sites on the Moroccan Atlantic coast [19, 20].

Diatoms and dinoflagellates are the main phytoplankton groups in this ecosystem. These two groups represent respectively 54 % and 42 % of the phytoplankton species richness in this zone.

Seasonal averages of phytoplankton specific richness recorded during the drought period (A15-W16) exceed those recorded during the flood period (A14-W15) (Fig. 4).

The maximum seasonal averages for SR of phytoplankton were recorded in winter 2016 with 54 ± 2.08 species in the Aglou station, 49 ± 2.08 species in the Sidi Boulfdail station and 50 ± 2.65 species in the Sidi Ifni station.

The analysis of variance revealed a significant difference ($p < 0.05$) between the seasonal means of the diatoms and dinoflagellates species richness at the three stations. This analysis did not reveal significant differences ($p > 0.05$) between seasonal averages of total phytoplankton species richness.

The average chlorophyll "a" concentrations recorded at the three stations during the dry period seasons (A15-W16) are higher than those recorded at these stations during the same flood period seasons (A14-W15). Maximum concentrations were recorded in the autumn of 2015 and minimum concentrations in the winter of 2015 (Fig. 4).

The maximum seasonal means for chlorophyll "a" were recorded in autumn 2015 with 0.90 ± 0.06 µg/l in Aglou, 0.91 ± 0.06 µg/l in Sidi Boulfdail and 0.87 ± 0.05 µg/l in Sidi Ifni. While the minimum seasonal averages were recorded in winter 2015 in Aglou and Sidi Boufdail stations with 0.72 ± 0.14 µg/l and 0.73 ± 0.11 µg/l respectively and in autumn 2014 in the Sidi Ifni station with 0.70 ± 0.09 µg/l.

The analysis of variance did not reveal any significant difference ($p > 0.05$) between the seasonal means of chlorophyll "a" concentration in the three stations.

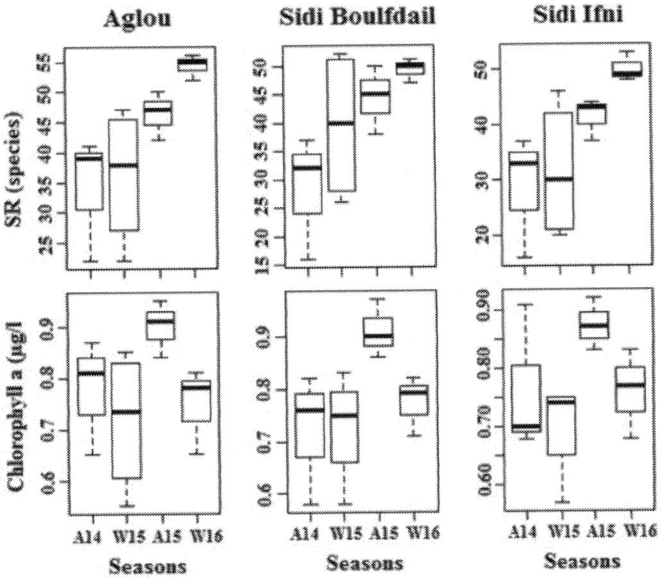

Fig. 4. Seasonal and interannual variability of the species richness (SR) and chlorophyll "a" in the three stations of the Aglou-Sidi Ifni coastline. The seasons are represented by their initials (A: autumn, W: winter) followed by two numbers indicating the year .

3.2.2 Relative abundance of different phytoplankton groups

The seasonal averages of total phytoplankton abundance recorded during the flood period (8.10^3 cell/l to 12.10^3 cell/l) are lower than those recorded during the drought period (32.10^3 cell/l to 52.10^3 cell/l).

Analysis of variance divulged a difference between seasonal averages of phytoplankton abundances in each of the three stations. This difference is very highly significant in the Aglou ($F = 9.97$, $p = 1.84 .10^{-5}$) and Sidi Ifni ($F = 5.028$, $p = 0.00067$) stations and highly significant in the Sidi Boulfdail station ($F = 5.909$, $p = 0.00181$).

The relative abundance of diatoms is higher during the fall and winter seasons of the flood year and exceeds that of the other phytoplankton groups. While the relative abundance of dinoflagellates is greatest during the fall and winter seasons of the drought year and exceeds that of other phytoplankton groups (Fig. 5).

The difference between the seasonal averages of relative diatom abundance is highly significant ($p < 0.01$) in the Aglou and Sidi Boulfdail stations and significant ($p < 0.05$) in the Sid Ifni station, while the difference between the seasonal means of the relative abundance of dinoflagellates is very highly significant ($p < 0.001$) in the Sidi Boulfdail station, highly significant ($p < 0.01$) in the Aglou station and significant ($p < 0.05$) in the Sidi Ifni station.

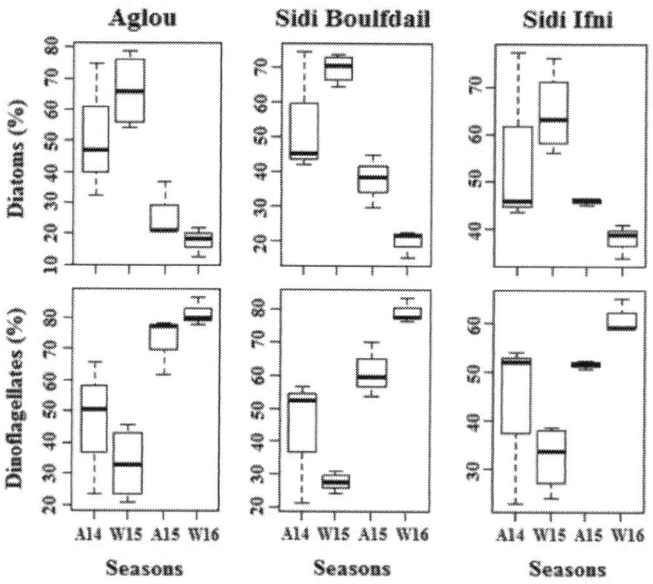

Fig. 5. Seasonal and inter-annual variation of the relative abundance of diatoms and dinoflagellates in the three stations of Aglou-Sidi Ifni littoral. The seasons are represented by their initials (A: autumn, W: winter) followed by two digits that indicate the year.

3.3 Physicochemical and phytoplankton typology of the Aglou-Sidi Ifni coastline

In order to visualize the relationship between climate change and the physicochemical and phytoplankton typology of the Aglou-Sidi Ifni coastline, a standardized PCA was carried of a data matrix containing the measurements of the above-mentioned environmental variables, the concentration of chlorophyll "a", the abundance of the main phytoplankton groups (diatoms and dinoflagellates) and the abundance and specific richness of the total phytoplankton.

The factorial design F1*F2 represents a total inertia of 78.23 % (Fig. 6). Salinity, water transparency, dinoflagellates density and phytoplankton species richness are positively correlated with each other and with the F1 axis. Nutrient salt concentration and diatom abundance are positively correlated with each other and negatively correlated with the F1 axis (Fig. 6).

Total phytoplankton density and temperature contribute strongly to the formation of the factorial axis F2 and are positively correlated with each other and with this axis. Water transparency and phytoplankton species richness are negatively correlated with the second factorial axis (Fig. 6).

The concentration of chlorophyll "a" is positively correlated with the two factorial axes and thus contributes to the formation of the two factorial axes of this PCA (Fig. 6).

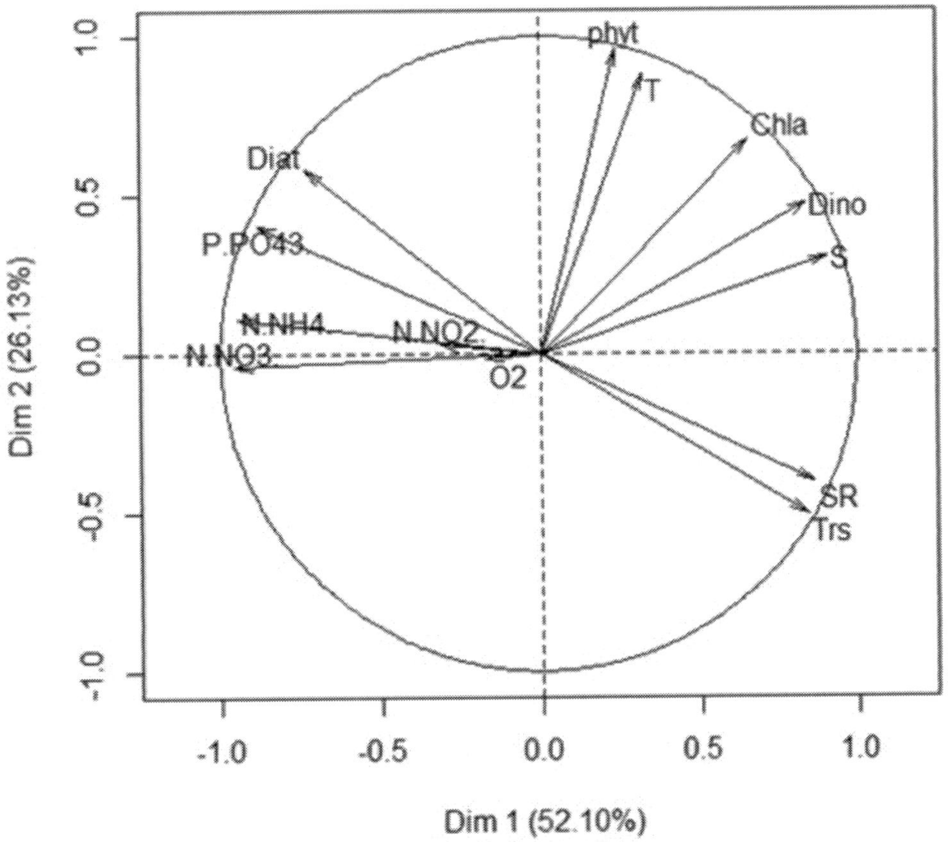

Fig. 6. Variables factor map (PCA): Temperature (T), Salinity (S), Dissolved oxygen (O_2), Transparency (Trs), Nutrients ($N.NO_3^-$, $N.NO_2^-$, $N.NH_4^+$ and $P.PO_4^{3-}$) Chlorophyll "a" (Chla), Total phytoplankton abundance (Phyt), Diatom abundance (Diat), Dinoflagellate abundance (Dino) and Phytoplankton species richness (SR).

The projection of the individuals (Stations-Seasons) on the factorial plane F1*F2 and the hierarchical classification allowed to highlight a spatiotemporal variability of the characteristics (physicochemical and phytoplankton) of the waters of the Aglou-Sidi Ifni coastline and a clear difference between the two periods of this study (Fig. 7).

The first grouping is formed by the samples taken at the three stations during the flood period (autumn 2014 and winter 2015). These samples are characterized by high nutrient contents, a relative abundance of diatoms higher than that of dinoflagellates, minimal temperatures and transparencies (Fig. 7).

The second grouping groups the samples collected at the three stations during the drought period (fall 2015 and winter 2016). These samples are characterized by a relative abundance of dinoflagellates that exceeds that of diatoms, high temperatures and clearances, and low nutrient concentrations. Chlorophyll "a" concentrations in the samples from this grouping are higher than those recorded in the samples from the first grouping.

Each of the two groups is made up of two subgroups, each of which represents the samples collected at the three stations during one season (flood or drought period). For each season, the samples taken in the Sidi Ifni station differ from those taken at the other two stations on the Aglou-Sidi Ifni coast (Fig. 7).

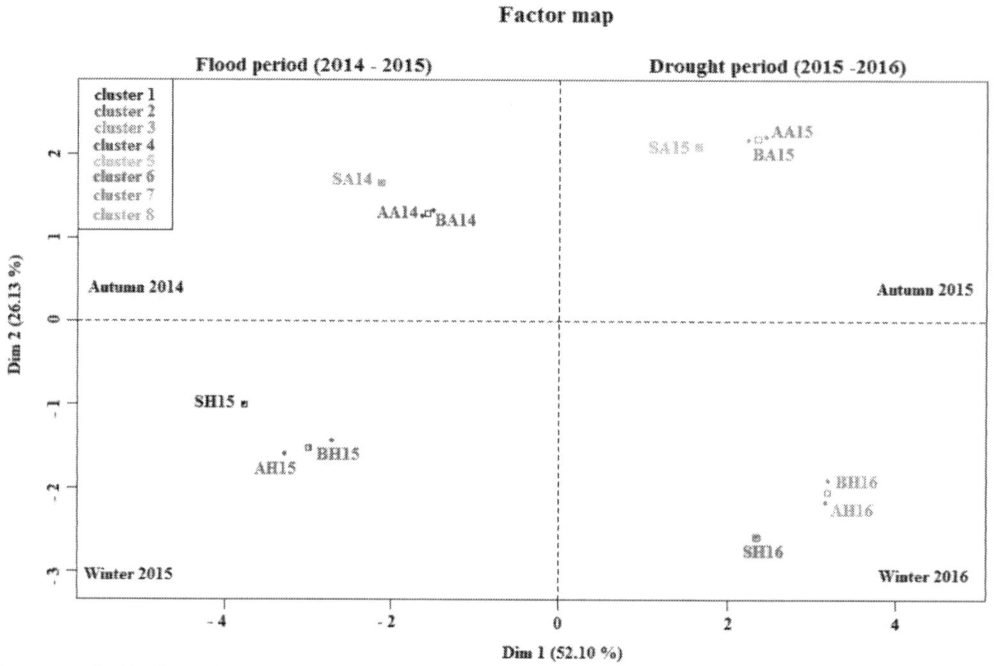

Fig. 7. Individuals projection on the factorial plane (F_1*F_2). Each individual is represented by the initial of the station (A: Aglou, B: Sidi Boulfdail and S: Sidi Ifni) followed by the initial of the season (A: autumn, W: winter) and by two digits that represent the year.

4 Discussion

4.1 Physicochemical characteristics of waters

The work carried out in the temperate and subtropical zones of the North Atlantic showed that the normal dynamics of water masses are naturally regulated by the alternation of the agitated and cold winter season, and the less agitated and hot summer season [21]. In Morocco, there are typically two rainy periods per year, one in autumn and the other in winter. The number of rainy days varies from 30 in the south of the country to approximately 70 in the north [22]. In the Guelmim-Oued Noun region, this rhythm seems to be modified by the hydro-meteorological and climatic hazards that this region has been undergoing over the last 20 years. In fact, precipitation in this region has shown great spatial and temporal variability [23]. The provinces of Guelmim and Sidi Ifni experienced bad weather conditions during the first period of this study (autumn 2014-winter 2015). These conditions generated high floods and strong agitation of the waters of the Aglou-Sidi Ifni coastline during this period. This led to a decrease in temperature, salinity and transparency and an increase in the concentration of dissolved oxygen and nutrient salts. The second period (autumn 2015-winter 2016) was on the contrary drier (no continental input) and the waters of this coastline were less agitated, warmer, more saline, more transparent, less oxygenated and less rich in nutrients.

The lowest seasonal temperature averages were recorded during the first period of this study and the highest during the second period. This warming was also reported in a preliminary study comparing the spring periods of two successive annual cycles (2013-2014) [24]. In the North Atlantic Ocean, the most pronounced warming since the 1960s has been recorded in the temperate and subtropical coastal zones [3]. The warming of water that accompanies the lack of continental water supply has a range of effects, both direct and indirect, on the physicochemical and biochemical processes occurring within the water body, as well as on aquatic biocenoses [25].

4.2 Phytoplankton Communities

Phytoplankton biomass is generally estimated by the concentration of chlorophyll a (Chla), which provides a simple and integrative measure of the response of the phytoplankton community to nutrient enrichment [26].

The seasonal mean chlorophyll "a" concentration varies between 0.70 ± 0.09 µg/l recorded at the Sidi Ifni station in autumn 2014, and 0.91 ± 0.06 µg/l recorded at the Sidi Boulfdail station in autumn 2015. The averages recorded during the drought period are higher than those recorded during the flood period. Seasonal averages of chlorophyll "a" concentration recorded in each of the three stations did not differ significantly from each other. This would be explained by the fact that environmental conditions are unfavourable for the photosynthetic activity of phytoplankton during the flood and dry periods of this study. In the eastern channel and southern bay of the North Sea, low chlorophyll "a" concentrations are recorded during the season of low light intensity and high chlorophyll "a" concentrations are recorded when light intensity increases [27].

The phytoplankton species richness recorded in the Aglou-Sidi Ifni coastline (233 taxa) is higher than that recorded at most sites on the Moroccan Atlantic coast [28-30]. Diatoms and dinoflagellates represent 54% and 42% of this species richness respectively. These two groups are the most abundant. The dominance of these two phytoplankton groups has also been reported in other ecosystems around the world [1, 31, 32].

The seasonal averages of total phytoplankton abundance recorded during the flood period (8.10^3 cell/l to 12.10^3 cell/l) are lower than those recorded during the drought period (32.10^3 cell/l to $52.\ 10^3$ cell/l). In the channel of Oualidia Lagoon the low phytoplankton cell concentrations ($1.64.10^3$ cell/l to $12.24.10^3$ cell/l) are recorded during the cool period and the maximum concentration ($1.19.10^5$ cell/l) is recorded during the autumn of the dry period [28]. The total phytoplankton abundance recorded in winter 2002 (2.10^3 cell/l) in the Bay of Agadir [29] remains lower than those recorded in Aglou-Sidi Ifni coastline during the two winter seasons of this study (8.10^3 cell/l during winter 2015 and 32.10^3 cell/l during winter 2016).

It should be noted that the increase in temperature and the decrease in nutrient concentration during the drought period may be one of the factors that favour the increase in the species richness and relative abundance of dinoflagellates and the decrease in those of diatoms in the Aglou-Sidi Ifni coastline during the drought period. The same phenomenon has been reported in different studies carried out on the Moroccan Atlantic coast [30] and the Mediterranean [33].

5 Conclusion

The climatic forcing that the province of Sidi Ifni underwent during the study period led to temporal modifications of the physicochemical characteristics of the waters of the Aglou-Sidi Ifni coastline (temperature, salinity, transparency, concentration of dissolved oxygen and nutritive salts) which would be at the origin of the temporal variability of the phytoplankton typology of this coastline. During floods and heavy swells, the water is more turbulent, more oxygenated, cooler, richer in nutrients and less transparent. This change is thought to be responsible for the decrease in species richness and relative abundance of dinoflagellates and the increase in those of diatoms. On the other hand, the period of severe drought and the weak swells is characterized by a stability of the water column with an increase in temperature, transparency and a decrease in the concentration of dissolved oxygen and nutrients, which may be the cause of the increase in the specific richness and relative abundance of dinoflagellates and the decrease in those of diatoms.

This study also disclosed a spatial variability in the typology of phytoplankton which would be linked to the spatial variation in the intensity of the currents (canary current and updraft) under the effect of the morphology of the coast in this zone which curves towards the west and carries Cape Aglou at its northern end, which makes the station of Sidi Ifni more exposed to these currents compared to the other two stations which are less exposed.

References

1. P. Bougis, Ecologie du plancton marin: Le phytoplancton I (Masson et scie, Editeurs, Paris, 1974)
2. S. Gobert, P. Lejeune, J. Romano, Température de l'Eau de Mer, in: Changement climatique et milieu marin en Corse (Report Card IUCN, 2018)
3. G. Beaugrand, E. Goberville, VertigO **8**, 14 (2010)
4. J.E. Cloern, Aquatic ecology **33**, (1999)
5. C. Sauvage, Ann. Phys. Globe Météo. Inst. Sci. Chérif **20**, (1963)
6. J. Alibou, *Impacts of climate change on water resources and wetlands in Morocco*, in Mediterranean Regional Round Table, 10-11 December 2002, Athens, Greece (2002)
7. A. Aminot, M. Chaussepied, Manuel des analyses chimiques en milieu marin (Centre national pour l'exploitation des océans, Paris, 1983)
8. C.J. Lorenzen, Limnol Oceanogr **12**, 343-346 (1967)
9. H. Utermöhl, Arch. Hydrobiol **9**,1-38 (1958)
10. F. M. Reid, J. Plankton. Res **5**, 235-252 (1983)
11. A. Sournia, Form and function in marine phytoplankton, Biol Rev **57**, 347-394 (1982).
12. A. Sournia, Atlas du phytoplancton marine: Cyanophcées dictyochophycées dinophycées raphidophycées (Editions du CNRS, France, 1986).
13. E. Balech, Helgol Mar Res **44**, 387 (1990)
14. E. Nézan, G. Piclet, H. Grossel, Guide pratique à l'usage des analystes du réseau national de surveillance phytoplancton (Ifremer, 1996)
15. G.R. Hasle, E.E. Syvertsen, K.A. Steidinger, K. Tangen, C.R. Tomas, Identifying marine diatoms and dinoflagellates (Elsevier, 1996)
16. G. Paulmier, DRV-RH-RST **14**, 428 (1997)
17. L. Bérard-Therriault, M. Poulin, L. Bossé, Guide d'identification du phytoplancton marin de l'estuaire et du golfe du Saint-Laurent: incluant également certains protozoaires (NRC Research Press, 1999)
18. R.C. Team, R: A language and environment for statistical computing (Vienna, Austria, 2013)
19. M. Fraikech, A. Berraho, M. Ramdani, A. Chafik, M. Serghini, A. Moukrim, Mar. Life **15**, 9 (2005)
20. L. Natij, Z. Damsiri, K. Khalil, M. Loudiki, O. Ettahiri, K. Elkalay, Int. J. Adv. Res **2**, 1022-1032 (2014)
21. S. Gobert, Le Plancton, in: Changement climatique en Corse (Report Card IUCN, 2018)

22. NU, Examen des performances environnementales-Maroc, in: Série des examens des performances environnementales (Comission economique des nations unies pour l'europe-bureau pour l'afrique du nord, New york et Genève, 2014)
23. F. Elame, R. Doukkali, Revue Marocaine des Sciences Agronomiques et Vétérinaires **6**, 124-131 (2018)
24. E.M. Gallouli, J. Aziko, H. Oulad Ali, M. El hafa, A. Aamiri, IJPSAT **4**, 7 (2016)
25. P. Balland, La sécheresse: conséquences sur le milieu, salubrité, environnement, La Houille Blanche **2**, 103-117 (1990)
26. J. Brodie, G. De'Ath, M. Devlin, M. Furnas, M. Wright, Mar Freshw Res **58**, 342-353 (2007)
27. A. Lefebvre, W. Perrouchet, Résultats de la mise en oeuvre des réseaux REPHY et SRN–Zones côtières de la Manche orientale et de la baie sud de la Mer du Nord–Bilan de l'année 2015 (RST.LER.BL, 2016)
28. A. Bennouna, O. Assobhei, B. Berland, J. El Attar, Mar. Life **10**, 3-18 (2000)
29. M. Fraikech, Variabilité spatio-temporelle des populations phytoplanctoniques dans la baie d'Agadir-Maroc (Faculté des sciences, Univ Ibn Zohr, Agadir, 2008)
30. A. Loumghari, R. Akallal, A. Mouradi, A. Mouradi, Afrique science **5**, 128-148 (2009)
31. L. Somoue, Structure des communautés planctoniques de l'écosystème pélagique de l'Atlantique sud marocain entre Cap Boujdor et Cap Blanc (26° 30′–21° N) (Fac Sc Ain chock,Univ Hassan II, Casablanca, 2004)
32. L. Lampert, Dynamique saisonnière et variabilité pigmentaire des populations phytoplanctoniques dans l'Atlantique Nord (Golfe de Gascogne) (Iuem, Univ Bretagne Occidentale, 2001)
33. M. Kamel, Etude de surveillance du phytoplancton nuisible sur la méditerranée marocaine dans la région de Tétouan (Restinga-kaâ sras) (IAV Hassan II, Rabat, 1997)

E3S Web of Conferences **183**, 01005 (2020) https://doi.org/10.1051/e3sconf/202018301005
I2CNP 2020

Maturity and spawning of Atlantic chub mackerel *Scomber colias* in M'diq Bay, Morocco

Mohamed Techetach[1*], *Hafid* Achtak[1], *Fatima* Rafiq[1], *Abdallah* Dahbi[1], *Rabia* Ajana[2], and *Younes* Saoud[2]

[1]Environment and Health Team, Department of Biology, Polydisciplinary Faculty of Safi, Cadi Ayyad University, Sidi Bouzid District, P.O. Box 4162, 46000, Safi, Morocco

[2]Applied Biology and Pathology Laboratory, Department of Biology, Faculty of Sciences, Abdelmalek Essaadi University, 93002, Tetouan, Morocco

Abstract. Knowledge of reproductive parameters is necessary to understand the ecology, the population dynamics and to enable rational management of fish of economic interest. This work is a contribution to the study of some aspects of the reproductive biology of the Atlantic mackerel *Scomber colias* (Gmelin, 1789) in the Mediterranean Moroccan coast. The study is based on samples taken from commercial catches in M'diq Bay. The spawning period was determined following both the monthly changes of the gonadosomatic index and the histological maturity stages. The Atlantic chub mackerel spawn between November and March, with maximum activity in December.

1 Introduction

The Atlantic chub mackerel *Scomber colias* (Gmelin, 1789) is a costal pelagic species. It is frequent over the continental slope from the surface to 300 m depth. The *S. Colias* is a gregarious and migratory species. It is widely distributed in the warm waters of the Atlantic Ocean, the Mediterranean Sea and the Black Sea. In the eastern Atlantic, chub mackerel occurs between the Bay of Biscay and South Africa [1]. Recently, genetic divergence and phenotypic variation has been demonstrated that *Scomber japonicus* and *Scomber colias* were distinct. Indeed, *S. japonicus* traditionally considered as a cosmopolitan species, their distribution is localized in the Indian and Pacific oceans [2, 3].

The Atlantic chub mackerel is one of the most important fishery resources in the Moroccan coasts, where is fished mainly by purse seiners. The total catch of *S. Colias* was 225 500 tons in Moroccan waters, which represent 19% of the whole landing of small pelagic fish [4]. Despite the socio-economic and ecological interest, researches on mackerel biology and ecology in the Moroccan coasts are very limited.

The optimal temperature for spawning activity of mackerel is 15-20 °C [1]. For the reproductive success, the strategy of mackerel is characterized by being gonochoric, iterparous, oviparous, multiple batch spwaner with indeterminate fecundity [5]. Knowledge of the characteristics of the reproductive biology of mackerel is an important basis for understanding the population dynamics and rational management of this species. However,

* Corresponding author: mtechetach@gmail.com

information on its reproductive biology is still scarce and fragmentary for the Mediterranean stock [5, 6, 7, 8, 9]. Thus, the main objective of the present study was to describe the spawning season of *S. colias*, and to conduct a comparative analysis with results previously obtained.

2 Materials and methods

Samples of Atlantic chub mackerel were collected in M'diq region (35°N, 5°W) from purse seiners catches. A total of 845 individuals were collected between January and December 2004 and specimens were obtained from weekly collections. For each fish, both total length (TL, 1 mm) and body weight (TW, 0.1 g) were measured. Sex determination was made by visual examination of the gonads. The gonads were removed, weighed (total weight, 0.1 mg), and ovaries were fixed in 10% buffered neutral formalin for subsequent histological processing. Sections of the ovaries (5 µm) were stained with Harris Haematoxylin and counterstained in eosin to determine maturity stages and to follow the stage of gonadic maturity. No difference was found in maturation stage in respect to ovary position [10].

Temporal patterns in female and male gonad development were followed using the gonadosomatic index (GSI) according to [11]:

$$\%GSI = Gonad\ weight\ (g)\ /\ Body\ weight\ (g) \times 100$$

The statistical analyses were made using the Statistica software v.6.0. Analysis of variance (ANOVA) was used to test the difference in GSI between males and females.

3 Results

3.1 Sample composition

Of the total fish samples examined, 53.61% were females and 46.39% were males. The length of fish ranged between 16.4 and 35.9 cm TL, while their weight varied from 31.6 to 489.7 g TW.

3.2 Seasonal fluctuations of gonadosomatic index

The monthly variation in the mean GSI is shown in Figure 1. The GSI fluctuated throughout the year and revealed an increase in reproductive activity in November until March. The maximum average GSI value was obtained in December for both sexes. However, the low values were recorded between April and October. The comparison between sexes was also informative. Females had GSI values significantly higher than those of males ($P < 0.05$). In fact, their GSI ranged from 0.01 to 4 compared to 0.005 to 4.22 for males. A similar monthly evolution in GSI was evident for females and males.

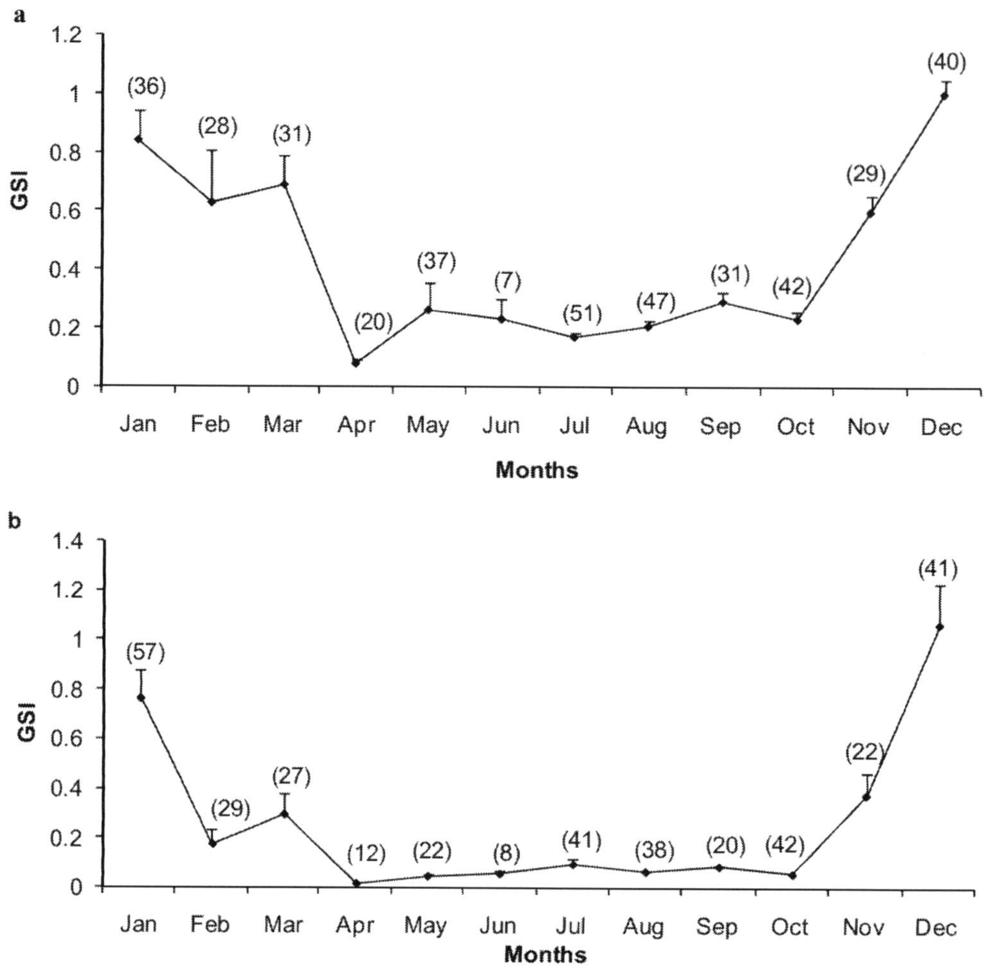

Fig. 1. Variation in gonadosomatic index (GSI) of *S. colias*: (a) Females; (b) Males. Vertical bars indicate standard deviations. Figures in parentheses indicate sample sizes.

3.3 Seasonal changes in the maturation of the ovaries

Histological classification of the ovaries was based on the developmental stage of the most advanced oocytes [12]. Four phases were assigned to the ovaries: (1) immature, (2) maturing, (3) spawning and (4) spent. The Atlantic chub mackerel has an asynchronous ovary containing oocytes at different stages of development.

The monthly distribution of the histological ovaries maturity stages of *S. colias* revealed the occurrence of immature individuals throughout the year. Females in the maturing and spawning stages were collected between November (72.41%) and March (20.83%) with a peak in December (100%) (Fig. 2).

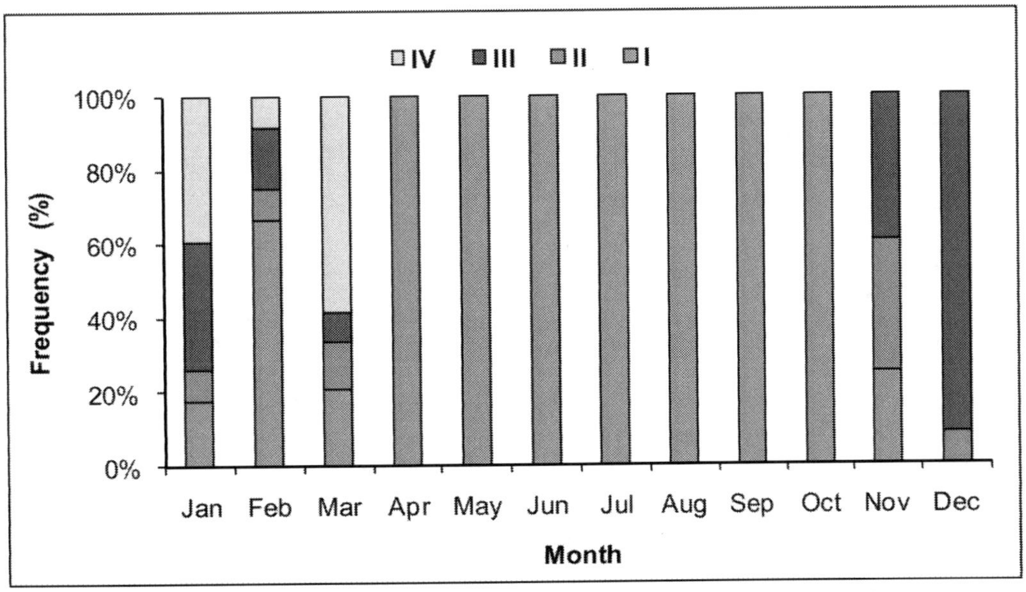

Fig. 2. Temporal variation in the distribution of the ovary maturity stages I-IV for *S. colias*.

4 Discussion

Scomber colias of M'diq Bay is gonochoric, without evidence of sexual dimorphism. The histological study showed that this species has an asynchronous ovarian organisation.

The present results on the monthly variation of the gonadosomatic index and ovarian maturity stages revealed that the spawning period of Atlantic chub mackerel in the study area extended from November to March and peaking in December. These results is in agreement with previous studies [13, 14, 15], but are discordant with other investigations. In fact, the Atlantic chub mackerel living in Adriatic Sea spawn during the spring and summer seasons, with peak spawning in June [16]. In Tunisian waters, a closer ecosystem, this species was shown to be biannually spawning in winter and in summer [17, 18]. Interestingly, in Moroccan North Atlantic coast, the spawning season was shown to occur in tow annual periods. The first period was from December to March, whereas the second period was concentrated in June-July [19]. These differences in spawning period between different regions (e.g. the Mediterranean and Atlantic coasts) might be related to the variation in ecological factors like water temperature and quality and abundance of nutriments [20]. According to [21], temperature appears to be the most determinant environmental factor influencing the reproduction of fish.

Higher GSI values were found in females than in males. Variation in GSI indicates the relative allocation of energy between somatic and gonadal growth [22]. This suggests that females have a higher energy allocation for gonadal development relative to males, such patterns also observed in other teleost fishes [23].

The Atlantic chub mackerel seem to have a prolonged spawning period in the Mediterranean Sea. This strategy is related probably to warming of the sea. Additionally, we must highlighted that success of any fish spawning depend of many environmental factors such as water temperature, food availability and upwelling indices. However, the ecological significance of such difference between populations needs more investigation including more geographical areas.

5 Conclusions

The information gathered in this study contributes to specify the reproductive period of Atlantic chub mackerel *S. colias*. This pelagic species has a long spawning season during the year (five months), which might affect the behavioural schools of fish as a whole. The results of this study could be fundamental and constitute a baseline to improve the assessment and management of this marine resource in the Moroccan Mediterranean coast.

References

1. B. B. Collette, C. E. Nauen, FAO Species Catalogue. Vol. 2. FAO Fish Synop **125**, 137 (1983)
2. C. Infante, E. Blanco, E. Zuasti, A Crespo , M. Manchado, Genetica **130**, 1–8 (2007)
3. G. Catanese, M. Manchado, C. Infante, Gene **452**, 35–43 (2010)
4. I.N.R.H, Rapport annuel de l'état des stocks et des pêcheries marocaines. DRH (2015)
5. M. Techetach, R. Ajana, Y. Saoud, J. Mar. Biol. Assoc. UK **99**, 957–962 (2019)
6. M. Tuggaç, CGPM 4. Débats et Doc Techn **21**, 145–159 (1957)
7. N. Demir, Rapports et Procès-verbaux des réunions **2**, 313–316 (1961)
8. R. Mužinic R, Pomorski zbornik **17**, 613–632 (1979)
9. S. Kiparissis, G. Tserpes, N. Tsimenidis, Belg. J. Zool **130**, 3–7 (2000)
10. T. Yamada, I. Aoaki, I. Mitani, Fisheries Research **38**, 83–89 (1998)
11. R. O. Anderson, S.J. Gutreuter, Fisheries Techniques. American Fisheries Society, Bethesda, 283–300 (1983)
12. S. K. Lowerre-Barbieri, K. Ganias, F. Saborido-Rey, H. Murua, J. R. Hunter, Mar. Coast. Fish **3**, 71–91 (2011)
13. B. Habashi, J. Wojciechowski, ICES CM **20**, 1–9 (1973)
14. J. M. Lorenzo, J. G. Pajuelo, Afr. J. Mar. Sci **17**, 275–280 (1996)
15. F. Wahbi, A. Errhif, O. Ettahiri, Symposium on Science and the Challenge of Managing Small Pelagic Fisheries on Shared Stocks in Northwest Africa, 11–14 March 2008, Casablanca, Morocco, 127–138 (2011)
16. V. Cikes Kec, B. Zorica, J. Appl. Ichthyol **28**, 12–21 (2012)
17. A. Hattour, Contribution à l'étude des poissons pélagiques des eaux tunisiennes. PhD thesis, Faculty of sciences of Tunis, Tunisia (2000)
18. H. Allaya, A. Hattour, G. Hajjej, M. Trabelsi, Afri. J. Biotechnol **12**, 3040–3048 (2013)
19. M. Techetach, J. A. Hernando-Casal, Y. Saoud, M. H. Benajiba, Cybium **34**, 159–165 (2010)
20. G. V. Nikolsky, The ecology of fishes. Academic Press, London (1963)
21. R. J. Wootton, Ecology of teleost fishes. Londres: Chapman and Hall (1990)
22. M. C. Yin, Fish ecology. Beijing, China Agriculture Press (1993)
23. V.M. Lemos, A. S. Varela Junior, G. Velasco, J. P. Vieira, Zoologia **28**, 603–609 (2011)

Challenges and Opportunities in the Operationalization of the Water-Environment-Energy-Food (WE^{2}F) Nexus: Case Study of the Upper Niger Basin and Inner Niger Delta, West Africa

Ousmane Seidou[1], *Fatoumata* Maiga[1], *Claudia* Ringler[2], *Spela* Kalcic[3], *Luca* Ferrini[4]

[1]University of Ottawa, Department of Civil Engineering, 161 Louis Pasteur, Ottawa (ON) K1N 6N5, Canada
[2] Environment and Production Technology Division, International Food Policy Research Institute, 2033 K Street, NW, Washington, DC 20006-1002, USA
[3] Freelance researcher and consultant, Brilejeva 3, 1117 Ljubljana, Slovenia / Unamontie 8c 22, 00610 Helsinki, Finland
[4] GIZ Office Niger, Avenue de l'Afrique, Porte N° 36, Niamey, Niger

Abstract. The ever-increasing demand for water, food, and energy is putting unsustainable pressure on natural resources worldwide, often leading to environmental degradation that, in turn, affect water, food, and energy security. The recognition of the complex interlinkages between multiple sectors has led to the creation of various holistic approaches to environmental decision making such as Integrated Natural Resources Management (INRM), Integrated Water Resources Management (IWRM), Virtual Water (VW), Water Footprint (WF) and lately the Food-Energy-Environment-Water nexus (WE2F). All these approaches aim to increase resource use efficiency and promote sustainability by increasing the cooperation between traditionally disjoint sectors, and mainly differ by the number and relative weights of the sectors included in their framework. They also suffer from the same face and the same barriers for implementation, some of which may never be fully overcome. The paper discusses the benefits of adopting a WE2F nexus approach in the Upper Niger Basin (UNB) and the Inner Niger Delta (IND), but also the multiple difficulties associated with its practical implementation. IWRM/WE2F initiatives in the UNB/IND such as the BAMGIRE project piloted by Wetlands International and funded by the Dutch Embassy in Mali to secure livelihoods and biodiversity in a changing environment, is taken as an example of partial success in the use of a nexus approach to watershed management. It was shown there are multiple barriers to the operational implementation of the WE2F. However, while a full understanding of all interlinkage between sectors may never be possible, data collection, scientific research and model development can improve our ability to understand the complex system in which we live, and hence take better decisions

1 Introduction

The United Nations expect the world's population to be increasing at least until 2050[1]. The population will peak at 11.2 billion people in 2100 according to the median projection, and the fastest growth will be occurring in Africa. Resource consumption per capita will also be increasing as people in developing countries are inexorably trending towards a western-style lifestyle, which require more energy, more water and has a greater environmental footprint. Rather than increasing, the availability of natural resources are actually decreasing as the result of overexploitation, environmental degradation and climate change. According to the UN International Panel on Resources, 90% of biodiversity loss and water stress in addition to half of the global greenhouse gas emission are caused by resource extraction and processing[2], despite the fact that the call for sustainability in resource extraction is getting louder and louder in the international discourse. Sustainability was promoted to the centre stage of the global agenda with the Sustainable Development Goals (SDG), put forward by the UN at the United Nations Conference on Sustainable Development in Rio de Janeiro in 2012. A key characteristic of the SDG is that the UN explicitly recognize that 'action in one area will affect outcomes in others, and that development must balance social, economic and environmental sustainability'[3].The need to account for positive and negative interactions between interventions and objectives in different sectors of the economy of the environment is not new, and has led to the creation of various holistic approaches to environmental decision making such as Integrated Natural Resources Management (INRM[4], [5]), Integrated Water Resources Management (IWRM[6]–[8]), Virtual Water (VW[9]–[11]), Water Footprint (WF[12], [13]) and lately various declinations of the Food-Energy-Environment-Water nexus (WE^{2}F[14], [15]).

All these frameworks aim to promote resource management that transcends sectors, disciplines and scales, and the main differences reside in the number and relative weights of disciplines/scales/sectors included. The WE^{2}F nexus focuses on the inter-linkages across three key sectors, namely the water, energy, food sector, and their interlinkages with environmental sustainability (Figure 1).

Given growing natural resource scarcity worldwide, these interlinkages continue to strengthen, making it increasingly unlikely that developments or investments in one of these sectors do not also affect one or more of the other sectors. The whole system is under direct and indirect divers of change such as population growth and climate change. The changes induced by these drives affect sustainability outcomes such as food security and environmental quality. Fig. 2 illustrates the complex linkages, dependencies and trade-offs between food, water and energy, the drivers of change and the environmental services that their provision depends on.**Fig. 2**.

https://doi.org/10.1051/e3sconf/202018302001

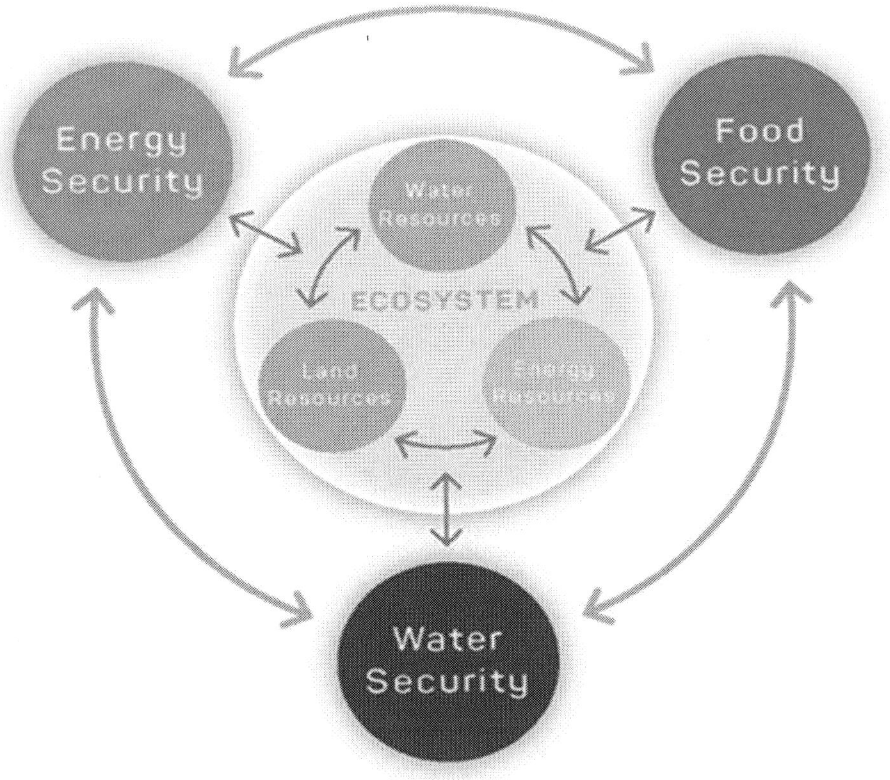

Fig. 1. WE2F diagram[15]

Nexus related frameworks that examine the link between two seemingly disjoint sectors date back to the '80s[16], but one of the earliest explicit references to the water-energy-food nexus can be found in background paper of the Bonn 2011 Conference on Water, Energy and Food Security Nexus[14]. The nexus was introduced as a novel approach that can enhance water, energy and food security by increasing efficiency, reducing trade-offs, building synergies and improving governance across sectors.

However, the ambitious aims of the nexus—the desire to capture multiple interdependencies across major sectors, across disciplines and across scales—could become its downfall[17]. Barriers to the implementation of the WE^{2}F include ambiguity in its definition, the strong assumption that integration (of sectors, disciplines, knowledge and stakeholders) is possible and desirable[18], the complexity of the natural system and the lack of understanding of the interdependencies and connections across the WE^{2}F systems. The operationalisation of the WE^{2}Fis also limited by the fact that comprehensive WE^{2}F nexus tool is yet to be developed[15]: existing tools are either conceptual, qualitative, and large scale, possibly oversimplifying the interactions between sectors, either quantitative and limited in space or number of sectors covered to manage complexity. Additional challenges come from the fact synergies and trade-offs involve monetary and non-monetary values, the mismatch between nexus boundaries and traditional management units, multiple spatial and temporal dimensions, heterogeneous procedures for various nexus dimensions[19]. All the above challenges can be found in the Upper Niger Basin (UNB), where food and energy security are seen as national priorities, but clash their development may clash with environmental sustainability already threatened by climate change, desertification and population growth. The UNB drains into the Inner Niger Delta (IND), a wetland of global importance that provides natural resources and livelihoods for over 10 % of Mali's population and 80% of Mali fish production. The IND also produces food for about 40% of the country's cattle {Citation}. The wetland is known to be extremely sensitive to flood regime in the IND, there is a widespread concern current, and planned dams and irrigation schemes in the IND could lead to a collapse of ecosystems in the IND, making it a perfect case study for a WE^2F analysis.

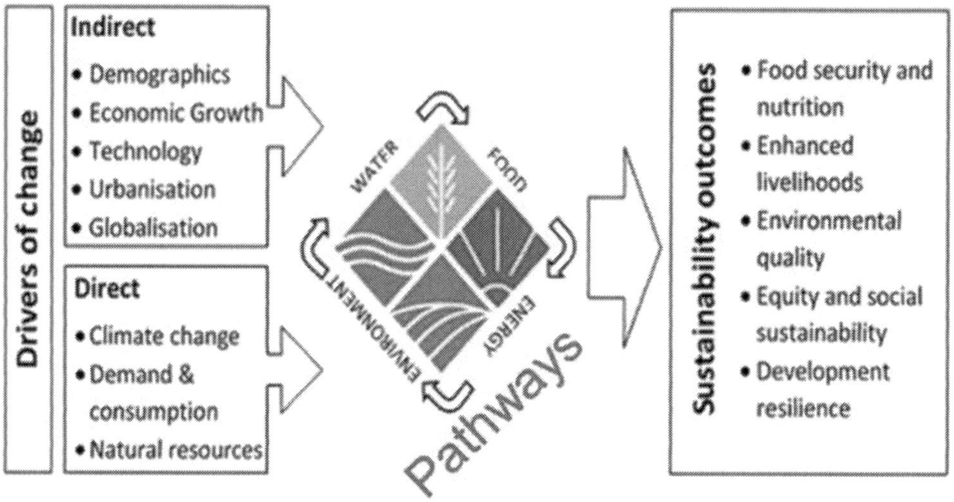

Fig. 2.Dependencies between drivers of change, the W^2EF and sustainability outcomes. Adapted from [20]

In 2016, the Embassy of the Kingdom of the Netherlands (EKN) in Mali commissioned Wetlands International to implement The BAMGIRE (BAM=Bamako; GIRE=Gestion Intégrée des Ressources en Eau=Integrated Water Management). The aim of the BAMGIRE is to secure livelihoods and biodiversity in a changing climate, and build the capacity in Mali and Guinea on water management with a special focus on sustaining essential ecosystem services in the IND and other upstream wetland ecosystems. While the project title refers to IWRM rather than the WE^2F, the project simultaneously considered hydropower production, food security, water security and environmental sustainability and hence qualifies as a W^2EF project. Wetlands International tasked international partnersto support Mali in the development of state of the art knowledge and technical tools that would help.This paper describes the efforts deployed between 2016 and 2020 within the BAMGIRE project to decipher the complex relationships between dam management, irrigation, climate change and ecosystem services in the IND and UNB.

2 The Upper Niger Basin and Inner Niger Delta

The Upper Niger Basin (UNB) and the Inner Niger Delta (IND) are part of the Niger River basin (Figure 3), the third longest river in Africa. The Niger River watershed has a total area of 2.262 million Km^2 and flows through 10 African countries starting in Guinea, where the river headwater is located, then flows 4100 km into the Atlantic Ocean in coastal Nigeria. The majority of the river is located in Nigeria, Mali, and Niger with respectively 25.7%, 25.5% and 24.8% of the total area. Over 1 million people depend on the Niger River resources. The Inner Niger Delta is a large floodplain located on the edge of the Sahara Desert (Mali) and is one of the world most productive temporary wetlands.The IND climate is dry most of the year and the wet season usually expands from July to September/October, leading more water availability. The communities surrounding the IND depend mainly on the agriculture and fishing culture. The wetland provides natural resources and livelihoods for over 10% of Mali's population and 80% of Mali fish production. The wetland also produces food for about 40% of the country's cattle[21], [22]

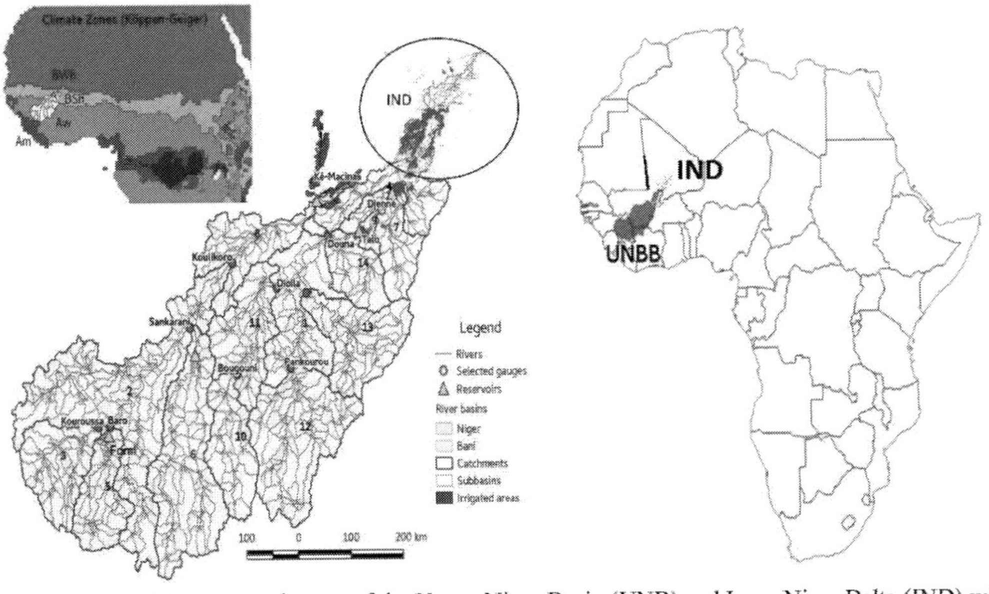

Fig. 3. Map of the case study area of the Upper Niger Basin (UNB) and Inner Niger Delta (IND) with catchments, sub-basins, discharge gauges, reservoirs, irrigated areas, and climate zones (adapted from [23])

The two main factors affecting the hydrological regime in the UNB and the IND are human-made structures, climate variability and climate change. Rainfall in the Sahel is highly variable and the region experiences alternating episodes of droughts and floods. As in all part of the Sahel, the river flows in Mali have drastically decreased in the 1970-2010 period compared to the 1950-1970 period. Decreasing precipitation and depleted groundwater reserves are thought to be the main causes, but dam operation and irrigation are also contributing factors. There are currently 5 major dams in operation in the UNB, and 4 other planned or in development[24]. Among existing reservoirs, the Selingué multipurpose dam, commissioned in 1981, has the largest impact on flows in the watershed. Besides producing hydropower, the dam has the mission to maintain a minimum discharge during the dry season for a large-scale irrigation scheme called Office du Niger (OdN). The OdN is located upstream of the IND and occupies around 100,000ha, for a potentially irrigable area of more than 1,000,000ha. The maximum inundation area shrinked from around 18,000 km^2between 1905 and 1960 to around 10,000 km^2 under the combined impacts of climate variability, irrigation withdrawals, and dam operation (Fig.3).

 https://doi.org/10.1051/e3sconf/202018302001

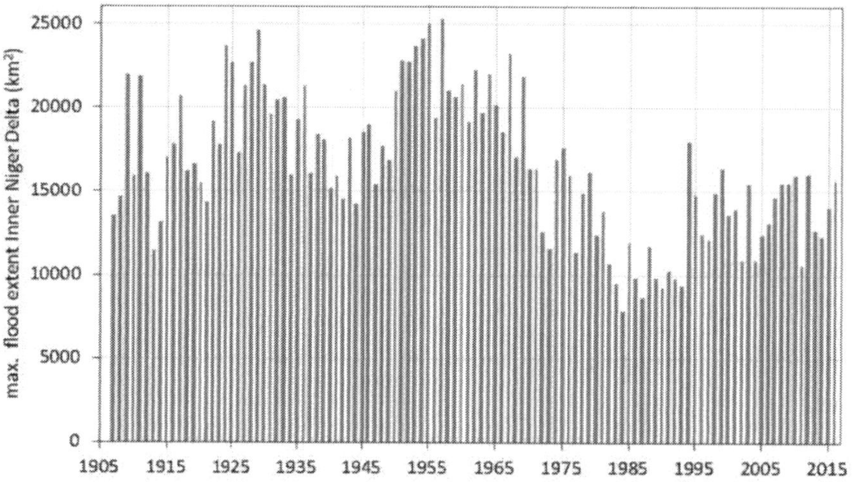

Fig. 4. Flood extent of the Inner Niger Delta[24]

Despite the already visible degradation of ecosystems in the IND, the Malian authorities are eager to expand the irrigated area in the future to maintain food security in the region. Furthermore, the Guinean government is going forward with the planned Fomi/Moussako dam to provide hydropower, with irrigation and flood control, creating even more uncertainty in the future of the IND. At this date, the combined effect of planned and existing dams and irrigation schemes on the flood regime in the IND is largely unknown. This is particularly worrisome as empirical data shows that resource productivity in the IND is highly correlated to the maximum flood extent[25].

3 IWRM/WE^{2}F nexus initiatives in the UNB and IND

The IND and UNB are part of the Niger River basin, which is shared by nine countries:Benin, Burkina Faso, Cameroon, Chad, Ivory Coast, Guinea, Mali, Niger, and Nigeria. The Niger Basin Authority (NBA) is an intergovernmental agency tasked with the promotion of the coordinated development of Niger Basin resources in a broad range of sectors: water, energy, agriculture, livestock, fisheries, forestry, mining, transport and communication. The NBA and its partners fully adhere to the concept of IWRM, and the NBA typically coordinates the studies and development of projects that are likely to have transboundary impacts. It collects and centralises hydrometric data throughout the watershed, and has commissioned the development of a water allocation model of the Niger basin using Mike Basin. Based on the model, the NBA has established minimum flow targets at the boundaries of each country. Potential projects are submitted to the NBA, which estimates the potential impacts using the Mike Basin model of the watershed. Unfortunately, the model has several limitations. It does not have a rainfall-runoff module and hence cannot account for climate variability/change; it cannot represent the complex hydrologic behaviour of the IND; it does not have an ecological module; an external firm developed it and the NBA does not have the in-house expertise to update it and represent the current configuration of the watershed. Aware of the concerns about the impacts of FOMI/Moussako, the NBA has recently launcheda study on advanced modelling of ecosystem services in the Inner Niger Delta (IND), which involves the development of a hydrodynamic model and the estimation of ecosystem services in the IND under various development and climate change scenarios, as well as Environmental flow (Eflow) estimation at key locations. Unfortunately, because of the lack of high-resolution elevation data for the IND, the simulated flood extents significantly deviate from observations.

Furthermore, ecosystem services were estimated using relatively imprecise correlations with water levels such as the one developed in [25] (Figure 5), leading to a huge uncertainty in the results. Therefore, as of today, despite its willingness to implement IWRM in the basin, the NBA does not have the tools and data to state with certainty how ecosystems in the IND will react to climate change and infrastructure development.

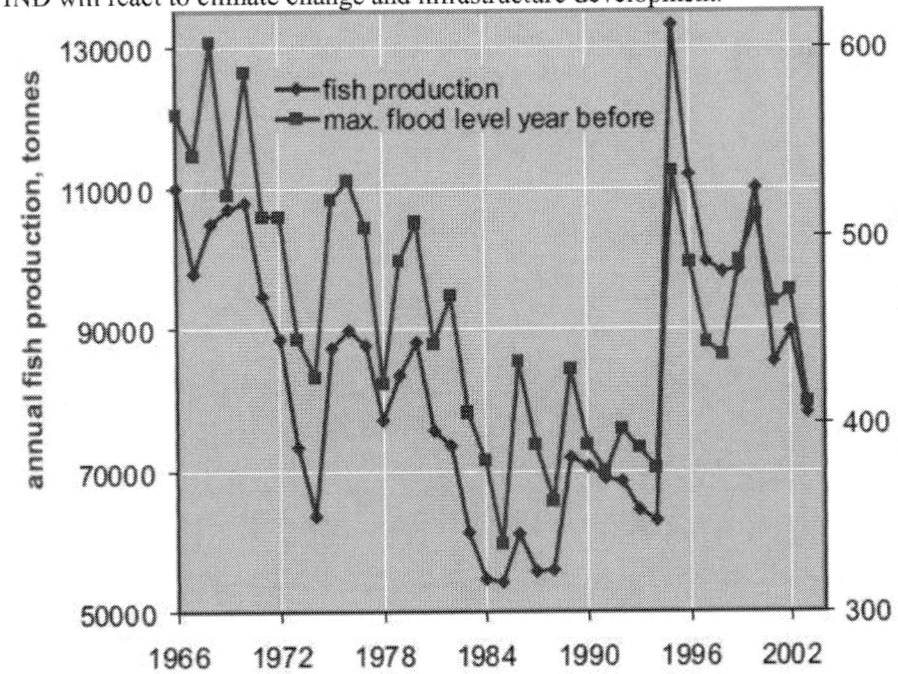

Fig. 5. Fluctuation in annual fish catch in the Inner Delta between 1966 and 2003 compared to the fluctuation in flood level (Akka, cm) in the year before[25]

A good understanding of the physical, environmental and economic system is not sufficient to achieve WE^2F. A continuous dialogue between stakeholders is fundamental to build public adhesion and political willingness to take decisions that may not align with the interests of individual actors. The EU Nexus Regional Dialogue Programme, funded by the European Union, was initiated to sustainably strengthen the political processes needed at the regional and national levels to meet the increasing demand for water, energy and food. It brings together stakeholders to develop concrete policy recommendations and action plans for future investments, with specific emphasis on multi-sector infrastructure and corresponding capacity development activities. There have been so far five regional dialogue regions: The Middle East and North Africa, America and the Caribbean, Southern Africa, and Central Asia and finally, the Niger Basin. As part of the Niger Basin Nexus Dialogue, the GIZ funded a study aiming to help the NBA use a Nexus angle to screen/prioritize intervention and infrastructure projects in the basin. A conceptual procedure was developed to score interventions and estimate trade-offs, but the lack of adequate data and appropriate modelling tools led to a lot of subjectivity in the assignment of scores, hence in the final prioritization of investments.

The BAMGIRE project was probably the most comprehensive attempt to understand the interlinkages between water, energy and sustainability in the UNB and IND. A continuous dialogue was established with the Malian and Guinean governmental services through a series of workshops dedicated to either two-ways knowledge exchange, capacity building or both. In parallel, knowledge products such as ecosystem services (ESS) mapping, hydrological (SWIM: Soil and Water Integrated Model) and hydraulic models were being developed (Figure 6). The ultimate goal was to develop a decision support system, called *Decidaid*.

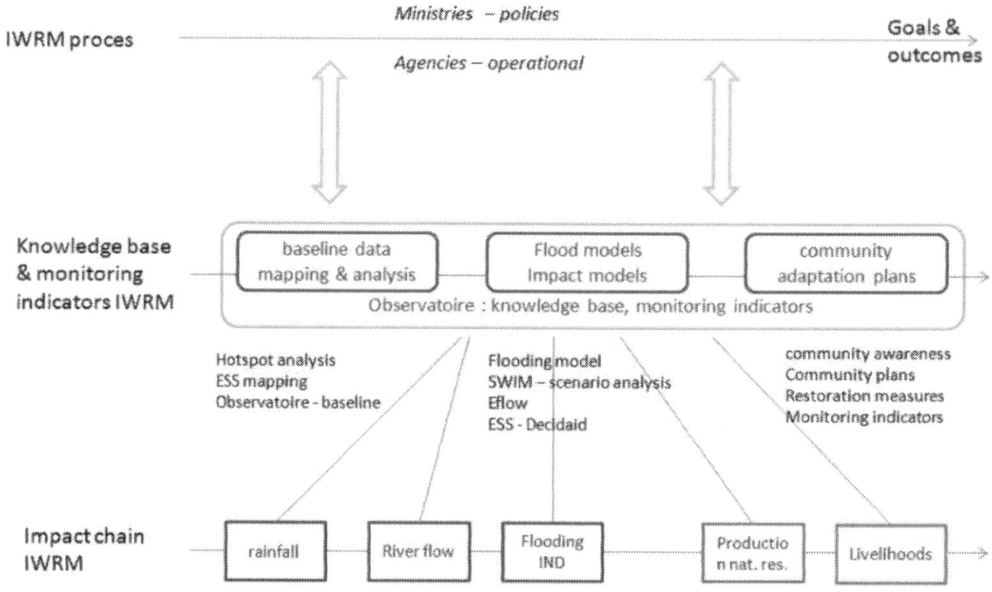

Fig. 6. Knowledge generation process in the BAMGIRE project. ESS

While data availability has been a challenge for all knowledge products generation, the two main challenges were the description of the flood regime in the IND and its links to ecosystem services. Flood model development of the IND is particularly challenging because of its complex network of streams and lakes, its large size, and the lack of precise topographic data. A first model was developed by coupling observed water levels with satellite imagery, before a full hydrodynamic was developed using a topographic data derived from secondary sources. Before the BAMGIRE project, the only available method to estimate ecosystem series in the IND was through oversimplified regressions using water levels at specific stations. These relations do not account for the actual ecological processes and may not hold in situations where the flood regime deviates from what have been observed in the past. A significant amount of resources was invested in developing aconceptual model called PROBFLO that demonstrate the causal risk pathways from identified sources (including anthropogenic and natural activities/events) to stressors (water quality, flow and habitat modifications for example), socio-ecological receptors in multiple habitats to endpoints[26], [27]. The impact pathways in PROBFLO are presented in Figure 7. According to the developers, PROBFLO is flexible enough to use any kind of relationships between drivers and impacts, from conceptual equation to empirical relations, or expert opinion. While the PROBFLO model constitute a significant achievement in the understanding of how climate and water management is likely to affect ecological endpoints, the lack of data to calibrate and validate the model leads to a low credibility in its output. A strong recommendation of the BAMGIRE project is the implementation of a monitoring plan that would allow a better calibration of the model, and an adaptive management plan that would be regularly updated according to new evidence and improvements in the understanding of the system.

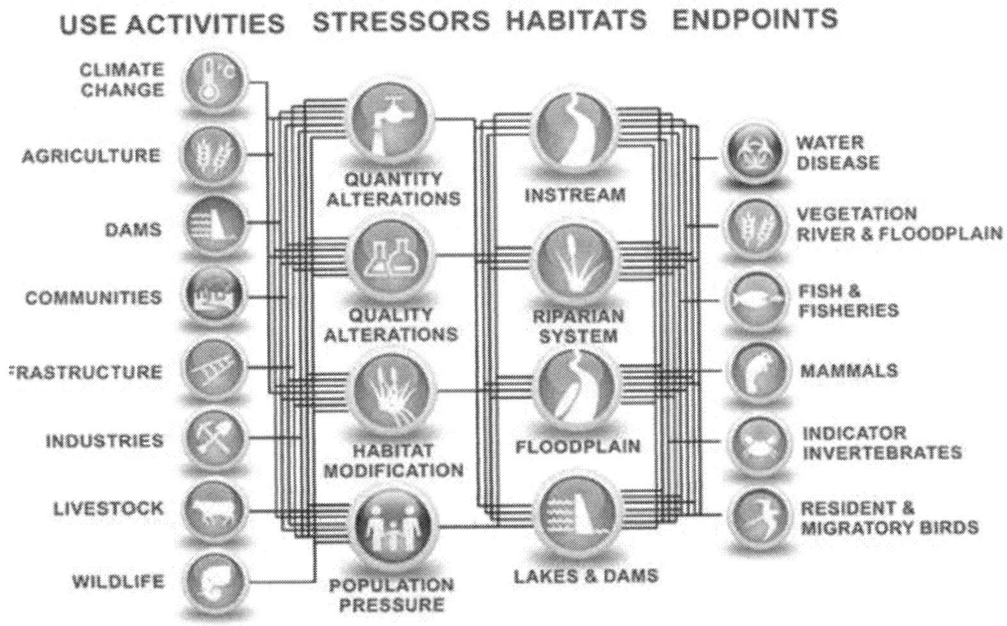

Fig. 7. Impact pathways in the PROBFLO model of the UNB and IND[26], [27]

The authors believe that despite significant progresses made in the BAMGIRE project, our knowledge of linkages between water, food, energy and environmental sustainability is still insufficient to estimate precisely trade-off between sectors. However, the project triggered valuable discussions between scientists of different disciplines, between stakeholders and highlighted the limits of our knowledge of the system, while improving our collective understanding of the challenges related to resources management in the IND and UNB. The project set the stage for more data collection, more tool development and more dialogues among actors. It is unlikely that a full understanding of the system will come anytime soon, but the outputs of the project made Mali and Guinea closer to a common understanding of how their infrastructure and management decisions would affect development objectives across sectors.

4. Conclusion

Just like previous holistic approaches to environmental decision making such as IWRM, the WE2F is a powerful tool for framing sustainability issues and foster discussions among stakeholders. It has the advantage over IWRM of (at least in theory) treat all sectors equally. However, there are multiple barriers to its implementation, such as complexity and data availability. It demonstrated that initiatives in the Upper Niger Basin related to WE2F nexus or other similar concepts such as IWRM were successful in stimulating inter stakeholder dialogues and tightening the political will implement a trans-sectorial and transnational management of resources in the watershed. The initiatives also led to significant knowledge generation and data collection, although considerable knowledge gaps remain, especially about flood dynamics in the IND and the relationships between the flood regime and ecosystem services. A full understanding of all interlinkage between sectors may never be possible, but data collection and model development can improve our ability to understand the system and take better decisions.

The authors thank the Wetlands International and Embassy of the Netherlands, Mali, for funding this research.

References

[1]United Nations, Department of Economic and Social Affairs, Population Division, *World Population Prospects 2019: Highlights.*, (United Nations, Department of Economic and Social Affairs, New York, USA,2019)

[2]International Resource Panel, *Global Resources Outlook 2019: Natural Resources for the Future We Want,* (United Nations Environment Programme, Economic Division, Nairobi, Kenya,2019)

[3]United Nations General Assembly, *Transforming our World: the 2030 Agenda for Sustainable Development,* (United Nations, New York, USA, 2015)

[4]M. Rohlmann, Integrated natural resource management? a question of property institutions?, GeoJournal**29**, 4, (1993)

[5]B. Douthwaite, J. M. Ekboir, S. J. Twomlow, and J. D. H. Keatinge, The concept of integrated natural resource management (INRM) and its implications for developing evaluation methods., in *Natural resource management in agriculture: methods for assessing economic and environmental impacts*, B. Shiferaw, H. A. Freeman, and S. M. Swinton, Eds. Wallingford: CABI, (2005)

[6]M. M. Rahaman, O. Varis, Integrated water resources management: evolution, prospects and future challenges, Sustain. Sci. Pract. Policy, **1**, 1, pp. 15-21, (2005)

[7]O. Graefe, Integrated Water Resources Management as a New Approach to Water Security, in *Eating, Drinking: Surviving*, P. Jackson, W. E. L. Spiess, and F. Sultana, Eds. Cham: Springer International Publishing, (2016)

[8]J. Hassing, Unesco, World Water Assessment Programme, *Integrated water resources management in action: dialogue paper*. Paris: UNESCO, (2009)

[9]J. A. Allan, Virtual Water - the Water, Food, and Trade Nexus. Useful Concept or Misleading Metaphor?, Water Int., **28**, 1, pp. 106-113, (2003)

[10]A. Y. Hoekstra, P. Q. Hung, Globalisation of water resources: international virtual water flows in relation to crop trade, Glob. Environ. Change, 15, 1, pp. 45-56 (2005)

[11] T. Oki, S. Kanae, Virtual water trade and world water resources, Water Sci. Technol., **49**,7. pp. 203-209, (2004)

[12] A. Y. Hoekstra, A. K. Chapagain, Water footprints of nations: Water use by people as a function of their consumption pattern, Water Resour. Manag., **21**, 1, pp 35-48, (2006)

[13] A. Y. Hoekstra, M. M. Mekonnen, The water footprint of humanity, Proc. Natl. Acad. Sci., **109**,9, pp. 3232–3237, (2012)

[14] H. Hoff, Understanding the Nexus. Background Paper for the Bonn2011 Conference: TheWater, Energy and Food Security Nexus, Stockholm Environment Institute, (2011)

[15] S. J., McGrane, M., Acuto, F., Artioli, P.-Y., Chen, R., Comber,J., Cottee, G., Farr-Wharton, N., Green, A., Helfgott, S., Larcom, J. A., McCann, P., O'Reilly, G., Salmoral, M., Scott, L. C., Todman, T., van Gevelt, X., Yan, Scaling the nexus: Towards integrated frameworks for analysing water, energy and food, Geogr. J., 185, 4, pp. 419-431,(2019)

[16] A. Endo, I. Tsurita, K. Burnett, P. M. Orencio, A review of the current state of research on the water, energy, and food nexus, J. Hydrol. Reg. Stud., **11**, (2017)

[17] H. Leck, D. Conway, M. Bradshaw, J. Rees, Tracing the Water-Energy-Food Nexus: Description, Theory and Practice: Tracing the Water-Energy-Food Nexus, Geogr. Compass, **9**, 8, (2015)

[18] R. Cairns A. Krzywoszynska, Anatomy of a buzzword: The emergence of 'the water-energy-food nexus' in UK natural resource debates, Environ. Sci. Policy, 64, (2016)

[19] J. Liu , H. Yang, C. Cudennec, A.K. Gain, H. Hoff, R. Lawford, J. Qi, L. de Strasser, P.T. Yillia, C. Zheng, Challenges in operationalizing the water–energy–food nexus, Hydrol. Sci. J., **62**, 11, (2017)

[20] S. Fan, The 'nexus' between food, energy and water security, (Crawford School of Public Policy, Canberra, 2014)

[21] S, Liersch, S. Fournet, H. Koch, *Assessment of Climate Change and Water Management Impacts on the Water Resources in the Upper Niger and Bani River basins,*(Wetlands International, 2018)

[22] Royal Haskoning, Altenburg & Wymenga, and GID., Etat des lieux du Delta Intérieur, -vers une vision commune de développement, (Ministère de l'environnement et de l'assainissement, République du Mali, 2010).

[23] S, Liersch, S. Fournet, H. Koch, A. G. Djibo, J. Reinhardt, J. Kortlandt, F.V. Weert, O. Seidou, E. Klop, C. Baker, F.F. Hatermann, Water resources planning in the Upper Niger River basin: Are there gaps between water demand and supply?, J. Hydrol. Reg. Stud., **21**, (2019).

[24] Observatory of Integrated Water Management in the Upper Niger basin and Inner Niger Delta — English. https://onisdin.info/en (accessed Jun. 28, 2020).

[25] L. Zwarts, P. van Beukering, B. Kone, *The Niger, a lifeline: effective water management in the Upper Niger Basin*. Lelystad: RIZA - Rijkswaterstaat, (2005)

[26] C. Dickens, G. O'Brien, R. Stassen, N. Eriyagama, M. Kleynhans, K. Rowntree, M. Graham, R.G. Vere, J. MacKenzie, E. Wymega, E. Mapedza, M. Brunet, M. Desai, J. Hean, *E-flows for the Upper Niger River and Inner Niger Delta: Synthesis Report*, (IWMI, 2019)

[27] C. O'Brien, Gordon, C. Dickens, E. Hines, V. Wepener, R. Stassen, L. Quayle, K. Fouchy, J. MacKenzie, P.M. Graham, W.G. Landis, A regional-scale ecological risk framework for environmental flow evaluations, Hydrol. Earth Syst. Sci., **22**, 2, (2018)

Review of water energy food nexus in Africa: Morocco and South Africa as case studies

Lahcen El Youssfi[1a]*, *Wesley* Doorsamy*[2], *Adil*Aghzar[1], *Sidi Imad* Cherkaoui[1], *Ihssan* Elouadi[3], *Alex* Godoy Faundez[4], and *Diego* Rivera Salazar[4,5]

[1] Research team of environment and natural resources management, B2DRN Laboratory, Khenifra Higher School of Technology, Sultan Moulay Slimane University, Morocco
[2] Institute for Intelligent Systems, University of Johannesburg, South Africa
[3] École Nationale Supérieure des mines de Rabat, ENSMR, Morocco
[4] CISGER, Facultad de Ingenieria, Universidad del Desarrolo, Chile
[5] CRHIAM, University of Conception, Chile
[*] Equally contributed to this work.

Abstract. The population of the African continent is significantly increasing which has given rise to societal, economic and environmental changes. This will generate an increase in demand for food, water and energy. Thus, strategies for sustainable management of natural resources need to be adopted urgently in order to match the future needs and to ensure sustainable development on the continent. The Water Energy Food (WEF) nexus is one the avenues for studying the current and future interactions and interdependencies of the WEF sectors under different scenarios and challenges in the effort towards establishing sustainable use of natural resources. Policies around the nexus approach are few or absent at the continental and countrylevels. This paper is a first review on the adoption of the Nexus in two model African countries, Morocco and South Africa. The review compares the water, energy and food sectors and the related national policies, plans and frameworks. The data related to the WEF Nexus revealed key difficulties of availability in both countries. A comparative analysis of the situation in both countries demonstrated differences in policies relating to the WEF Nexus across the sectors: water, energy, and agriculture. Additionally, the paper explores the relevance of technology-assisted frameworks for the WEF nexus and also analyzed risks to agriculture and food security in the context of the reviewed countries.

Introduction

Africa's population is rapidly increasing and along with this is an increasing demand for food, water and energy. This places a tremendous strain on natural resources: soil, water and natural energy sources, especially for food production. The situation is further compounded by extreme societal, economic and environmental changes. Thus, African countries need to account for these changes in their long-term sustainable development strategies. The nexus of water, energy and food is seemingly becoming more critical for developing new adaptation and mitigation strategies to reduce the negative consequences of over-exploitation of natural resources and thus provide a more sustainable and resilient food production systems.

The use of precision technology in food production systems is one of the promising solutions for limiting and reducing over-exploitation of natural resources and pollution of the production ecosystem. In fact, innovative research and development programs and initiatives that focus on the role of technology in large scales and agricultural technology in small scales are required in addressing nexus challenges across Africa. The ability of the African countries to develop and to use low-cost, accessible and affordable agricultural technologies is thus a critical point of interest. This coincides with resolving the challenge of promoting small scale farmers so as to provide them with access to technology that is currently

[a]Corresponding author: l.elyoussfilahcen@usms.ma

not available/accessible to them. This could potentially support economic growth through enhancing small-scale farming, as well as promoting sustainability in the vast arable lands of Africa.

According to the OECD, a 55% increase in water demand by 2050 is predicted, and the FAO projects that food production will need to increase by 70% by 2050, and the IEA (US Energy Information Administration) expects energy demand will increase by 40% in 2030 [1].The future of the nexus will undoubtedly become more complex in all directions. Through considering agriculture as the most important economic activity in Africa as it provides food, not only for its population, but also for other populations on other continents, and the water and energy consumption thereof, optimizing water and energy in agriculture and food production is found to be a necessary focus.

Taking the aforementioned into consideration, the Nexus is studied and adopted because it's an evidentiary-based approach with which to develop scenarios, as well as to design and appraise potential options, and a means for supporting multi-stakeholder dialogue. This paper is aimed at drawing an overview on the Water Energy Food Nexus and its reflections on the sustainability of the concerned resources in agriculture and on food security with particular focus on Morocco and South Africa. The relevant sectors were reviewed in the context of these countries, primarily focusing on their interactions. The potential impact of the development and the use of technology-assisted frameworks to support the nexus approach is briefly explored. The detailed review and comparative analysis of the studied countries were then presented.

1 Overview on Water Energy Food Nexus

1.1. The nexus

The nexus corresponds to the central link or the important connection associating different elements and parts of a system. The Water Energy Food Nexus (WEF Nexus) corresponds to the connections and the interdependencies of water, energy and food as a system or as three unique systems [2].The study of the WEF Nexus aimed to define the different components and aspects of the interdependencies of these systems. The interdependencies are multidimensional and very complex ~~are difficult~~ to study. In fact, the nexus is treated from different angles and aspects in order to desegregate the interdependencies. On the other hand, the interactions of the Nexus with other systems are also considered while dealing with Nexus challenges.

Water is utilized in agriculture for irrigation and food processing. Water is also essential for energy generation, whether directly in the case of hydropower or indirectly in coal mining. Energy is utilized in the pumping and distribution of water and for the mechanization of various agricultural activities. The nexus approach to resource sector management seeks to understand the linkages, dependencies, and trade-offs associated with the core elements within the nexus under assessment [3].

Many connections can be evaluated in interaction with the WEF Nexus; for instance, climate change, food security, water management, energy security and others.

1.2. Water Energy Food nexus and resource security

The WEF Nexus is important because of the importance of food, energy and water as resources and systems. Those resources are not only pillars of the economy of a nation but critically important for its security and not only the natural resources security. The increase of population increases the pressure on the available resources of food, water and energy; and countries exports and/or imports one or two or all the resources in order to maintain equilibrium and to satisfy domestic demand on those resources. In this way countries have to face the challenge of ensuring resource security. In addition, the sustainability of resources is associated key challenge for the different countries. This process is also influenced by external factors such as political, geopolitical, international relations and others. In this sense, food, water and energy security form the basis of a self-sufficient economy of a country [4]. In fact, resources security can be evaluated separately or maybe overviewed based on an integrated evaluation of different resources via the nexus and in interaction with others.

1.3. Water Energy Food nexus and food security

Food security has several indicators or pillars: availability, accessibility and utilization. These pillars are related and influenced by many factors. Food availability relates to the supply of food through production, processing distribution, and exchange. Food production is determined by a variety of factors including land ownership and use; crop, water and soil management; crop management. Food accessibility is related to affordability, purchasing power, transport and other factors. Utilization of food is related to food quality and safety, health and sanitation conditions, clean water and other factors. In fact, food security is influenced by complex systems and various factors; however, food production related factors can be the most important one; because it's the most important source of food at the local food then comes trade and imported/exported food. An efficient food production system requires a robust agricultural system and strong economic growth. But, a robust agriculture includes energy and industrial crops, not just food and it's also water consuming especially in irrigated areas. Agriculture is responsible for an important freshwater overuse. According to UNESCO, irrigated agriculture is actually responsible for about more than 70% of freshwater abstractions by humans.

Energy is required for the entire food system including food production, harvesting, transport, processing, packaging, and marketing. This has received limited attention in the water-energy-food nexus discourse so far, but is an important factor that needs to be considered [1]. And it's also important to mention that when there is an increase in agricultural productivity, the agricultural sector is supposedto become a more significant energy consumer. Going forward in this process, it is clear that energy will be a fundamental input to ensure universal food security [5].

Food production further affects the water and energy sectors through degradation of land, changes in overflow, disturbance of groundwater release, water quality, accessibility of water and land for different purposes [6]. In this way, an important pillar of food security is perturbed which is the food production ecosystem.

There is uncertainty in the food and agricultural sector with regards to policies and plans that are directly linked to the WEF nexus. As population increases a substantial increase in agricultural productivity is required. However, action plans do not consider the fact that the expansion requires more land, water and energy resources. Consequently, there is a need for changing and revisiting existing water and agricultural policies in order to prevent the impacts of increasing agricultural productivity on water and energy resources [3, 4].

A healthy WEF nexus requires the development of a sustainable agricultural policy through efficient land use and good planning, good linking between food production systems to ecosystems, innovative redesigning of waste management policy as well as new formulation of a proficient water quality strategy [3]. Several initiatives and policies were implemented in different part of the continent and worldwide which are environmentally friendly but there is still a need for more ones and more coordinated and impactful ones.

1.4. Water Energy Food nexus and climate

The exposure of the nexus components to climate is gaining more and more importance, especially under the risk of climate change. This aspect concerns the evaluation of the interaction between characteristics of the climate system, inter-annual rainfall variability for example, and the dependence of a country on climate-sensitive economic activities such as agriculture and sectors such as drinking water and energy production from hydroelectric sources [2].

Climate plays an important role in determining medium-term water availability, potential agricultural production and some components of energy production and demand. Climate variability drives fluctuations in water–energy–food elements with secondary effects across the whole nexus. Exposure to climate variability and climate change are high across nexus sectors that include substantial areas of economic activity in southern Africa and there is strong evidence of the effects of individual climate events [2].

1.5. Technology-assisted Frameworks for Water Energy Food nexus

The role of innovative technologies towards achieving sustainability is well recognized. Effective management of water and energy resources - particularly with competing large-scale usage models - implores technological intervention. For example, energy and water sectors are amongst the most significant consumers of others as a resource, globally. While there is certainly a need for basic research for transforming fundamental approaches in this regard, applied research and technological developments can assist with disrupting current practices relating to these resources - i.e. generation, transport, consumption etc. [7].

Recently, there have been tremendous technological advancements – particularly in the area of agriculture. From the Internet of Things (IoT) to enhanced Earth observation, recent technological advancements provide new avenues of better understanding, monitoring and regulating the nexus interactions. The complexity of the WEF nexus provides a challenge, in that, the interactions themselves are difficult to model, monitor and regulate. Thus, technology is fast becoming a critical enabler for development and implementation of WEF nexus frameworks, especially when scalability, flexibility and robustness are sought after features.

Modelling WEF interactions is highly complex and context specific [8], and there has been much work involving computer-based modelling of various issues within the sphere of the WEF nexus. There are several different modelling approaches put forward that vary according to modeling priorities/goals, scale and interdependencies of interest, as summarized in [9]. For example, [10] proposes a computational decision framework for modelling and analysing WEF nexus particularly in risk adverse environments. Another recent example of computational modelling in this area is presented by [11] which deal specifically with optimizing land use - in a WEF nexus framework - for agriculture application.

Technological advancements are also aiding in overcoming some of the key challenges towards implementation and utilization of the aforementioned frameworks. [12] discusses some of these key challenges: limited observability within and across WEF systems; observation data management and access (protocols) - when observation data are available - are generally inadequate. Although there is inadequate sharing of regional and global observation data is inadequate, the widespread availability of pervasive computing and IoT with the cost-effective wireless sensor technologies provides the opportunity for generating useful observation data to furnish and operationalize these WEF frameworks. An example of this is remote sensing technology - inter alia satellite-based remote sensing, aerial (manned or unmanned) vehicle, ground vehicles, ground-based sensor networks, advanced metering - which enables cheaper, robust and high-spatiotemporal resolution data collection [13].

2 Reflections on food security in Morocco and South Africa

2.1. South Africa

2.1.1. Agriculture in SA: Overview

South Africa's agriculture sector contributes by about 2.5% to its Gross Domestic Product (GDP), and has seen an average decline since 1993 [14, 15, 16]. Although the agriculture sector itself makes a relatively small contribution to the country's GDP, the agri-food system accounts for a substantially larger GDP contribution at approximately 14% [14], which has increased over the years due to supermarket expansions and further separation between the food consumer and producer. Generally, the total number of commercial farms in South Africa has decreased over the years, but farms have increased in average size and their usage of technology. This has led to a steady decline in manual labor resulting in jobs losses in both casual and permanent type work categories. The agriculture economy in South Africa is market-oriented with a highly diversified production portfolio including all major grains – excluding rice – subtropical fruits, oilseeds, sugar, citrus, wine, several vegetables, cattle, diary, hogs, sheep, poultry and eggs [17].

In addition to several major challenges that South African agriculture faces, the government recognizes two of the foremost threats to food security as climate change and sustainable resource management [14]. This is further compounded by growing population and declining farm yields. And the declining yields are owing to water and energy constraints as well as unsustainable land-use practices.

2.1.2. Risks to agricultural food production

- *Water quality and availability*

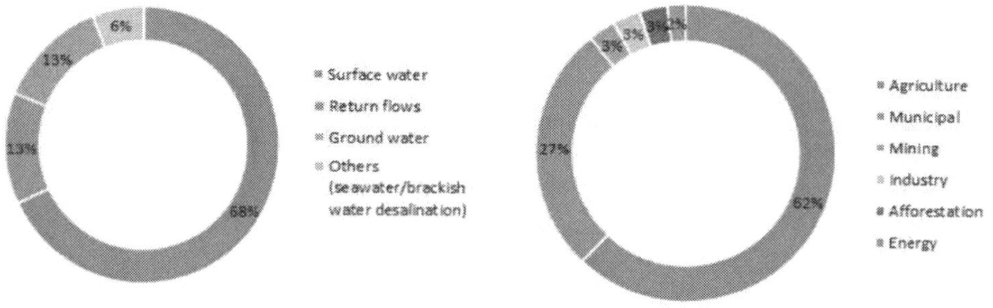

Fig.1. Water sources and usage by sector in South Africa [18]

South Africa has a semi-arid climate and is amongst the driest countries in the world ranking 30th in this regard [19]. This means that the water scarcity problem due to increasing demand, and drought resulting from climate change exacerbate the general water resource challenge in the country. In recent years, due to these constraints, the agricultural yields – particularly major grains – have been inexorably at the mercy of drought. Agriculture is by far the majority user of water in South Africa accounting for approximately 62% of the total water usage. The figure below shows the fluctuating agricultural yields (as contribution in ZAR millions to SA GDP) which were directly linked to droughts in the 2015 season, rainfall in the 2016/2017 and drought once again in the 2017/2018 season [20]. This drought period continued into 2019, and was found to be the worst in the past 100 years in the country's history [16].

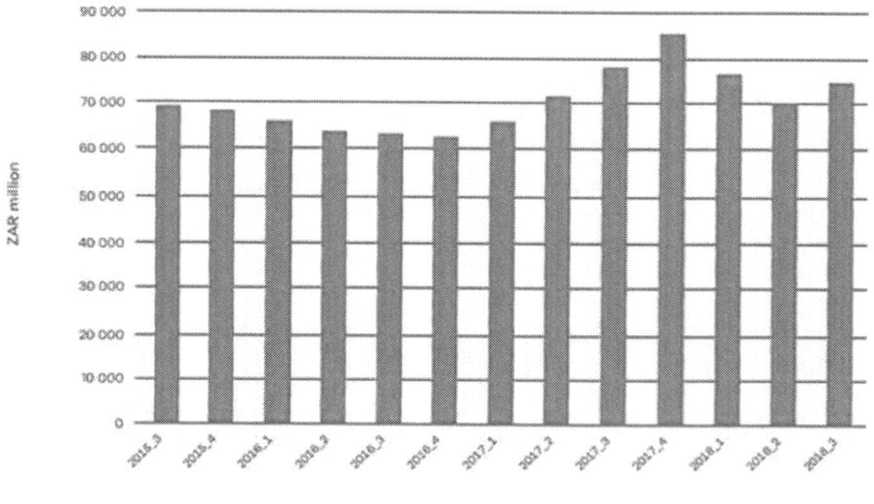

Fig.2. South African GDP from agriculture per quarter[20]

- *Electricity availability, reliability and cost*

The energy infrastructure of any country forms a crucial component of the country's economic lifeblood. In South Africa, the vast majority of the nation's energy is provided through one state-owned utility. In terms of energy sources, coal is the main source of electricity generation accounting for around 83% of the total mix - with nuclear, gas and pumped storage forming the other significant components. Since the dawn of democracy in the country, there has been a steep rise in demand, particularly with the noted increase in industrialization and massive electrification programmes for providing access to citizens via the National Development Plan.

Together with the steady increase in the country's demand, the lag in establishment of new generation capacity coupled with ageing generation fleets have resulted in an energy crisis [21]. This energy crisis has culminated in rolling blackouts and a destabilized grid which has occurred intensely over different periods beginning in 2007 and is still ongoing. This has had major impacts on all sectors of the economy. According to the World Bank, the Co2 emissions reached a value of 9 million tons per capita which comes mainly from consumption of energy, electricity production and industry.

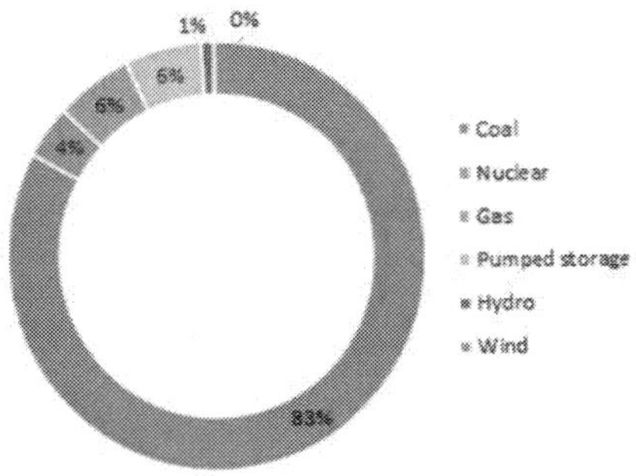

Fig.3. Energy generation mix in South Africa [22]

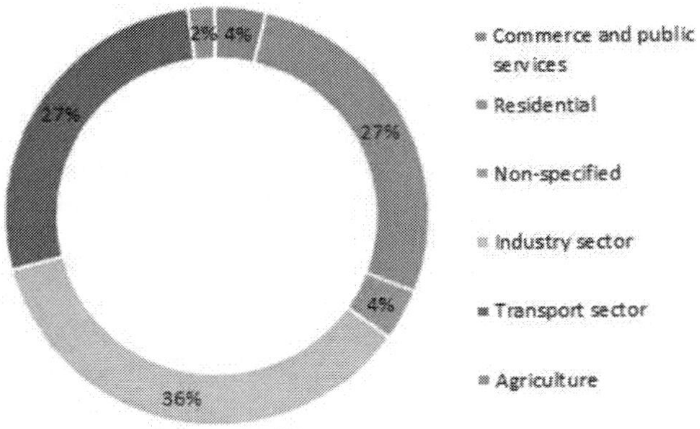

Fig.4. Energy consumption by sector in South African [22]

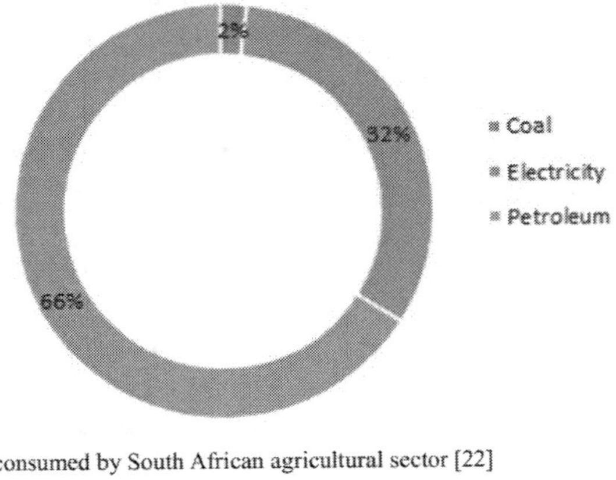

Fig.5. Types and proportions of fuel consumed by South African agricultural sector [22]

Although the Agricultural sector depends mainly on petroleum products, it also is significantly dependent on the electricity grid and is thus one of the affected sectors. Irrigated farms that use pumps for irrigation water have been especially hard hit with rising energy costs. It should also be highlighted that not only is the agricultural sector affected by the reliability and increasing costs of electricity in the country, but the food security itself is at risk with the entire food supply chain being affected.

2.1.3. Policy and Regulation

- *Food*

South African agricultural exports reached a record high in 2018 at $10.6-billion, and fell by 8% in 2019 largely due to the aforementioned drought [24]. These negative impacts on the agricultural sector are also compounded by the energy crisis and rising electricity and fuel costs. Furthermore, there has been recently declining investment in the sector due to policy uncertainty with respect to land reform and agriculture, and poor progression in the informal food production sector due to hegemony of the industrial players together with policy that largely favors them [25]. As suggested in [25], the food-policy challenges of food security, poverty, hunger and inequality in South Africa require coordinated responses across the government as water, energy and economic factors deeply influence the sector.

- *Water*

The water scarcity in South Africa, coupled with the estimated population growth and subsequent demand means that a shortfall in supply is expected. Thus, there have been significant infrastructural investments in the country in recent years with plans to improve its capacity. Water management in South Africa is a national competency as declared under the National Water Act, Act No 36 of 1998. The Department of Water and Sanitation is the ministry that is primarily responsible for water in South Africa. As Agriculture is by far the largest consumer of water in South Africa, there is a huge investment opportunity in water resources and infrastructure in this sector [18].

The forecasted scenarios for water demand have been revised in recent years and show improvements, but the revised projections are sufficient to warrant significant concern [26]. According to forecasts and analysis presented [26], it is found that only through markedly improved system efficiencies can water balance be achieved and vulnerabilities arising from other influences such as international commodity shocks. Although there have been some policy and regulation revisions in recent times with respect to water, there still exists gaps which require attention in order to close investment gaps for the necessary infrastructure improvements and to improve system efficiencies.

- *Energy*

Energy policy in a democratic South Africa has historically reflected the country's recognition for transition into more sustainable growth and development. In its White Paper on Energy Policy [27] more than twenty years ago, the government recognized the critical importance for renewable energy developments including wind, solar and hydro-power. However, despite further emphasis on this sustainable energy development including revisions to policy such as the White Paper on Renewable Energy [28] and the more recent Integrated Resource Plan [23], the adoption of these policies in terms of implementation has been slow and inadequate. Multiple regulation and policy failures have resulted in further deepening of the energy crisis resulting in huge economic losses and sustained negative impacts on all sectors including the agricultural and food production sector. The true impact of the energy crisis and policy failures is not well understood as the interrelated issues of the need for widening access, improving reliability, extending sustainable generation versus the infrastructure costs, impacts on the environment and sectors of the economy have not been adequately dealt with in studies that inform the policy and regulation.

2.1.4. WEF nexusframeworks

Energy and water play an immense role in the context of any country. As has been described, South Africa suffers from the water and energy crises having both short term and long-term effects on the country's agricultural sector and its food security. It should also be highlighted that the policies - and the scenario modeling that informs those policies - are generally done in isolation. This means that compounding or cascading effects of these strongly related issues are not sufficiently accounted for and in some cases not even considered. The risks are thus not adequately evaluated, which means that most plans are revisited because the actions do not prove to have the expected impacts. Although there is awareness of climate change and the nexus on food security, much more work is needed into transitioning this into knowledge that informs policy and regulation, which can lead to improved sustainable development strategies for the country [4].

2.2. Morocco:

2.2.1. Agriculture in Morocco: Overview

Agriculture is one of the pillars of the Moroccan economy. The agricultural production contributes by 15-20 % of the country's GDP and agro-food industry insures 5 % of the national GDP while it covers 30 %ofexport earnings. The sector is very important for job creation and employs about 40% of the country's workforce depending on the productivity of the year. The cultivated area in Morocco is around 87 million hectares of which 16 million hectares are irrigated. More than 90% of the country's agricultural production depends on rain and the outputs vary from a year to another. Agriculture products are diverse and permit the supply of the local market and also export. Thus, the main agricultural products are wheat, sugar beet, citrus fruits, tomatoes, potatoes, olives, and olive oil in addition to eggs, meat and milk. However, products such as grains (wheat, barley, maize), sugar, tea and coffee constitute the main food product depending on importation. Grain consumption is characteristic of the food system in Morocco and the national production depends on rainfall and the country imports on average more than 40 % of its needs. On the other hand, Morocco is known to be an exporter of fish and fruits and vegetables mainly to Europe[29]. The main challenge for the Moroccan agriculture is climate variation and the dependency of the sector to rainfall with an over exploitation of water resources for irrigation (use of more than 85 % of mobilized water resources).

2.2.2. Risks to agricultural food production

- *Water quality and availability*

Currently, Morocco reached "water stress" status according to the United Nations Economic Commission for Africa (UNECA) corresponding to 1000 m^3 to 1700 m^3 of available freshwater per person per year. In the future and by 2025, the country's water availability will face more difficulties under "water scarcity" state with predictions of less than 1000 m^3 of available freshwater per person per year. This situation is resulting in an increase of the water demand from different sectors and from climate instability. Predictions estimate, according to the Food and Agriculture Organization, that Morocco will experience both a decrease of 13 to 30% in precipitation and an increase of almost 3 degrees Celsius (2.3 to 2.9 °C) in annual temperatures by 2050. This impacts directly the recharge of ground water. Agriculture production consequently decreases significantly with a high variation over the years. 7.2 million hectares of agricultural land representing 83% are not irrigated and yields of major crops rely on rainfall. Significant variations of yields are due to the high variability of rainfall and a high frequency of droughts.

This situation is critical and will be more complicated for ensuring water and food security. In fact, water use efficiency should be increased via integrated water resources management strategies.

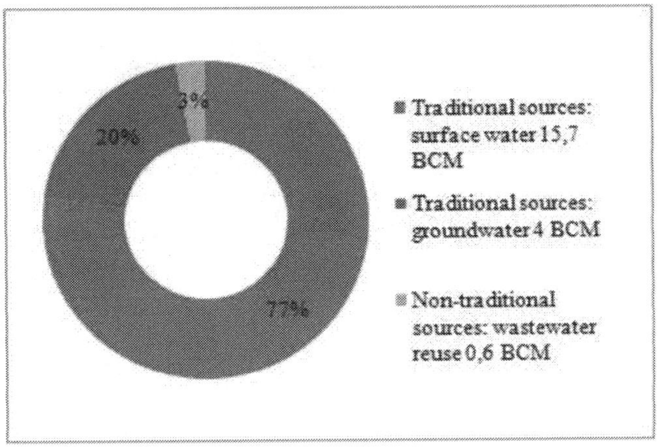

Fig.6. Water Resources in Morocco in billions of cubic meters[30]

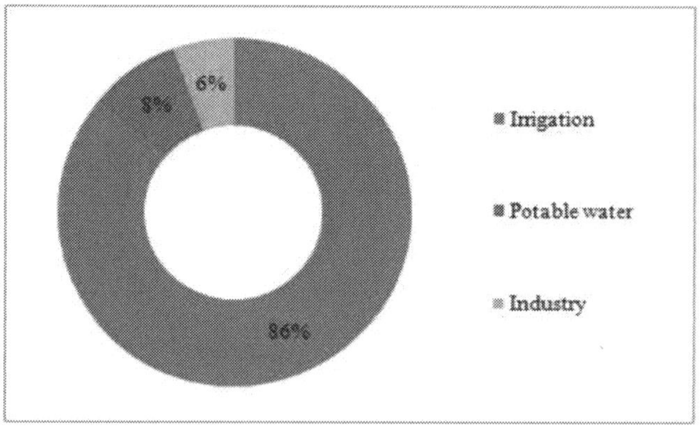

Fig.7. Water utilization in Morocco (%) [30]

Irrigation consumes more than 85 % of mobilized water (surface and groundwater) in Morocco and this shows the dependency of the agriculture sector on irrigation to insure production and the important pressure of agriculture activities on water resources.

Another big challenge for the water sector in Morocco is the increasing level of urbanization. Thus, urbanization will have a significant influence on water resources availability in the future. 60% Population of Morocco was residing in urban areas in 2015 and the future projections expect that this percentage will reach 74% by 2050. In fact, population growth and rapid urbanization, economic developments will make the competition for water resources across all sectors very complicated. Consequently, this will lead to a depletion of traditional groundwater sources. Meanwhile, water sources will be more and more exposed to contamination through changes in land use changes, weak and difficult solid waste storm water and flooding management, insufficient wastewater treatment, and weak infrastructure especially for places with informal urban expansion. Climate change is adding more uncertainty and vulnerability to these constraints. If water management was not balanced, majority of cities in morocco will be facing a water-insecure future [30].

- *Energy availability, reliability and cost*

Morocco's energy sector depends primarily on oil and coal imports that have heavy environmental and financial costs. The production of oil and coal is not present in the country and a small amount of gas is produced and it's oriented to a limited industrial use. The majority of energy needs of the country is coming from oil (67.6 %) and coal (16.1%). This corresponds to 95 % of the total primary energy supply (TPES). Electricity is also imported with an increasing cadence during the last decade in order to satisfy the demand in peak periods [31].

The consumption of energy in Morocco increased during the recent years following the economic growth (an average annual growth of 4 %), population rise and urbanization expansion [31].

Transport is the most consuming sector of energy in Morocco (35.9%) followed by the residential sector (24.9%), industry sector (23.6%), commercial and public services 8.0% and agriculture and forestry activities (7.6%). Consumption in transport and agriculture sectors increased significantly during the last few years while the consumption in residential and industry sectors decreased. However, it remained stable for commercial and public services [31]. Agriculture production activities rely on non-renewable sources of energy (diesel and gas, figure 8) as a mainsource of energy than and electricity as the second source of energy [31].According to the World Bank, the CO_2 emissions reached a value of 1.8 million tons per capita which comes mainly from consumption of energy, electricity production and industry.

https://doi.org/10.1051/e3sconf/202018302002

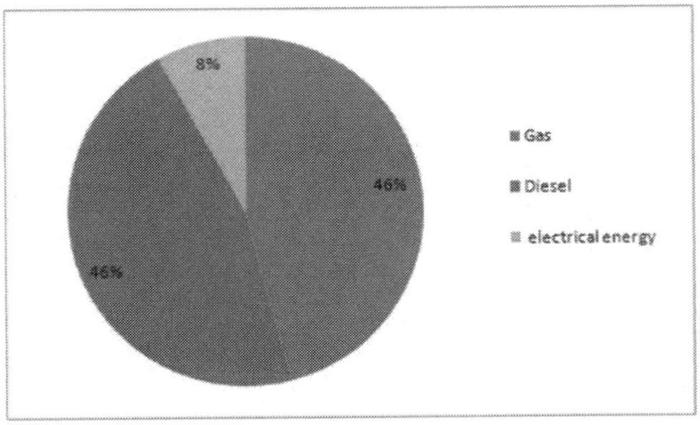

Fig.8. Distribution and types of source of energy consumed by Moroccan agricultural sector [32].

Agriculture sector consumes a significant share of national energy, mainly in irrigation systems (pumping, water distribution), tractors and engines, dryers and livestock buildings (bovine milking blocks, air conditioning and heating systems) in poultry farming, feeding equipment). On the other hand, there is indirect energy consumption resulting from the use of inputs (fertilizers, livestock feeds) and the energy impact of investments in buildings and equipment. In addition to its impact on the environmental quality of rural areas, energy consumption has a significant impact on the economic competitiveness of the Moroccan agricultural sector, severely increasing the operating expenses for producers [32]. Those activities rely mainly on the use of diesel (46 %) and liquid gas (46%) in addition to a small percentage of electrical energy (8%) [32].

Separately, food processing (with tobacco plant processing) consumes 11% of the total consumption of the industrial sector.

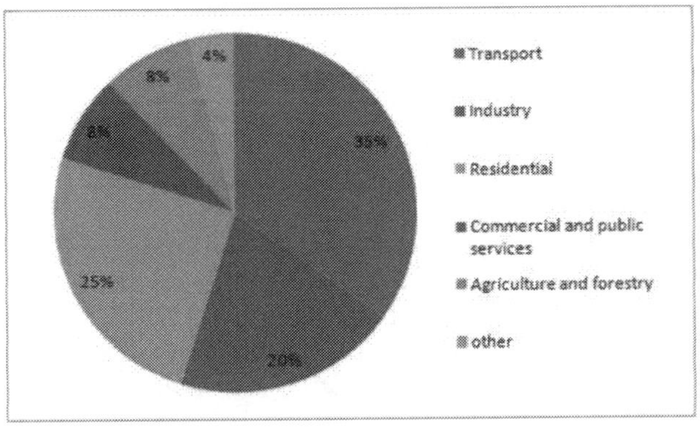

Fig.9. Energy consumption by sector in Morocco (share in TFC)[31]

Taking into consideration the dependency of the Moroccan energy sector of oil and coal imports (Figure 10) and the increasing consumption, the Moroccan government is trying to change this situation by developing renewable energy technologies in order to diversify and to enrich the future energy mix of the country. In fact, Morocco set up a strategic plan to produce more 42 % of its energy needs from renewable energy by 2020 and more than 50 % (20% solar, 20% wind, 12% hydro) by 2030. Currently, the percentage of renewables is around 34 % among the mixture of electricity [32].

 https://doi.org/10.1051/e3sconf/202018302002

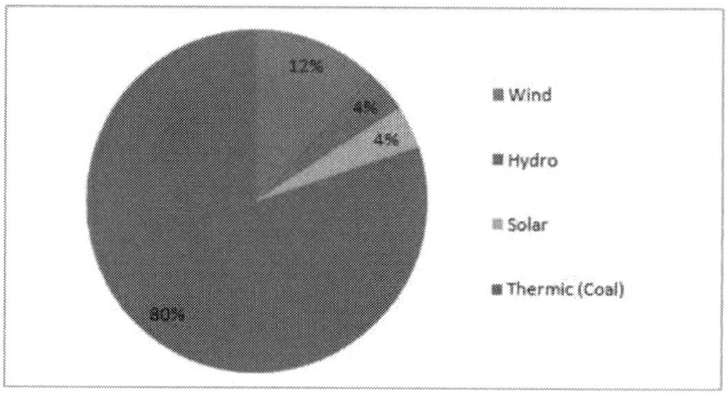

Fig.10. Energy generation mix in Morocco [31]

2.2.3. *Policy and Regulation*

- *Food:*

Morocco is known as an exporter of fresh fruits and vegetables especially tomato and oranges. The main external market is the European Union because of the historical and geographical factors. According to that, the food policy of Morocco depends on the EU with whom a pricing system is applied between the 2 partners based on bilateral agreements. Those agreements include the one related to fisheries and fish exportation to the EU countries.

Food production, distribution and control is under the responsibility of the Ministry of Agriculture, Fisheries and forestry. The agricultural production, distribution and packaging professionals are organized in associations and federations specific for the main agricultural productions. On the other hand, agriculture producers are represented by the Agriculture Chamber which is an elected body that ensures the defense of their interests and to contribute in the development of their activities and their organizations. The import and the production of seeds, plants and animals, the use of pesticides, the control of quality of food in the market is the responsibility of the national office of food safety: Office Nationale de Sécurité et Salubrité Alimentaire(ONSSA).The development of agriculture production and valuation of products projects are managed by the Agency for agriculture development. The organization of the food sector and agriculture production changed significantly after the adoption of the Moroccan Green Plan in 2008. This plan aims to double the contribution of agriculture production in the national GDP by 2020.

The Green Morocco Plan (Plan MarocVert, 2008-2020) is a national strategy adopted and implemented to make food systems more sustainable. This strategic plan aimed at the improvement of food security and the promotion of resilience, sustainability and competitiveness of the agriculture sector [33].

The Green Morocco Plan impact was a significant increase of production of certain crops, mainly fruits and vegetables, milk and meats but the problem of commercializing the excess of production occurred. In fact, the future strategy of the food production in Morocco will focus on developing and boosting the food processing sector. A new governmental strategy for agriculture development is under development and named 'Generation Green' that aims to foster the agriculture sector productivity and business for a more added value and job creation.

- *Water*

Morocco's water policy changed in time in order to face the climate constraints and the increasing and important pressure on water resources. Three important phases characterized the past of Moroccan strategies in the water sector. The first one corresponds to the post-independence era in the sixties where the focus was done on water offer management. During this phase, the hydraulic infrastructure was developed by adopting dams' construction policy in order to ensure water security and currently more than 150 dams of different sizes are available in Morocco with storage capacity of 18 Billion m^3. On the other hand, the development of large irrigated perimeters occurred for the enhancement of the agriculture sector. In a parallel with that, water sanitation development was occurring in urban areas.

During the second period, the Moroccan water policy changed significantly and was marked by the adoption in 1995 after adopting the law 10-95. These law articles aim to optimize and to organize water use in the different sectors and to provide equal access to water for the different territories of the country and it insure water security as limited and public resources. The regulations of water production and distribution, the improvement of water development and use in agriculture and the prevention of water pollution and illegal water use are other principles of water law.

The third period started at the beginning of this century with the evolution of water demand and marked by the adoption and the start of implementation of the National Water Plan in 2009. This plan focuses on the role of complementary water management actions for addressing water problems and for the achievement of a coordinated management of supply and demand, while trying to ensure equity in water distribution in rural and urban territories. In fact, this plan encouraged water valuation and water economy. Thus, the National Programme for Water Economy in irrigation (Programme National de l'Économie de l'Eau en Irrigation PNEEI) was adopted for conversion of irrigation systems to drip and pressurized ones.

Non-conventional water resources started to be mobilized during this third period with the adoption of water desalination of seawater and demineralization of brackish water. This phase also was axed on maintaining the quality of water and the protection of the resources against pollution with adoption of the National Plan of Sanitation (Plan National de l'Assainissement Liquide PNA) and the National Programme of household solid waste (Programme National des Déchets Solides Ménagers PNDM).

The water sector in Morocco is characterized by the intervention of many stakeholders. The ministry of infrastructure, transport, logistics and water is the main department for the water sector. Previously, a secretariat of state was dedicated to water and was removed from the last structure of the Moroccan government. The Supreme Council for Water and Climate and the Interministerial Commission for Water are the highest consultative and coordination instances in all water related issues. Planning and decision making for water sector includes the participation of the department of Agriculture, Fisheries and Forestry department, the internal affairs department, the health department and the economy and finance department in addition to the department of water.

Water management includes the intervention of the River basin agencies (RBAs), considered as the most important institutional instance in water resources management; The regional offices for agriculture enhancement are the mean institution for management, distribution and use in agriculture that are controlling large scale irrigated perimeters in the country. Water use and management for drinkable water services and sanitation sector is mainly controlled by the National Office for Electricity and Potable Water, Private companies and Local municipalities. The sector is characterized by the presence of water users association, especially in the agriculture sector but also for drinkable water, especially in rural areas.

- *Energy*

The energy sector belongs to the Ministry of Energy, Mines and Environment with also the presence of an important institution involved in the energy sector. The mean public institution in energy sector are : (1) The National Agency for Electricity and Water (Office National de l'Électricité et de l'Eau potable, ONEE) is a government-owned vertically integrated utility, (2) The Moroccan Agency for Sustainable Energy (MASEN) which is a limited company than includes many public shareholders responsible for the deployment of renewable energy, (3) the Institute for Research into Solar and Renewable Energies (Institut de Recherche en Énergie Solaire et Énergies Nouvelles, IRESEN) is an agency for energy research, development and innovation (RD&I) created in 2011 in order to support the national energy strategy through applied research and innovation and (4) The Moroccan Agency for Energy Efficiency (AMEE Agence Marocaine pour l'Efficacité Énergétique) [34].

The presence of such an institutional organization helped in fostering the change in the national strategies for energy management and development.

In the last decade Morocco has been adopting new legislative text, policies and strategies in the energy sector in order to respond to the future challenges of the energy sectors and related issues. The government adopted the National Priority Action Plan for Energy from 2008-2012 and in 2009 the National Energy Strategy was launched. In 2013 a national debate was launched by the Moroccan Agency for Energy Efficiency for adopting a new national Energy Efficiency Strategy for 2030 in order to significantly reduce the energy consumption in different sectors. The draft of this strategy was presented by the government but its adoption is not official yet. An action plan was prepared in parallel with this strategy in order to implement a series of actions in the different regions of the country with the participation of the different ministries (transport, housing, interior, industry, agriculture and finance) with responsible ministry and agencies. A positive economic, environmental and societal impact is expected as a result of this action plan in the energy sector [30].

On the other hand, a Law 47-09 on energy efficiency was adopted in September 2011 for creating a legal framework for improving energy efficiency across the different economic sectors.

Initiatives in the agriculture sector for reducing its energy consumption bill and its carbon footprint are conducted by the Ministry of Agriculture, forestry and rural development and the Moroccan Agency for energy efficiency (AMEE). The AMEE intervenes at the level of control energy consumption: develop alternative energy, through a program dedicated to promoting solar pumping, and on second level for reducing energy consumption, through a program aiming the optimization of energy performance of farms, launched in partnership with Crédit Agricole du Maroc which is a public bank [32].10 pilot farms were identified for testing the control and reduction of energy consumption and the results were promising and a guide of the best practices of energy efficiency and the use of renewables in Moroccan agriculture production sites. On the other hand, subsidies of solar pumping equipment are scheduled by the Ministry of agriculture in the short term for the replacement of diesel and liquid gas for water pumping.

2.2.4. WEF nexusframeworks

Water scarcityin Morocco is becoming very serious in recent years. In fact, alternative solutions to freshwater under stress needs to be found and adopted urgently. Desalination of seawater is one of the promising solutions. Moroccan has a very long coastal line, nearly 2,000 km (Atlantic Ocean and the Mediterranean Sea), offering an ample sea and ocean water to desalinate. However, important investments are required for water treatment into the high cost of energy required for running the treatment plant. The desalination plants can satisfy or replace the pressure on groundwater and surface water used mainly for irrigation of crops (More than 85 % of mobilized freshwater resources). This will contribute significantly in the reduction of variation of rainfall over the years and the potential climate change on crop production and food security in the country. Thus, Morocco has to find a solution to its dependency on oil and coal imports for energy production and invest in integrated renewable energy production and water desalination plants. The water energy nexus needs to be studied in order to evaluate the impacts of such a kind of solution and different future scenarios and the most sustainable solution for water. One of the initiatives on the nexus concerns a Water-Energy Nexus platform which was planned to be launched in 2019 by IRESEN and will be dedicated to water treatment, water desalination, and the linkages between the energy and water sectors [30]. This platform is an example of solutions that may help in disseminating technical reports and exposing successful experiences and best practices related to the water energy nexus and that will impact automatically the agriculture sector and food security of the country.

2.3. General overview on food security in both countries:

Cereal import dependency ratio and the value of food imports in total merchandise exports [35] are two important FAO indicators that can help analyzing food security in a specific country. They are used to evaluate partially the stability criteria as one the food security pillars in both countries Morocco and South Africa. The third indicator that was compared is the average value of food production which is on the criteria used to evaluate availability as also a pillar of food security.

2.3.1. Cereal import dependency ratio

The most important cereals are Maize, rice and wheat. They constitute the most widely cultivated crops in the world and the foundation of world food security. On the other hand, they are critical to food security especially in the developing regions. In the Southern Africa region, they make up half the calorie supply [36]. Thus, this ration was chosen to compare the situation of food security in Morocco and South Africa based on this ratio.

The analysis of the available data for those indexes (Fig.11) shows that Morocco has a higher ratio for cereals import dependency compared to the ratio of South Africa and compared to the world ratio that is about 16 percent [35]and the one of the developing countries that is 15.5 percent [35]. However, the South African ratio is still low compared to these two values. Negative values reveal Morocco's dependency on cereals imports because the cereals production in the country is mainly practiced in rain-fed areas and still relies on rainfall which is directly related to water resources and water statusof the country. Variation in the Moroccan is due the introduction of new varieties more tolerant to drought and a change in the cultivation system (zero-tillage cultivation, fertilization) in addition to climate variation that is the main factor of this variation.

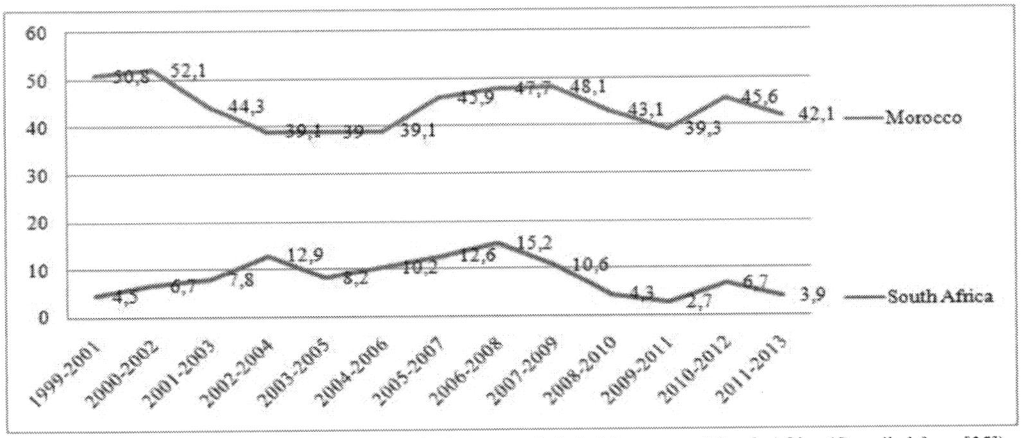

Fig.11. Evolution of Cereal import dependency ratio (%) for 3 years periods in Morocco and South Africa (Compiled from [35])

2.3.2. *Value of food imports in total merchandise exports*

The value of food imports in total merchandise exports indicator provides a measure of vulnerability and captures the adequacy of foreign exchange reserves to pay for food imports, which has implications for national food security depending on production and trade patterns. The analysis of the available data for this indicator shows a higher value for Morocco indicator in comparison with its value for South Africa (Fig.12) and to the average value for the World that is around 5 % [35]. This shows that this indicator South Africa is not higher and the export value of goods permits to cover the food imports. This shows how important the foreign trade balance and its direct connection and influence on food security that is represented here by the value of food export.

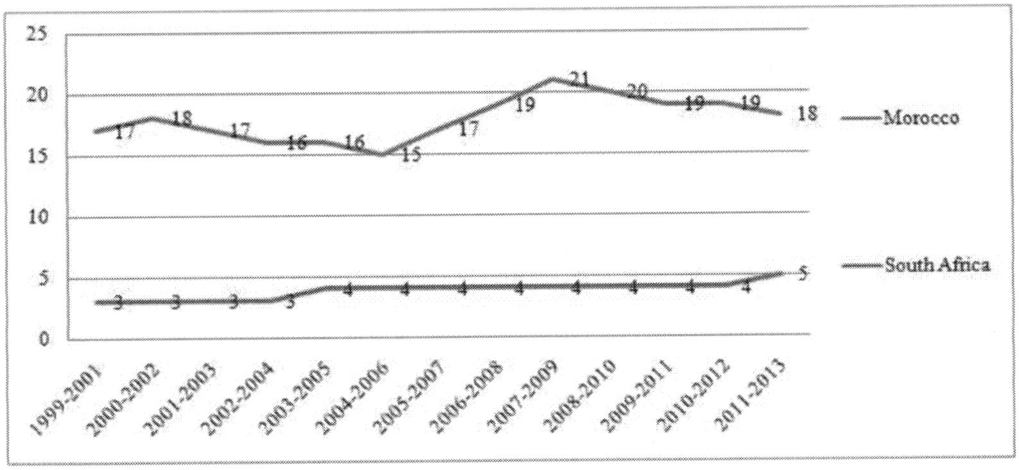

Fig.12. Evolution of Value of food imports in total merchandise exports (%) for 3 years periods in Morocco and South Africa (Compiled from [35])

2.3.3. *Average value of food production*

The average value of food production provides a cross country comparable measure of the relative economic size of the food production sector in both countries. The comparison of this value for Morocco and South Africa showed an aligned evolution across the years with slightly higher values for Morocco between 2008 and 2016 period (Fig.13). However, the values for both countries are lower the average value for food production in the world which is around 283 (constant 2004-

2006 I$/cap), and very low in comparison of the average of the higher income countries that have a value around 480 (constant 2004-2006 I$/cap) for the same periods of years [35].

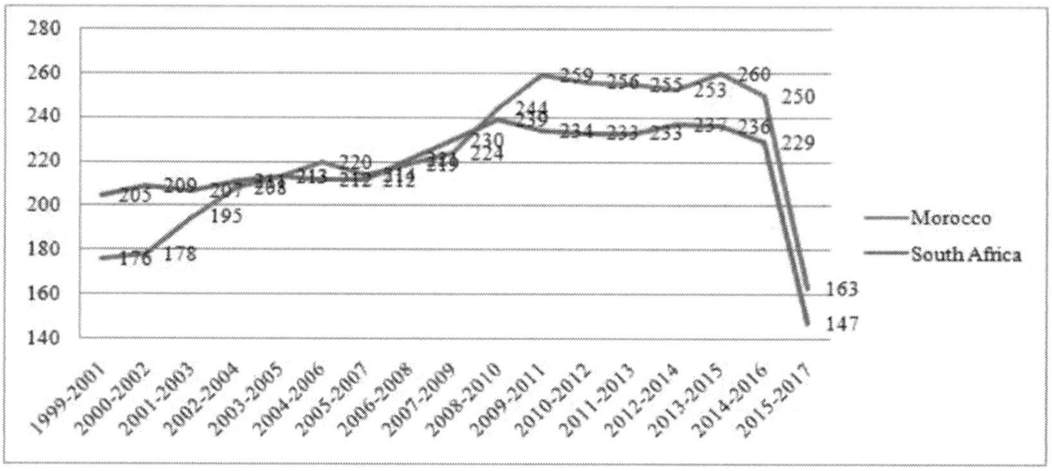

Fig.13. Evolution of the average value of food production (constant 2004-2006 I$/cap) (3-year average) for Morocco and South Africa (Compiled from [35])

Conclusion

This review paper aims to compare the statute of the Water Energy and Food nexus in Morocco and South Africa and describe the basic frameworks in both countries. It is found that the existing frameworks are inadequate and do not sufficiently reflect, or not all in most instances, the interconnection between the different components of the Nexus - especially for the agricultural sector and with respect to food security. Although the water and energy systems are well developed in both countries, there is a need to develop more complex and complete WEF Nexus frameworks for both countries.

The research of WEF Nexus policies showed that there is a lack of policies that take into consideration all the nexus components. In fact, one of the key findings in this paper is that policies are largely formed in "isolation", and there is a lack of consideration of interdependencies - i.e. WEF Nexus in both countries. For this reason, it is thus impracticable to even begin to make meaningful comparisons and direct analyses of risks for water, energy and food components. There is overwhelming evidence to indicate as the complexity of these systems grows, assessments and analyses in practice are best made via the WEF Nexus. The complexity of gathering data and modeling these complex interconnections is also recognized and thus technology-assisted WEF nexus frameworks are suggested in this paper as a means toward overcoming challenges with implementation. Finally, it is recommended that models for the evaluation of the WEF Nexus be adopted as test approaches of the policy makers in Morocco and South Africa, which is expected to significantly contribute in informing and facilitating the decision making process.

References

1. H. Bellfield. *Water, Energy and Food Security Nexus in Latin America and the Caribbean*. (IUCN/IWA/GIZ, 2015)
2. D. Conway, E. A. Van Garderen, D. Deryng, S. Dorling, T.Krueger, W. Landman, B. Lankford, Nat Clim Chang **9** (2015)
3. G. Simpson, T. Mabhaudhi, J. Badenhorst, M. Mohammed, T. Motongera, A. Senzanje, G. Jewitt *Assessing the State of the Water-Energy-Food (WEF) Nexus in South Africa* (2018)

4. S. Carter, M. Gulati, *Climate change, the Food Energy Water Nexus and food security in South Africa. Understanding the Food Energy Water Nexus*. WWF-SA, South Africa, (2014)
5. R. Ferroukhi, D. Nagpal, A. Lopez-Peña, T. Hodges, R. Mohtar, B. Daher, M. Keulertz, *Renewable energy in the water, energy, and food nexus*, (IRENA, 2015)
6. M. Sajid, I. Muhammad, R. Muhammad, B. Saqib, Y. Gutha.*A Report on Water, Energy and Food Relationship*.Int.J. of Nutr.,**3** (2019)
7. J.J. Urban, Joule, **1** (2017)
8. B. Daher, R. H. Mohtar, H. Lee, A. Assi, *"Modeling the water-energy food nexus," in Water-Energy-Food Nexus: Principles and Practices*, eds P. Abdul Salam, S. Shrestha, V. Prasad Pandey, and A. K. Anal (Washington, DC: John Wiley and Sons, Inc., 2017)
9. C. Zhang,X.ChenY., Li,W. Ding, G. Fu, J.ofclean.prod., **195** (2018)
10. R. Govindan, T. Al-Ansari, Renew. Sust. Energ. Rev., **112** (2019)
11. Y. Nie, S.Avraamidou, X. Xiao, E.N.Pistikopoulos, J. Li, Y.Zeng, F. Song, J. Yu, M.Zhu, Total Environ., **659** (2019)
12. E. Eftelioglu, E. Jiang, X. Tang, S. Shekhar2017. *The nexus of food, energy, and water resources: Visions and challenges in spatial computing*, in Advances in geocomputation, Springer, Cham (2017)
13. K.T. Sanders, S.F. Masri, J.clean. prod., **117** (2016)
14. Department of Agriculture, Forestry and Fisheries (DAFF),(Strategic Plan. Republic of South Africa, 2015)
15. Bureau of Food and Agricultural Policy (BFAP), (BFAP Baseline: Agricultural Outlook (2018–2027), 2014).
16. Parliamentary Monitoring Group (PMG) .*Drought update: Provinces' state of readiness for 2019/20 planting season*, National Assembly Committee on Agriculture, Land Reform and Rural Development, Meeting Summary, (2019) https://pmg.org.za/committee-meeting/29261/
17. IBP Inc.,*South Africa Investment and Business Guide. Volume 1 Strategic and Practical Information. Lulu.com: Wo*rld Business and Investment Library, (2015).
18. Reddick J., R. Kruger, *Water market intelligence report*. (Green Cape, 2019).
19. I. Kuschke, A. Cassim, *Sustainable Agriculture: Market Intelligence Report 2019*, GreenCape (2019)
20. Department of Agriculture*, Forestry and Fisheries (DAFF, (*Economic Review of the South African Agriculture 2017/18. Republicof South Africa, *2018).*
21. K. Ratshomo, R.Nembahe, *South African energy sector report*. Department of Energy – Republicof South Africa, 2018)
22. Department of Energy (DoE), *The South African Energy Sector Report 2019*, (Republic of South Africa, 2019)
23. Department of Energy (DoE), *Integrated Resource Plan*, (Republicof South Africa, 2019)
24. W. Sihlobo,(Understanding South Africa's agriculture trade patterns, Daily Maverick, 2020) https://www.dailymaverick.co.za/opinionista/2020-02-17-understanding-south-africas-agriculture-trade-patterns/#gsc.tab=0
25. T. von Bormann (WWF-South Africa, 2019)
26. Z. Donnenfeld, C. Crookes, S.Hedden. *A delicate balance: Water scarcity in South Africa*, (ISS Southern Africa Report, 2018)
27. DME, *White paper on the energy policy of the Republic of South Africa* (Department of Minerals and Energ, 1998). http://www.energy.gov.za/files/policies/whitepaper_energypolicy_1998.pdf
28. Department of Minerals and Energy (DME) (White paper on renewable energy, Republic of South Africa, 2003)
29. Agence pour le DéveloppmentAgricole (ADA), (investor's guide in the agricultural sector in Morocco, 2015)http://www.agriculture.gov.ma/sites/default/files/investors_guide_in_the_agricultural_sector_in_morocco.pdf

30. World Bank Group (WBG) (Managing Urban Water Scarcity in Morocco, 2017) https://openknowledge.worldbank.org/handle/10986/29190
31. International Energy Institution (IEA)(Energy policies beyond IEA countries: Morocco, 2019)
32. AMEE Agence Marocaine pour l'Efficacité Énergétique, Website, (2019) http://www.amee.ma/index.php?option=com_content&view=article&id=138&Itemid=232&lang=fr
33. World Food Programme WFP, Morocco country strategic plan (2019–2021), Executive Board document, Annual session Rome, 10–14 June 2019)
34. MMEME, Moroccan Ministry of Energy, Mines and Environment Website, (2019) http://www.mem.gov.ma/SitePages/Ministere/Organigramme.aspx#
35. FAO. FAOSTAT, (AccessedMay 15, 2020) http://www.fao.org/faostat/en/#data/FS.
36. FAO, (Save and Grow: Maize, Rice and Wheat: A Guide to Sustainable Crop Production, 2016). http://www.fao.org/3/a-i4009e.pdf

Spatial and temporal rainfall variability and erosivity: Case of the Issen watershed, SW-Morocco

Mohamed Ait Haddou[1*], *Belkacem* Kabbachi[1], *Ali* Aydda[1], *Hicham* Gougni[1], and *Youssef* Bouchriti[1]

[1] Geosciences and Environment Team (EGE), Department of Geology, Faculty of Sciences, Ibn Zohr University, Agadir, Morocco

Abstract. The Issen basin is a dry area affected by climate variability and desertification. It is located in high spot of Moroccan flora biodiversity that includes argan tree ecosystem as a part of the Western High Atlas (WHA). The objective of this work is to assess rainfall variability and erosivity as determining factors of soil erosion and vegetation cover degradation. The results reveal an arid to superior semi-arid climate where annual rainfall range from 200 mm in the centre of the Argana corridor, to 500 mm in the snow-covered northern foothills. The analysis of the 37 years of rainfall data indicate a coefficient of variation Cvis about67%. The erosivity index presents a high variability range from 34 to more than 81 MJ.mm/ha.h.yr in the wettest areas, indicating considerable soil erosion. In addition, the basin suffering from severe periods of drought due to the occurrence of the continuous decrease of the accumulated precipitation. Eventually, the rainfall variability and erosivity can affect ecosystem function and services (ecosystem degradation, water availability, etc).

1 Introduction

The Moroccan country characterized by three different climatic systems; the humid Mediterranean and humid Atlantic climates in the north and the arid Saharan climate in the south [1]. During the last decades, it was affected by a longest dry period which characterized by rainfall decrease and temperatures increase [2]. In fact, 53% of the total country was affected [3]. The rainfall decrease affects the vegetation cover growth, especially, in the Moroccan High Atlas mountains [4].

The ecosystems and agriculture activities in the semi-arid areas principally depend on the water availability [3]. In fact, the analysis of the rainfall data is crucial for assessing current environmental constraints and for optimal management of the limited water resources [5].In general, the climate change amplify existing risks and generate new ones for natural and human systems [6].

The Issen basin is the largest sub-basin in the right bank of the Souss river which covers an area of 1302 km² (fig. 1). The Issen river is the most important tributary descending from the Western High Atlas (WHA), and it is a part of the Argana corridor and its annexed

*Correspondingauthor:medaithaddou@gmail.com

valleys considered as a suitable area to characterize rainfall instability in the High Atlas Mountains [7]. Geologically, the basin cover all Permo-Triassic red sandstone, detrital sandstone and conglomeratic silty sandstone Formations, which including at least 2500 m of fluvial, lacustrine and clastic deposits and coarse to fine floodplains [8]. Based on the national soil map database (EuDASM) [9], the red mountain forest soils and the brown and red lightly forested podzolic soils with very heterogeneous vegetation coverare developed on the Triassic Formations. The depression between the ancient massif and the western plateaus creates a natural passageway for the air currents which developing a particular and complex climatic conditions [10].

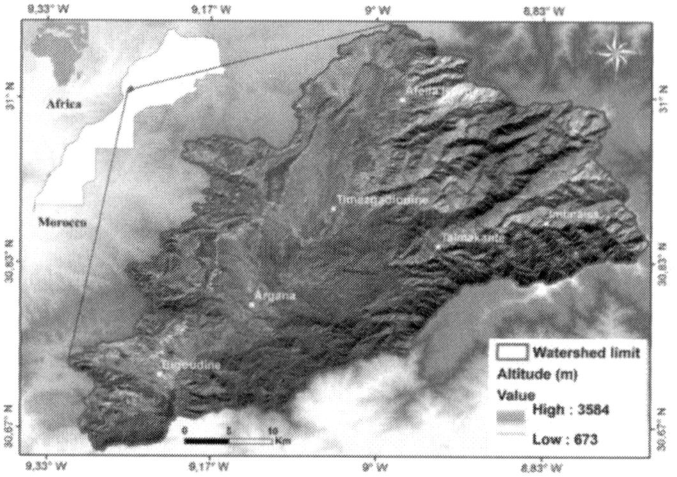

Fig. 1. Study area location.

Topographically, the study area is a sloped land (the average slope is about 33%). In addition, the study area is a fragile environment influenced by the pedoclimatic hazards and the frequency of atmospheric disturbances, which affectingthe vegetation distribution (Fig. 2). Also, the study area is characterized by the presence of the argan steppes. Due to the climatic aggressiveness andthe high anthropogenic pressure, the High Atlas mountains and the Issen basin as a part of it suffering from water erosion risks, including soil degradation and flooding [11, 12].

The vegetation cover in the Issen basin includes the argan tree (*Arganiaspinosa*) ecosystems, which is considered as a natural, national and universal heritage, associated with endemic or rare plants serving several applications; aromatic, medicinal, melliferous and pastoral [13].

 https://doi.org/10.1051/e3sconf/202018302003

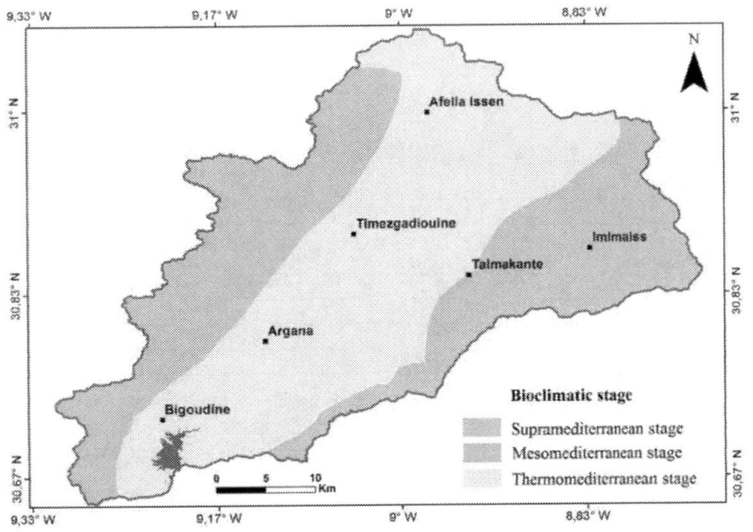

Fig.2. Map shows the bioclimatic zones in the Issen basin [14].

The present study aims to analyse the contemporary spatial and temporal variability of rainfall and its associated risk responsible of the soil degradation. Practically, we calculated simple and combined climatic indices commonly used in Mediterranean countries as a useful metric parameters to describe the climate variability. In fact, the climate variability may induces perenity of the remarkable endemic flora species and the water resources stored in the dams that ensuring the drinking water supply for the Agadir metropolis (9.5 Mm^3) and the agricultural irrigation for the Issen perimeter (13,000 ha) [15]. Moreover, we mapped the spatial distribution of rainfall and the erosivity in order to determine the susceptible areas to water erosion which will be helpful to manage the erosion risks.

2 Materials and methods

2.1 Data availability

Climatic data were obtained from the regional climate dataset compiled by the Souss Massaand Tensift Hydraulic Basin Agencies (ABH) and the stations under control of the South-West and High Atlas HCEFLCD (Tab. 1 and 2). The collected rainfall data covering the period of 37 to 50 years. In this study, we analyzed the data covering a satisfactory number of years without missing data.

Table 1. Characteristics of the rainfall stations in the Issen basin and adjacent zones. (ABH, DREFLCD)

Station	Lng. °E X	Lat.°N Y	Altitude Z (m)	Distance to the ocean (Km)	Annual Rainfall (mm)
A.Elmoumen dam	-9.190	30.670	635	49	402.05
Aguenza	-9.153	30.751	720	57	253.47
Argana	-9.122	30.778	771	61	349.66
Amsoul	-9.069	30.842	871	70	193.77
Tizguine	-9.120	30.918	1329	67	236.63
Iloudjane	-8.801	31.183	789	98	332.15
Aghbar WHA	-8.367	30.880	1750	130	537.40
TiziOu-Maacho	-8.968	31.058	1580	84	465.48
Ikakern	-8.975	30.832	1100	79	222.73
Timlilt	-9.139	31.033	1350	67	344.00
Dkhila dam	-9.285	30.569	232	44	307.67
Ain Asmama	-9.286	30.773	1600	50	473.72
Immouzzer	-9.400	30.670	1199	27	540.78

Table 2. Statistical characteristics of the precipitations (mm) in the Issen basin and the adjacent zones (ABH, DREFLCD). A.R: Annual rainfall, S.Div: Standard deviation, Cv: Coefficient of variance, and Q: Quartel

	A.Elmomen dam	Agunza	Argana	Amsoul	Tizguine	Iloudjane	Aghbar HAO	Tizi-Ou-Maacho	Ikakern	Timlilt	Dkhila dam	AinAsmama	Immouzzer
Sep.	11.57	4.48	9.96	5.16	13.75	10.45	20.50	12.33	7.43	24.6	5.07	25.12	9.16
Oct.	27.52	21.99	27.29	19.26	17.92	35.79	50.10	41.34	24.29	33.83	23.65	40.57	62.78
Nov.	78.20	46.53	54.52	31.43	29.83	36.13	73.90	59.23	52.16	25.89	57.95	82.52	85.34
Dec.	98.97	54.98	84.40	31.58	52.75	37.12	81.90	76.40	22.21	67.36	72.86	75.13	99.07
Jan.	40.47	29.12	45.76	29.36	33.25	41.82	105.90	55.30	34.67	31.69	36.16	76.52	89.17
Feb.	65.18	36.12	61.90	25.57	40.79	39.54	58.20	66.88	24.62	42.91	41.89	71.46	83.17
Mar.	48.90	36.09	41.87	27.00	20.46	46.02	88.90	66.18	33.63	40.21	46.60	53.96	63.21
Apr.	19.65	12.90	15.55	11.85	15.21	44.38	29.70	54.58	15.76	50.36	13.86	32.69	36.19
May	8.37	4.99	5.04	5.27	2.33	20.85	15.90	17.11	2.90	16.07	7.46	8.67	10.42
Jun.	0.99	3.00	1.13	1.07	2.42	11.17	4.70	4.20	0.67	2	0.48	2.73	1.58
Jul.	0.78	0.74	1.20	2.38	0.67	3.03	2.80	1.18	0.18	0.29	0.08	0.98	0.13
Agu	1.44	2.80	1.05	3.84	7.25	5.86	4.90	10.75	4.22	8.80	1.63	3.36	0.57
A.R.	402.06	253.74	349.66	193.77	236.63	332.15	537.4	465.48	222.73	344	307.67	473.72	540.78
Min	33.8	45	70.3	85	167	145.1	134	195.7	69	219.2	38.7	154.2	154.4
Max	1474.7	946	1801	485	314	519.1	1220.5	1109.2	635.5	512	928.2	977.1	1055.6
S.Div	263.9	195.03	332.65	100.98	60.37	100.94	304.01	330.078	155.17	78.13	220.83	224.34	248.31
Cv (%)	65.00	76.86	95.14	52.11	25.51	30.39	56.57	70.91	69.67	22.71	71.77	47.36	45.92
Q1	140.60	118.00	144.30	134.33	167.50	244.40	422.43	202.23	96.20	288.29	144.13	279.20	290.15
Q2	361.05	223.00	268.40	201.00	236.63	332.15	564.45	322.75	214.15	351.21	255.45	446.90	401.00
Q3	474.40	315.50	400.63	269.75	301.50	244.40	819.27	676.78	282.25	401.60	255.45	555.85	585.75

https://doi.org/10.1051/e3sconf/202018302003

The temperature is essential data because it is influences the spatial distribution of the vegetation. Table 2 shows the maxima (M), minima (m) and monthly means (Tm) of temperatures in °C of the three stations.

Table 2. Thermal characteristics of the Abdelmomen, Argana and Immouzzerstations.

Station	Para-meter	Ja.	Fe.	Ma.	Ap.	Ma.	Jun	Jul.	Ag.	Sp.	Oc.	Nv.	De.	Mean
Abdelm. dam (1)	M	25.4	22.8	24.2	28.5	28.6	34.2	39	38.4	33	32.1	29.4	27	30.22
	M	9.5	11.8	13	15.7	16.7	20.6	25.4	25.6	21.8	19	16	11	17.18
	Tm	17.5	17.3	18.6	22.1	22.7	27.4	32.2	32	27.4	25.6	22.7	19	24.25
Argana (3)	M	18.8	21.2	24.4	27.3	29.9	32.8	40.6	41.8	36.6	30.6	24.7	19.6	29.03
	m	1.9	3.4	5.5	7.3	9.4	14.9	15	16	13.3	10.4	6.8	2.8	8.89
	Tm	10.4	12.3	15	17.3	19.7	23.9	27.8	28.9	25	20.5	15.8	11.2	21.85
Imm. (13)	M	14	16	19	21	23	29	34	32	27	22	18	16	22.58
	m	5	6	8	9	10	14	21	19	16	12	9	7	11.33
	Tm	9.5	11	13.5	15	16.5	21.5	27.5	25.5	21.5	17	13.5	11.5	19.5

2.2 Data preprocessing

First, based on the monthly and annual averages of the collected rainfall data the missed values was deduced in order to complete datasets. Secondly, the monthly and annual averages were also used to calculate the climate indices (seasonal regime, monthly drought, deviation-to-mean index (Em), Standardized Precipitation Index (SPI) and rainfall erosivity (R-factor)). Finally, all the annual averages of precipitation and R-factor values were integrated and analyzed in a GIS tool. In fact, mapping rainfall distribution and erosion risk areas was achieved using interpolation krigging method.

2.3 Rainfall variabilityand trends

The rainfall deficit, the extent of the drought and its intensity were evaluated using thefollowingindices:

2.3.1Monthly drought

The monthly drought have been assessed based on the total rainfall (P) and the average monthly temperature (T). [16] reported that a month be dry if $P \leq 2T$. In other hand, [17] indicated that if $P < 3T$ or $P < 4T$ the month is dry.

2.3.2 Deviation-to-mean index (Em)

Emis the most used index to estimate the rainfall deficit [18,19] (equation 1). It is positive for wet years and negative for dry years.

$$Em = Xi - Xm \quad (1)$$

Where: Xi is the cumulative rainfall for year i, and Xm is the mean annual precipitation

2.3.3 Standardized Precipitation Index (SPI)

The Standardized Precipitation Index (SPI) was developed by Mckee et al. [20]. The SPI have been used to verify humid periods as well as dry periods according to equation 2.

$$SPI = (Xi - Xm) / Si \quad (2)$$

Where: Xi is the cumulative rainfall for year i, Xm is the mean annual precipitation, and Si is the standard deviation of the time series annual rainfall.
Effectively, several studies have been used the SPI and defined the different classes of drought severity [5,18,19,21]. The negative annual values indicate a dryperiod and positive values indicate a humid period (Tab. 3).

Table 3. Drought degree according toSPI by [5,18,21].

Drought degree (Class)	SPI Value
Extreme humid	SPI >2
Severe humid	1.5 <SPI<2
Moderate humid	1<SPI<1.5
Near normal	-1<SPI<1
Moderate drought	-1<SPI<-1.5
Severe drought	-1.5 <SPI<-2
Extreme drought	SPI < - 2

2.4 Rainfall Erosivity R-Factor

The rainfall erosivityis critical factor to assess the soil erosion risks [22].The rainfall erosivityis the product of the energy kinetics of rainy event and its maximum intensity in 30 minutes (IE30) [23].Wischmeier& Smith [24] provided a complete description of the rainfall erosivity. Arnoldus [25] proposed the Modified Fournier Index (IFM) which combines both the monthly average rainfall for all months of the year (Rm) and the annual average (Xm) (equation 3).

$$R \simeq IFM = \sum_{i=1}^{12} (\text{Rm}^2) / Xm \quad (3)$$

In this study, due to do not have sufficiently rainfall stations we used linear interpolation to generate the erosivity map. In general, the interpolation method guarantees the minimum variance topredict unknown values.

3 Results and discussion

3.1 Rainfall variability and trend

3.1.1 Annual and seasonal average

The isohyets of the Issen basin range between 100 and 700 mm (Fig. 3). The analysis of the obtained map reveals a high spatial irregularity of the total precipitation. In general, therainfall increasesfrom the center of the Argana corridor to its borders in relationship with altitude.In addition, the rainfall is lower and relatively homogeneous in the center of the

 https://doi.org/10.1051/e3sconf/202018302003

basin. The rainfall ranges from 100 to 350 mm/year in the Timzgadiouine and Argana basin because they are well protected by the high plateaus of the Ida-ou-Bouzia and Ida-ou-Tanana which brock up the oceanic effects.Furthermore, due to its high altitude, the ancient WHA massif (the snow-covered northern foothills and the Ida-Ou-Tanan plateaus) receives an average precipitation more than 400 mm/year.

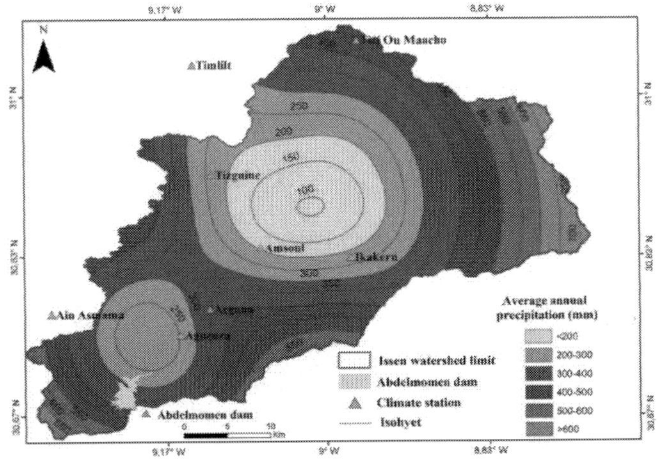

Fig. 3. Spatial distribution of the average annual precipitation in the Issen basin.

During the period 1980-2017 (fig. 4a), the arithmetic mean of rainfall is 302 mm/year, associated with a coefficient of variation Cv = 51% during the 37 years, which reflect a great fluctuation and variability of rainfall. The most remarkable peaks are at 86/87, 95/96 and 09/10 which represent the wettest years, while the peakat2011/12 represents the driest year. The whole basinreceives an average of 141 mm at winter and half at autumn and spring. However, dry season is summer during it the basin receives the minimum amount of the annual precipitation (Fig. 4b).
The spatial and temporal precipitation variability may affects vegetation development, which increasing soil erosion and then environment degradation [26].

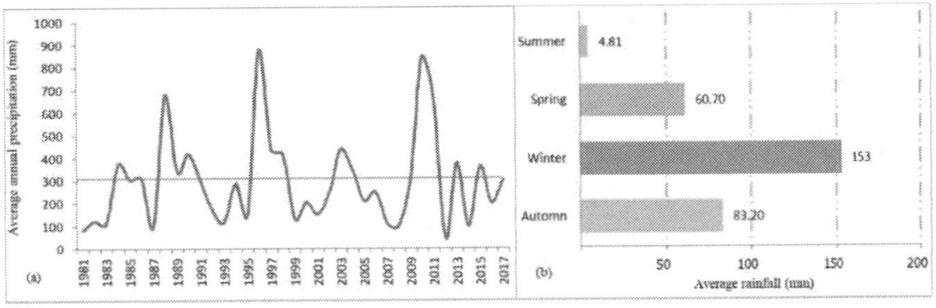

Fig. 4. (a) Interannual variation of rainfall from 1980 to 2017, (b) distribution of average seasonal precipitation in the Issen basin

3.1.2 Seasonal regime and monthly drought

Table 4 resumes the seasonal regime in the Issen basin. The winter season is the wettest, which represents 50% of the annual total. The maximum winter rainfall is about 271 and 246 m observed in the Immouzzer and Aghbar stations, respectively. The summer season is the driest, which represents only 1 to 6% of the annual total.

Table 4. Seasonal regime recorded in the studied stations

Stations	**Precipitation (mm)**	**Automn S-O-N M2 (mm)**	**Winter D-J-F M1 (mm)**	**Spring M-A-M m1 (mm)**	**Summer J-Jl-A m2 (mm)**	**Seasonal regime**
Immouzzer	540.78 (100%)	157.28 (29%)	271.41 (50%)	109.82 (20%)	2.3 (0.5%)	WASS
Dkhila dam	307.69	86.67(28)	150.91 (49)	67.92(22)	2.19 (1)	WASS
Abdelm. dam	402.04	117.29(29)	204.62(51)	76.92 (19)	3.21 (1)	WASS
Ain Asmama	473.72	148.21 (31)	223.11 (47)	95.32 (20)	7.07 (2)	WASS
Agunza	253.74	73.00 (29)	120.22 (47)	53.98 (21)	6.54 (3)	WASS
Argana	349.67	91.77 (26)	192.06 (55)	62.46 (18)	3.38 (1)	WASS
Ikakern	222.74	83.88 (38)	81.5 (37)	52.29 (23)	5.07(2)	WASS
Amsoul	193.77	55.85(29)	86.51 (45)	44.12 (23)	7.29 (4)	WASS
Tizguine	236.63	61.50 (26)	126.79 (54)	38 (16)	10.34 (4)	WASS
AghbarWHA	537.4	144.5 (27)	246 (48)	134.5 (25)	12.4 (2)	WASS
Timlilt	344.01	84.32 (25)	141.96 (41)	106.64 (31)	11.09 (3)	**WSAS**
T.o.Maacho	465.48	112.90 (24)	198.58 (43)	137.87 (30)	16.13 (3)	**WSAS**
Iloudjane	332.16	82.37 (25)	118.48 (36)	111.25 (33)	20.06 (6)	**WSAS**

N.B.: Values in brackets indicate seasonal rainfall rates (%).

The seasonal regime was defined to classify the seasons in descending order of precipitation.The seasonal rainfall regime in the majority of the basin is of the WASS type (Winter-Automn-spring and summer). All the stations representing a summer deficit (m2) and a winter maximum (M1) are subject to the "Soussien" currents in the south-west. The only exceptions are the very northern stations, Timlilt, T.O.Maacho and Iloudjane, that are WSAS type (winter-spring-automn and summer), which show a second maximum (M2) in spring as opposed to autumn and are subject to the "Haouzien" currents of the north-east (Fig. 5). The transition between the two types of regimes occurs in the high basin of Issen at the upstream of the Tizguine station.

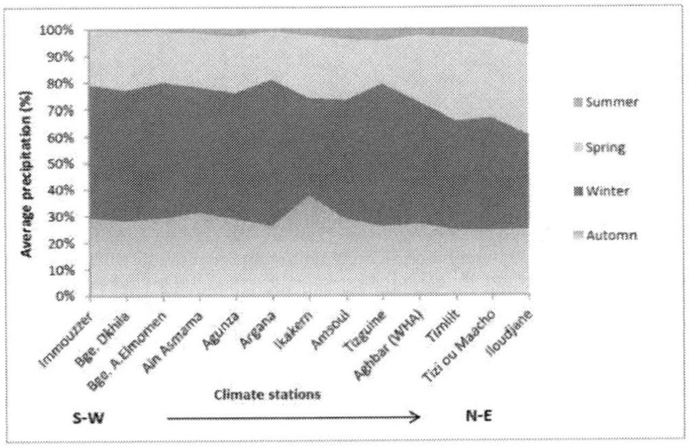

Fig. 5. Seasonal rainfall variation from SW to NE of the Issen river basin.

The ombrothermal diagrams for three stations show that the whole basin is characterized by a drought season lasting 5 to 7 months and a thermal amplitude characterizing semi-continental conditions (cold winter - very hot summer). According to the Embergerclimagram, the study stations are located in the arid bioclimatic zone with a temperate winter and in the semi-arid bioclimatic zone with a cold to worm winter.

3.1.3 Deviation-to-mean index (Em)

The results show very unequal spatial and temporal variation of the drought frequency, normal and wet years. The stations of Tizi-Ou-Maacho, Ikakern, Tizguine(upstream part) and Dkhila(downstream part) present an average of drought years of more than 55%, which indicates a tendency to a dryness. The stations of Argana, Abdelmomen, Aguenza, Ain asmama and Immouzzer recorded annual averages of normal rainfall over 20 years. However, a limited number of the stations receivesannual totals above the habitual average with a frequency of wet years over 35%. We notice that more than 40% of the years are droughty (tab.5). The long period of drought in the study area may affect the vegetation growth and development.

Table 5. Frequency of drought, normal and humid years during 1980-2017

Stations	**"DroughtYear" (%)**	**" NormalYear" (%)**	**" HumidYear" (%)**
Immouzzer	19.75	62.96	17.28
BgeDkhila	57.14	7.14	35.71
Bg. A.elmoumen	15.15	75.76	9.09
Ain Asmama	18.18	64.94	16.88
Aguenza	20.59	73.53	5.88
Argana	12.94	81.18	5.88
Amsoul	27.03	59.46	13.51
Ikakern	58.62	10.34	31.03
Tizguine	56.25	25.00	18.75
Aghbar	50.00	10.71	39.29
TiziOuMaacho	66.67	6.67	26.67

Timlilt	51.43	17.14	31.43
Iloudjane	34.62	53.85	11.54
Mean	**40.70%**	**45.72%**	**21.91%**

3.1.4 Standardized Precipitation Index (SPI)

Table 6 shows the results of the SPI. For all decennia, normal conditions prevailed in more than 60% of cases over drought and moisture conditions. The interval period 2010-2017 is still the driest, with moderate or severe drought manifesting in 37.46% of the cases and in 18 to 27% of the cases over the other interval periods 1980-89, 1990-99 and 2000-09. The common drought years are 2000/01 and 2001/02. We also observe that during the 37 years, 26% of the years are fairly, dry and 13% are humid. For all stations the rainfall continues to be deficient proving moisture deficit. Labbaci [18] reported a cumulative deficit of 3927.25 mm during a mean of 25 years in the Site of Biological and Ecological Interest (SBEI) of Ain Asmama(equivalent to an annual deficit of 133mm for the dry period).

Table 6. Frequencies (% of stations/years) of the Standardised Precipitation Index (SPI)

Droughtdegree	**Moderatedrouhgtperiod**			**Severedroughtperiod**	**1980-2017**
	1980-1989	**1990-1999**	**2000-2009**	**2010-2017**	
Extremehumid	0.17	0.20	0	12.4	3.19
Severehumid	9.01	1.00	1.12	0	2.78
Moderatehumid	10.05	9.13	9	0	7.04
Near normal	61.59	70.35	69.33	50.15	62.64
Moderatedrouhgt	19.18	18.23	18.20	25.32	20.23
Severedrought	0	1.10	2.31	12.00	4.81
Extremedrought	0	0	0	0	0
Total	100	100	100	100	100

Figure 6 shows that the Issen Basin has suffered from dry episodes during 1981, 1986, 1992, 1998, 2002, 2006, 2012 and 2014-2017. The analysis of these results indicates a higher frequency of drought years in comparison to the humid years. In addition, the annual droughts occur once or twice every 10 years. In other hands, the drought that lasting 5 years may return after 360 years [27].

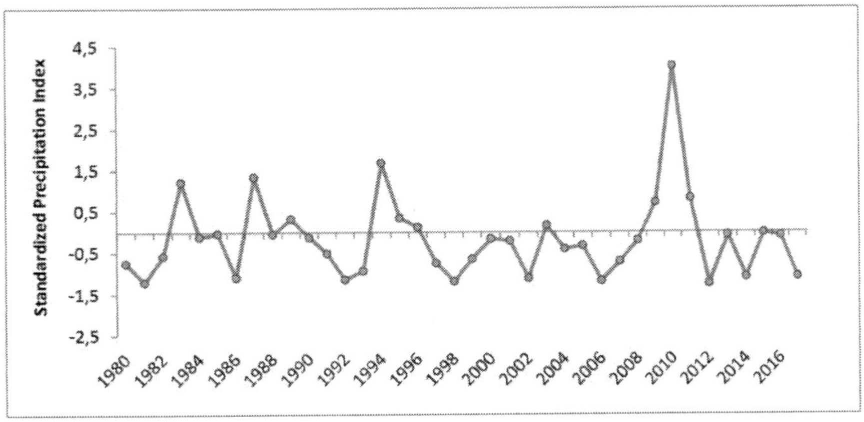

Fig. 6. Standardised Precipitation Index (SPI) in the Issen basin during the period 1980-2017

3.2 Rainfall ErosivityR-Factor

The R-factor considering the role of precipitation for soil erosion modelling: The obtained results show an ascending gradation of values with topography (Fig.6& tab.6). The erosivity of rainfall naturally progresses from downstream to upstream of the Issen basin. R values vary between 34 and 80.89 MJ.mm/ha.h.yr with an average of 47 MJ.mm/ha.h.yr. 79 % of the total area of theIssen basinhave R values over 41 MJ.mm/ha.h.yr which contributes to soil loss exceeding the tolerance threshold (10.4 t/ha/yr)[28]. The low values of R (21 to 40 MJ.mm/ha.h.yr) are situated in the center of the Argana corridor which includes the two topographic basins of Bigoudine and Timzgadiouine, subject to arid climate. The high values (>61 MJ.mm/ha.h.yr) are attributed to the plateaus region and Ida-Mahmoud valley, under strong oceanic influence or at high altitudes.
A recent study based on pollen in the marine sediment of the southwest Morocco reported the impact of the climate change and anthropogenic activities on the argan tree ecosystem and soil degradation [1].

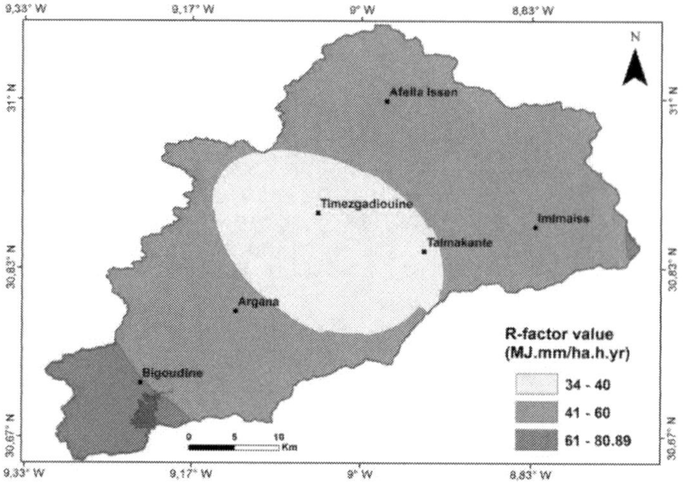

Fig. 7. Spatial distribution of rainfall erosivity (R-factor)

Table7. Conceptual scale for assessing the Fournier index [29]

Class	Soil Loss (t/ha/yr)	Fournier Index (MJ.mm/ha.h.yr)	Erosion risk	Area (%)
1	< 5	< 20	Very Low	0
2	5 – 12	21 - 40	Low	21.3
3	12 – 50	41 - 60	Moderate	69.6
4	50 – 100	61 - 80	Severe	09.1
5	100 – 200	81 - 100	Very Severe	0
6	> 200	> 100	Extremely Severe	0

4 Conclusions and recommendations

The study investigated the indicators of drought and climatic aggressiveness in the Issen basin in order to characterize the spatial and temporal variability of precipitation and evaluate the rainfall erosivity. In fact, the record data collected from 13stations over the study area covering the period 1980-2017 were analyzed using a GIS tools. Thefindings revealed that almost all the studied stations recorded a rainfall below 500 mm with an arithmetic mean of 302 mm over the whole basin. The record data of 10 stations (WASStype) reflecting a high aridity and inter-annual and seasonal variability in the study area. In detail, the aridity extends from the beginning of April to the end of October and the dry season duration is about six to seven months per year. Furthermore, the dry periods have been more frequent and remarkable in recent years than humid periods.The analysis of precipitation variabilityusing standardized precipitation index (SPI) indicates moisture deficit.In fact, 70% of the study area is affected by moderate rainfall erosivity which is responsible of ecosystem degradation.

The occurrence of a succession of drought, climate events and the rainfall aggressiveness, especially, during the last decade threaten water security, the resilience of the argan trees ecosystem (southwest part of the basin), the quality of the soil and the agricultural sustainability in the study area. In fact, the soil erosion problem can be solved by constructing other dam in the upstream part of the basin which is the most part affected by water erosion. The action will also conserve drinking and irrigation water resources. In addition, it is recommended to regenerate the natural forests such as the argan forest, which is will be helpful to reduce the climate vulnerability. Moreover, extending and serving the researches continuity in the whole area of the Western High Atlas (WHA) can be helpful to figure out the precipitation trend.

References

[1] X. Zhao, L. Dupontet,& al., Recent climatic and anthropogenic impacts on endemic species insouthwestern Morocco. Quaternary Science Reviews Volume **221**, 1 October 2019, 105889 (2019)https://doi.org/10.1016/j.quascirev.2019.105889

[2] L. Stour, &A. Agoumi, Sécheresseclimatique au Marocdurant les dernièresdécennies [Climatic Drought in Morocco During the Last Decades]. Hydroécol. Appl., **16**: 215–232. (2009)https://doi.org/10.1051/hydro/2009003

[3] M. Hssaisoune, L. Bouchaou, A. Sifeddine, I. Bouimetarhan, & A. Chehbouni, Moroccan Groundwater Resources and Evolution with Global Climate Changes. Geosciences, **10**(2), 81. (2020)https://doi.org/10.3390/geosciences10020081

[4] S. Hammi, V. Simonneaux, M. Alifriqui, L. Auclair,& N. Montes, Evolution des recouvrementsforestiers et de l'Occupation des sols entre 1964 et 2002 dans la haute vallée des Ait Bouguemez (Haut Atlas Central, Maroc). Impact des modes de gestion. Science etchangementsplanétaires/Sécheresse, **18**, 271-277. (2007) https://doi.org/10.1684/sec.2007.0101

[5] C. Faye,A. A. Sow, &J. B. Ndong, Étude des sècheressespluviométriquesethydrologiques en Afriquetropicale: caractérisation et cartographie de la sècheresse par indices dans le haut bassin du fleuveSénégal. Physio-Géo. Géographie physique etenvironnement, (Volume **9**), 17-35. (2015)https://doi.org/10.4000/physio-geo.4388

[6] GIEC,Changementsclimatiques, «Rapport de synthèse» : Contribution des Groupes de travail I, II et III au cinquième Rapport d'évaluation du Grouped'expertsintergouvernementalsurl'évolution du climat.GIEC, Genève, Suisse, 161p, (2014)

[7] M. Alifriqui, O. M'hirit, R. Michalet,& J. P. Peltier, Variabilité des précipitationsdans le Haut Atlas Occidental marocain. Le climat, **13**(1), 11-28. (1995)

[8] P.E. Olsen, D.V. Kent, M. Touhami, & J. Puffer, Cyclo-magneto and bio-stratigraphic constraints on the duration of the CAMP event and its relationship to the Triassic–Jurassic boundary. In: Hames, W. E., McHone, G., Renne, P. R. &Ruppel, C. (eds) The Central Atlantic Magmatic Province: Insights from Fragments of Pangea. AGU, Geophysical Monographs, **136**, 7–32, (2003)https://doi.org/10.1029/136GM02

[9] P. Panagos, A. Jones, C. Bosco, & P.S Kumar, European digital archive on soil maps (EuDASM): preserving important soil data for public free access. International Journal of Digital Earth **4**, 434–443, (2011)https://doi.org/10.1080/17538947.2011.596580

[10] B. Jaafar, Etude phytoécologiqueetmorphobotaniqued'une enclave sèche du Haut Atlas occidental Marocain (couloir d'Argana). Thèsedoctorat national, Univ. Cadi Ayyad, Fac. Sci. Semlalia, Marrakech 241p. (1994)

[11] E.A. El Morjani, M. SeifEnnasr,VA. Elmouden, S. Idbraim, B. Bouaakaz, &A. Saad, Flood Hazard Mapping and Modelling using GIS applied to the Souss River watershed. R. ChoukrAllah et al. (eds.), the Souss-Massa River Basin, Morocco. «Chapter part of the series The Handbook of Environmental Chemistry pp 1-30», DOI 10.1007/698_2016, Springer International Publishing Switzerland. ungauged Mediterranean catchments. Energy Procedia**18**: 934-943, (2016)

[12] B. Bouaakkaz, Z.E.A. El Morjani, L. Bouchaou, & H. Elhimri, Flood risk management in the Souss watershed, E3S Web of Conferences 37, 04005 (2018), published by (EDP Sciences, 2018)https://doi.org/10.1051/e3sconf/20183704005

[13] A. Benabid, & Y. Melhaoui, Écosystèmesnaturels à arganier (Arganiaspinosa), patrimoine national etuniversel: Bio-écologie, phytosociologie, phytodynamique et ethnobotanique; Restauration et certification. Actes du 1er congrès international de l'Arganier (ages 39-47). Agadir, Maroc, (2011)

[14] A. Benabid, Floreetécosystèmes du Maroc. Evaluation etpréservation de la biodiversité. Ibis Press, Paris. (2000)

[15] ABHSM, Optimisation de la gestion de ressources en eau dans le bassin du Souss Massa. Colloquesur la gestion des ressources en eau, mai (2005)

[16] Fr.Bagnouls&H. Gaussen, Les climatsbiologiquesetleur classification. Annales de géographie. Vol. **66**. No. 355. Armand Colin, (1957)

[17] F. Lebourgeois,& C. Piedallu,.Appréhender le niveau de sécheressedans le cadre des étudesstationnelleset de la gestionforestière à partird'indicesbioclimatiques. Revue forestièrefrançaise. (2005)

[18] A. Labbaci, Thèsedoctorat national, Univ. IbnZohr, Fac. Sci. Agadir, p.237, (2013)

[19] I. Jouilil, K. Bitar, H. Salama,&al., Sécheresseméteorologique au bassinhydraulique OUM ER RBIA durant les dernièresdécennies. LARHYSS Journal P-ISSN 1112-3680/E-ISSN 2602-7828, 2013, no **12**. (2013)

[20] T.B. Mckee, N.J. Doesken,& J. Kleist, The relationship of drought frequency and duration to time scale. Actes de la 8th Conference on Applied Climatology (Anaheim, Californie), p. 179-184. (1993)

[21] A.Ali,&T.Lebel, The Sahelian standardized rainfall index revisited. International Journal of Climatolology, vol. **29**, p. 1705-1714.(2009)https://doi.org/10.1002/joc.1832

[22] K.Meusburger, A.Steel,P.Panagos, L.Montanarella, &C.Alewell, Spatial and temporal variability of rainfall erosivity factor for Switzerland. Hydrology and Earth System Sciences, **16**, 167-177. (2012) https://doi.org/10.5194/hess-16-167-2012

[23] L.C. Brown&G.R.Foster, Storm erosivity using idealizedintensity distributions, Transactions of the Asae, **30**, 379–386, (1987).

[24] W. H. Wischmeier,& D. D. Smith, Predicting rainfall erosion losses: A guide to conservation planning. In: USDA Agriculture Handbook 537. Washington, DC: U.S.A, Government Printing Office. (1978)

[25] H.M.J. Arnoldus, An approximation of the rainfall factor in the Universal Soil Loss Equation. In M. deBoot and D. Gabriels (eds), Assessment of erosion. Wiley, chichester. (1980)

[26] R. Cheddadi, M. Nourelbait,O.Bouaissa, J.Tabel, A. Rhoujjati, J. A.López-Sáez,... ,& H. Lamb, A history of human impact on Moroccan mountain landscapes. African Archaeological Review, **32**(2), 233-248. (2015) https://doi.org/10.1007/s10437-015-9186-

[27] A. AGOUMI, Ressources en eau et bassins versants du Maroc : 50 ans de développements (1955-2005), pp. 61. (2006)

[28] Y.Ostovari,A.A. Moosavi, &H.R. Pourghasemi, Soil loss tolerance in calcareous soils of a semiarid region: evaluation, prediction, and influential parameters. Land Degradation & Development.1–12. (2020)https://doi.org/10.1002/ldr.3597

[29] K. Oduro-Afriyie, Rainfall erosivitymapforGhana. Geoderma**74**, 161–166. (1996)

Contributions of remote sensing in the diachronic study of the spatial and temporal evolution of the Ahmed El Hansali dam water reservoir from 2002-03 to 2018-19.

*Tarik*El Orfi[1,2*], *Mohamed*El Ghachi[1] and *Sébastien*Lebaut[2]

[1] "LDRH" Laboratory, Sultan Moulay Slimane University, B.P: 524, Beni Mellal, Morocco
[2] "LOTERR" Research Unit, Lorraine University, 57000 Metz, France

Abstract. The OumErRbiabasin is one of the watersheds with the largest number of hydraulic infrastructures in Morocco. These hydraulic structuressupply water for drinking, industrial and agricultural uses. The Ahmed El Hansali dam is a 740 Mm^3 reservoir located near Zaouyat Cheikh andhave an active storage of473 Mm^3. The succession of dry years in the OumErRbiabasin has had a negative impact on the water resource and has caused a remarkable decrease in the reservoir of the Ahmed el Hansali dam. In this paper, the MNDWI (Modified Normalized Difference Water Index) from Landsat 5-TM, Landsat 7-ETM, and Landsat 8-OLI satellite images was used to estimate the spatial and temporal fluctuations of the volumes of water stored in the reservoir between hydrological years 2002-03 and 2018-19. Results show that the volumes estimated by remote sensing reasonably match the volumes estimated by the OumErRbia Hydraulic Basin Agency (OERHBA)using recorded water levels and reservoir storage curve for years 2002-03 and 2013-14; the determination coefficient R^2 exceeds 0.90. The mapping of the extent of the dam's impoundment has shown a very significant decreasein the flooded area level during dry years.

1 Introduction

Water is a natural resource that is essential for human well-being and sustainable socio-economic development [1]. In Morocco, water management is even more vital because of the arid to semi-arid climate and limited water availability. Over the past three decades, Morocco has suffered five periods of severe drought [2], which have exacerbated several problems and underlines the seriousness of water scarcity in the country.

Storage is often advocated as a solution to droughts, as it allows us to build up reserves with the water available in rainy periods (winter) and mobilize them during the dry period [3]. To meet growing water needs, Morocco has built an extensive network of hydraulic infrastructure spread over the entire hydrographic network [4]. Morocco's water resources policy has long been oriented towards the construction of dams [5]. The late Majesty

*Corresponding author:elorfitarik@gmail.com

Hassan-II initiated the development of large dams as early as 1967, the main objective being ensuring food security and guaranteeing drinking water supply to the population [6]. Morocco's various development plans have always given priority to the agricultural sector, and large agriculturalhydraulicinfrastructures have always beenperceivedby Moroccan officials as the pillar of economic development[7].

Rational management of water resources must ensure continuous availability in quantity and quality. In this respect, the Master Plan for Integrated Water Resources Management (MPIWRM) was proposed to plan water resources development based on a prospective and concerted vision that takes into account, on the one hand, the evolution of resources and, on the other hand, the real water needs [8]. With the largest irrigated perimeters in the Maghreb, some reaching 100000 ha and more than 100 large dams, Moroccois the "country of large hydraulics"[9]. OumErRbia watershed contains several dams, including the Ahmed El Hansalidam "AEH'd" (formerly Dchar El Oued dam). Itis located near the Zaouyat Cheikh centre, and 50 km from KasbaTadla, and was commissioned in 2002 [6].

Remote sensing shows a great performance in the monitoring of water resources although there are difficulties related to their identification.Day-to-day dam management requires the knowledge of the amount of stored water as well, and the volume of expected inflows in the reservoir to optimally allocate water to local and downstream users. For several decades, an increasing number of satellites are used for the observation of the spatial and temporal variability of several compartments of continental waters (open water, snow cover, soil moisture, etc.) [10]. This study is a contribution to the use of Landsat images in the temporal monitoring of continental water surfaces. It also evaluates the possibility of calculating the volumes related to these surfaces by coupling them to a multisource Digital Elevation Model.

2 Material and Methods

2.1 Study area

The upper watershed of the OumErRbiariver (3380km²) is part of the central plateau and the Middle Atlas. It is the upstream part of the large watershed of Oum Er Rbiariver. It is bounded by the Moulouya watershed to the east, the Sebou watershed to the north, and the Bouragreg watershed to the west. The climate station located after AEH'd is used as an outlet (Fig. 1).

Six climate stations are managed by the Agency of the Hydraulic Basin of OumErRbia "OERHBA". The climate is Mediterranean with rainy and mild winters and hot and dry summers.

The annual variability of rainfall between 1974-75 and 2016-17 for three stations representative of the study area shows a decrease in the annual mean from upstream to downstream(Tab. 1)[11]:

Table 1. Principal characteristics of the three representative stations

Station	Elevation	Mean (mm)	Min		Max	
			R (mm)	Year	R (mm)	Year
Tamchachat	1685m	692	302	99-00	1314	10-11
Tarhat	886m	570	290	94-95	1117	95-96
Taghzoute	708m	508	260	06-07	857	95-96

Hydrologically, the OumErRbia originates from the Fellat wadi, which receives inputs from a group of sources known asthe "sources of OumErRbia". It is only from the confluence of Fellat river with Bouldjiriver that the river is called OumErRbia [12]. The flow of the OumErRbia river is perennial and the module at the Tarhat station is of the order of $16m^3/s$.

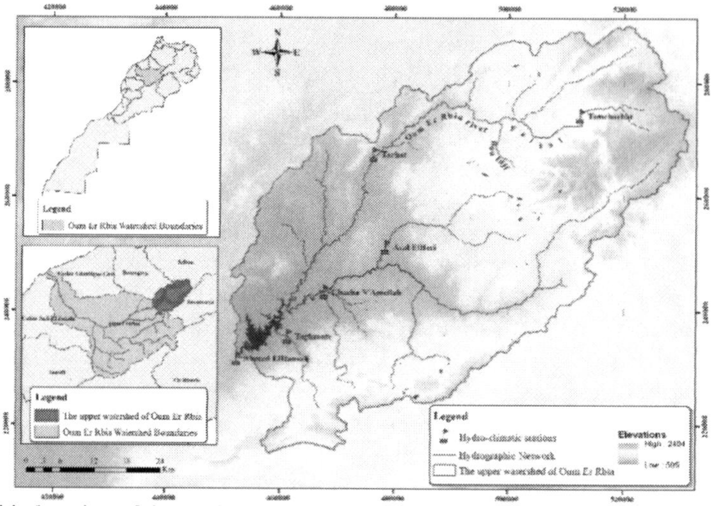

Fig. 1. Geographic location of the study area.

The OumErRbiawatershed is home of 15 dams that provide water for nearly 323000ha of agricultural land distributed between the Tadla, Tessaout, Haouz and Doukkala perimeters, as well as the production of about 350 Mm^3/year ofdrinking water and industrial water supply. The AEH'd(Image 1), commissioned in 2002, is a structure allowing the creation of a reservoir of 740 Mm^3, to regulate 473 Mm^3 and to obtain a waterfall varying between 51 and 82 m for the production of electric power.

Image. 1. Ahmed El Hansali Dam "http://www.sgtm-maroc.com/"

2.2 Lake extent estimation and validation

The temporal evolution of water volumes of the Ahmed El Hansali lake reservoir is done using two sources of information: the extent of the lake is estimated at a monthly time-scalefrom Landsat images. The overlay of these boundaries with a multi-source DEM allows the calculation of the volume of water in the dam reservoir. This study adopts an operational geomatics approach (Fig. 2) that is used to delineate and follow the lake's extent and volume using remotely sensed data.

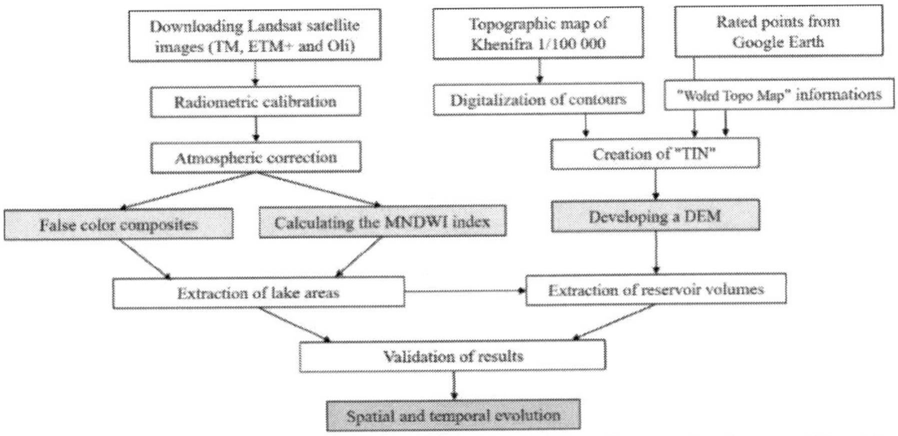

Fig. 2. Flowchart of the methodology for the calculation of the surfaces and volumes of the Ahmed El HansaliLake reservoir.

2.3 Data

Landsat satellite images were downloaded since 2000 from the NASA website (https://landsat.gsfc.nasa.gov/). They were preprocessed to have reflectance-calibrated images for the whole series. Radiometric calibration was performed to overcome differences between instrument generations (Tab. 2). The influence of the state of the atmosphere at the time of shooting was removed by an atmospheric correction. Both corrections were performed using the "Radiometric Calibration" and "Dark Subtraction" functions implemented in the ENVI software.

Table 2. Principal characteristics of Landsat satellites (https://landsat.gsfc.nasa.gov/); B: Blue, G: Green, R: Red, NIR: Near Infrared, SWIR: Short Wavelength Infrared, TIR: Thermal Infrared, Pan: Panchromatic, Aer: Aerosol, Cir: Cirrus

Satellites		Landsat-5	Landsat-7	Landsat-8
Sensors		TM	ETM+	OLI - TIRS
Mission period		1984-2013	1999- operational	2013- operational
Spectral Band	Band 1 :	0,45-0,52 µm (B)	0,441-0,514 µm (B)	0,435-0,451 µm (Aer)
	Band 2 :	0,52-0,60 µm (G)	0,519-0,601 µm (G)	0,452-0,512 µm (B)
	Band 3 :	0,63-0,69 µm (R)	0,631-0,692 µm (R)	0,533-0,590 µm (G)
	Band 4 :	0,76-0,90 µm (NIR)	0,772-0,898 µm (NIR)	0,636-0,673 µm (R)
	Band 5 :	1,55-1,75 µm (SWIR1)	1,547-1,74 µm (SWIR1)	0,851-0,879 µm (NIR)
	Band 6 :	10,4-12,5 µm (TIR)	10,31-12,36 µm (TIR)	1,566-1,651 µm (SWIR1)
	Band 7 :	2,08-2,35 µm (SWIR2)	2,064-2,345 µm (SWIR2)	2,107-2,294 µm (SWIR2)
	Band 8 :		0,515-0,896 µm (Pan)	0,503-0,676 µm (Pan)
	Band 9 :			1,363 - 1,384 µm (Cir)
	Band 10 :			10,60-11,19 µm (TIR 1)
	Band 11 :			11,50-12,51 µm (TIR 2)
Resolution		General : 30 m TIR :120 m	General : 30 m Pan : 15 m TIR : 60 m	General : 30 m Pan : 15 m TIR : 100 m

"OERHBA" providedtime series total volumescalculated using recorded water levels and reservoir storage curvebetween 2002-03 and 2013-14.

2.4 MNDWI Index

Water is characterized by a relatively higher reflectance in the visible than in the near and middle infrared. The absorption of solar radiation in the near infrared band is complete. Several water detection indices are based on this phenomenon such as the "NDWI"

(Normalized difference water index) introduced by McFeeters[13]. This index is used to delineate surface water by the ratio of green to near-infrared reflectance, where ρ is the reflectance.

$$NDWI = \frac{\rho_{Green} - \rho_{NIR}}{\rho_{Green} + \rho_{NIR}} \quad (1)$$

In 2006 Xu replaced NIR reflectance with middle infrared reflectance and suggested changing the NDWI nomenclature to MNDWI "Modified Normalized Difference Water Index" [14]. He observed that MNDWI is more accurate in delineating wetlands than NDWI.

$$MNDWI = \frac{\rho_{Green} - \rho_{MIR}}{\rho_{Green} + \rho_{MIR}} \quad (2)$$

We choose to work with the MNDWI index because several researchers[14–19] have shown that it is the most accurate in discriminating between dry and wet surfaces. A discrimination threshold of zero was set for all treated scenes, however, multispectral images acquired at different times always have different characteristics [16].

2.5 Lake topography

To calculate the volume of the lake, it is necessary to know the topography below the surface of the lake. Since the reservoir was commissioned in 2002, the topography of the lake had to be based on previous topographic data. The only source of such homogeneous data is the 1/100000 topographic map of Khenifra (InstitutGéographique National - France, 1972). Digitized contour lines are the basis of the pre-dam topographic information. To increase the accuracy of the final DEM, data from other sources have been merged into a TIN file: elevation points extracted from "Google Earth"and "the world topo map"layer (ESRI). From the latter the DEM with 84m spatial resolution was extracted.

3 Results and Discussion

3.1 Evolution of the surface area of AEH'd Lake

It is difficult to quantify the uncertainty associated with the lake boundaries determined by the MNDWI. An estimate can be made by overlaying our results with high spatial resolution Google Earth imagery (Fig. 3).

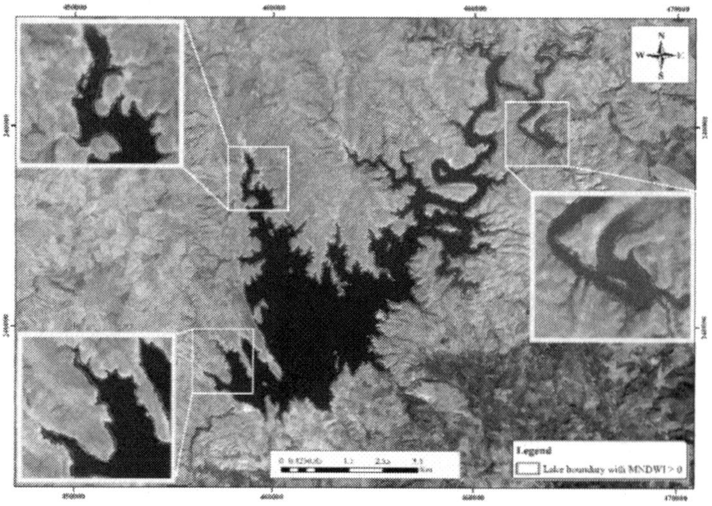

Fig. 3. Overlay of the extent as of 12/24/2013 on a Google Earth image as of 12/31/2013.

On the example of December 2013, the good superposition of the two inundation extents is obvious except on the most pronounced festoons of the shoreline. The MNDWI index is therefore a good predictor of water surfaces in the AEH'd, and the estimated water are reasonable.

During the period of filling of the dam, the surface area of the AEH'd lake increased until June 2004, when it reached 24 km² (Figs. 4 and 5). Then, and in accordance with the results of the analysis of rainfall series in the study area [11], the increase and retraction of the lake surface follow the annual fluctuations in rainfall. Until 2009, the end of the dry period that began in 1979, the average surface of the lake was 13 km² (Figs. 4 and 5). From 2009, another homogeneous wetter series begins for which the average surface area of the lake is 22 km², with a maximum in February 2009, 27.5 km², following the very heavy rainfall of autumn/early winter 2008/2009. However, since 2018 rainfall has been falling again, bringing the surface of the lake closer to that of the early 2000s.

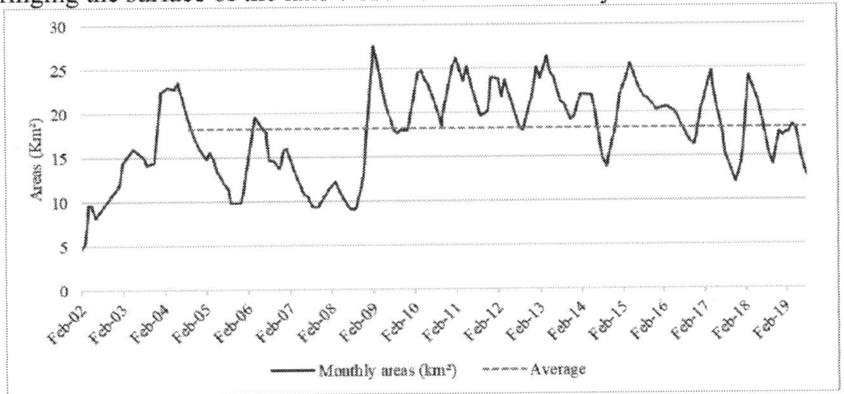

Fig. 4. Evolution of the surface area of AEH'd Lake between 2002-03 and 2018-19.

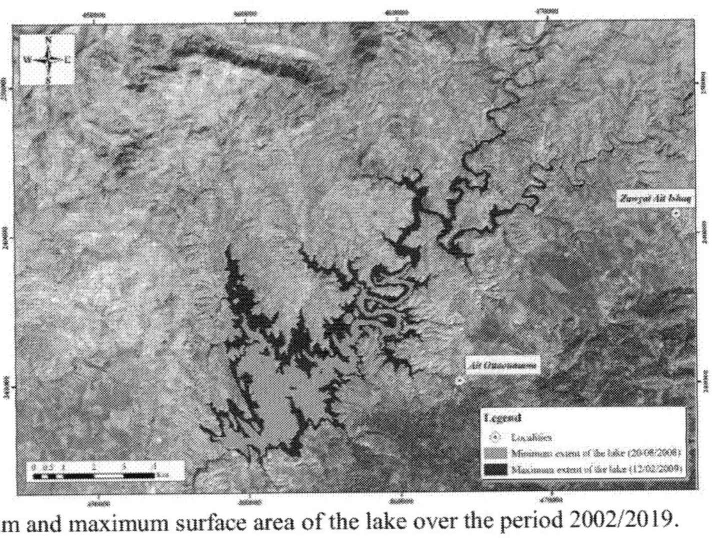

Fig. 5. Minimum and maximum surface area of the lake over the period 2002/2019.

There is also an average annual expansion/retraction cycle of around 8 km².

On a monthly scale, on average, the filling of the lake begins in November and ends in April (21 km²). From May the reservoir is emptied until October (15 km²) to ensure the 3 types of uses, the main one being agricultural since the Tadla plain located downstream is the first irrigated perimeter in Morocco (Fig. 6).

 https://doi.org/10.1051/e3sconf/202018302004

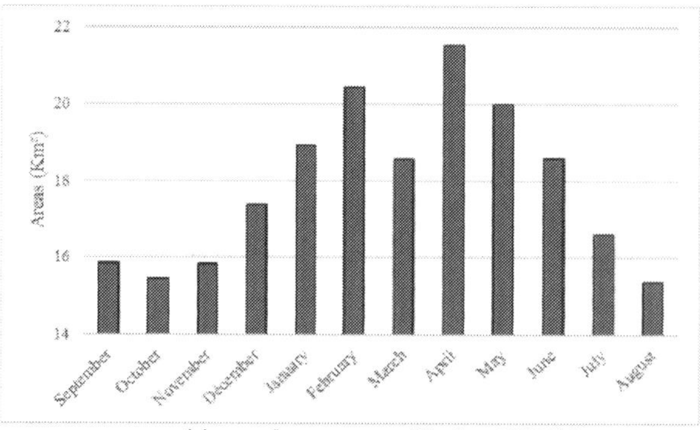

Fig. 6. Evolution of the average monthly surface areas of Lake AEH'd.

3.2 Evolution of the volume of AEH'd lake between 2002-03 and 2018-19

The calculation of the lake water volume is based on the pre-dam construction topography described from the 84 m spatial resolution multisource DEM. The greater the slope of the lake floor, the greater the uncertainty in the volume of water. Slopes, for 82% of the surface are less than 15% and only 2% are greater than 25% (fig.8).Thus, the system of distribution in the lake is contributes to a low uncertainty on the calculation of the volume; especially since the steepest slopes are located in the upstream part of the lake where it is narrow (fig. 7): the distance from one side to the other of about 300m maximum.

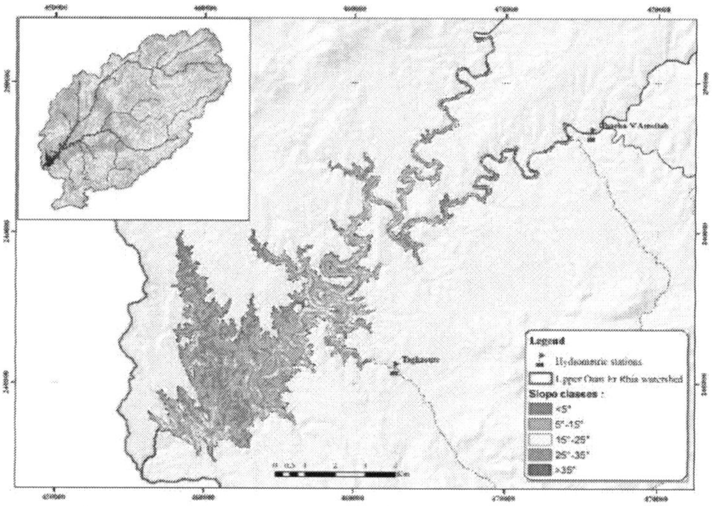

Fig. 7. Slope classes at the maximum extension of the lake.

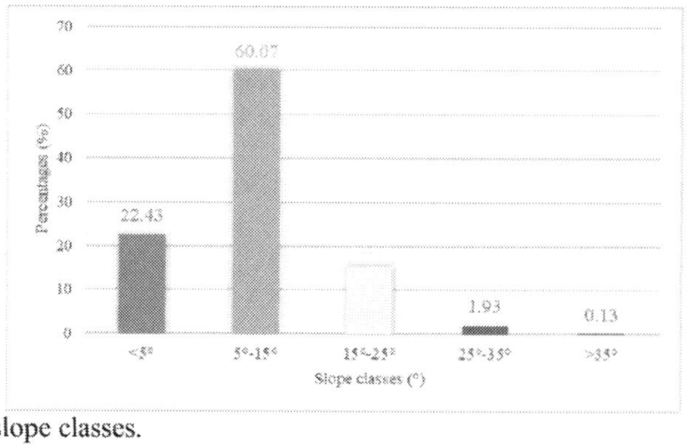

Fig. 8. Histogram of slope classes.

The validation of water volumes calculated from remote sensing uses the volumes of water measured and provided by the "OERHBA" over the period 2002/2014. Whether on an annual or monthly scale, the results are particularly good with determination coefficients (R^2) of 0.95 and 0.93 respectively (Figs. 9 and 10).

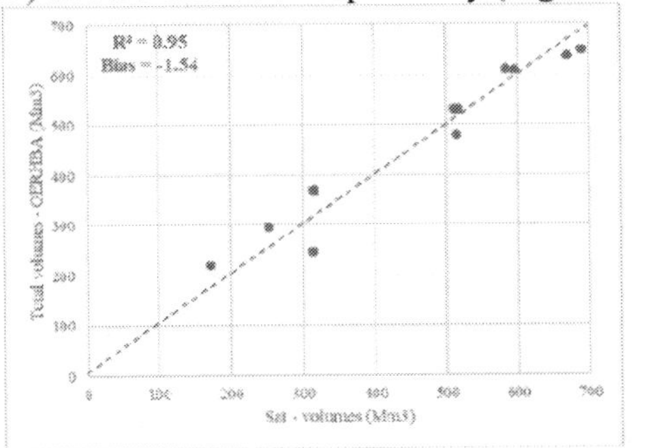

Fig. 9. Annual correlation between Sat-Vol and OERHBA-Vol.

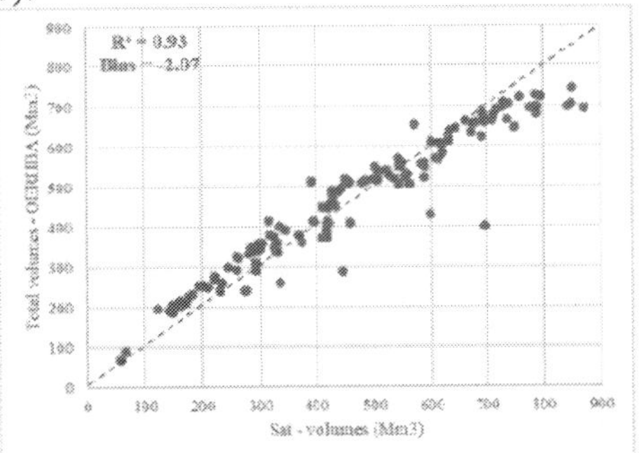

Fig. 10. Monthly correlation between Sat-Vol and OERHBA-Vol.

Although low, the relative bias shows that our methodology leads to an underestimation of the order of 15% for volumes below 500 Mm^3 and an overestimation of the order of 10% for higher values.

On an annual scale, after the filling of the dam, the end of the dry period is accompanied by an almost continuous emptying of the lake until it reaches 171.5 Mm^3 in 2008/2009 (fig.11). From 2009 onwards, the maximum reserves are reached and are of the order of 650 Mm^3, dropping below the 500 Mm^3 threshold in the last two years.

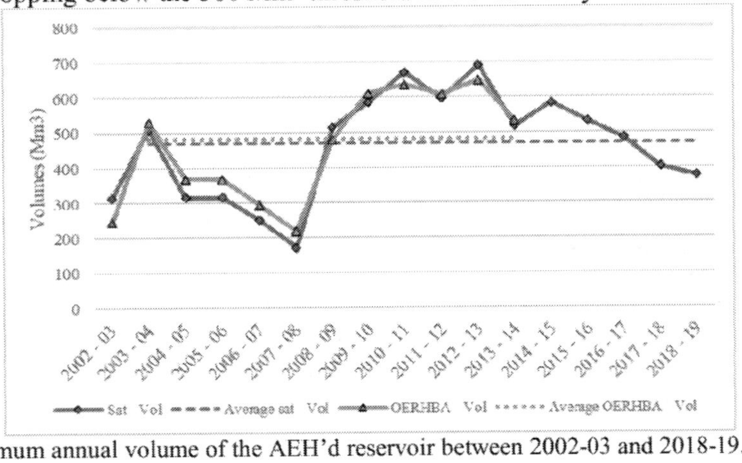

Fig. 11. Maximum annual volume of the AEH'd reservoir between 2002-03 and 2018-19.

On a monthly scale, on average we note that the quantities emptied during the dry period are of the order of 250 Mm^3.But for the driest periods it is more like 350 Mm^3 that are used. These results underline the merits of the mega-hydraulic infrastructure in this sector.

In accordance with the evolution of the surface area, the volumes of the reservoir are lower than the average (467Mm^3 for Sat - volumes and 471Mm^3 for the volumes provided by the "OERHBA") until August 2008 with a minimum for satellite data equal to 144Mm^3. Since 2009, volumes have generally been above average except in recent years where there has been a remarkable decrease in volume. (Fig12).

The period of filling of the dam begins in December with an increase in volume until April, which records an average volume equal to 620Mm^3. During summer, the volume of the reservoir decreases until it reaches 350Mm^3 in August (Fig13).

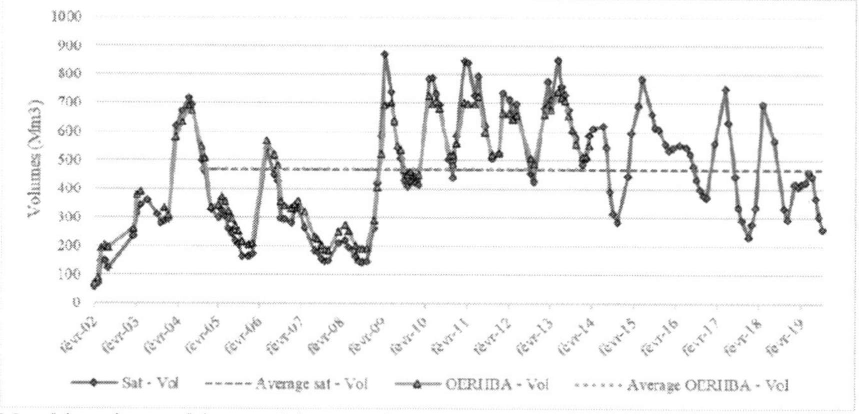

Fig. 12. Monthly volume of the AEH'd reserve between 2002-03 and 2018-19.

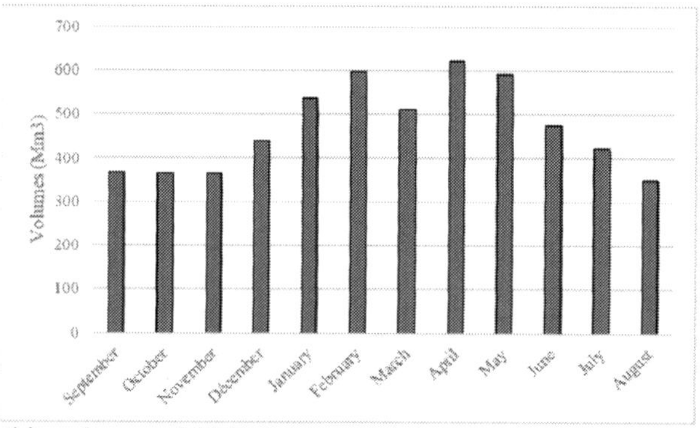

Fig. 13. Average monthly volumes of AEH'd reserves.

This paper demonstrated that remote sensing can be a valuable tool for water resources management. Flooded areas and stored volumes were estimated with a reasonable level of uncertainty. These uncertainties can be reduced using two approaches:

- Increasing the DEM accuracy by using pairs of stereoscopic photographs which are known to have been produced at the scale of Morocco during aerial missions dating from the early 1960s. In this case the spatial resolution of the DEM would be of the order of a metre [20].
- Increasing the resolution of optical images. An increase in the spatial resolution of the surface water layer could be tried by merging the channels used with the panchromatic channel. Thus, the spatial resolution would increase from 30m to 15m.

Once these improvements, the OERHBA will have a cost-effective alternative way of monitoring inundated areas and water stored in the dam for a more effective management of the system.

4 Conclusion

The application of the MNDWI index on Landsat images has enabled us to study the spatio-temporal evolution of the extent of the AEH'd lake and the associated water volumes from

2002/2003 to 2018/2019. Results show that there is a significant retreat of the lake area in dry years up to 2008, followed by a remarkable spatial increase in wet years. During this period, the water reserve varied between 200Mm^3 and 800Mm^3. These changes were shown to be consistent with regional climate trends. Optical images are hence valuable sources of information for the operational management of surface water reserves.

References

1. A. E. Boukhari, M. Naimi, M. Chikhaoui, and D. Raclot, Rev. Mar. Sci. Agron. Vét**7**, 288 (2019)
2. A. Lahlimi Alami, *Agriculture 2030 : Quels Avenirs Pour Le Maroc 2030 ?* (Haut Commissariat au Plan, n.d.), p. 108
3. F. Habets, The Conversation (2019)
4. M. Chahboune, A.S **10**, 199 (2014)
5. B. Abdellaoui, A. Merzouk, M. Aberkan, and J. Albergel, Revue des sciences de l'eau **15**, 737 (2005)
6. *Brochure de l'Agence Du Bassin Hydraulique de l'Oum Er Rbia* (n.d.), p. 28
7. A. Benhadi, 275 (1975)
8. *Projet de Plan Directeur d'aménagement Intégré Des Ressoureces En Eau Du Bassin de l'Oum Er Rbia et Des Bassins Côtiers Atlantiques* (2012), p. 19
9. A. Laouina, *Gestion durable des ressources naturelles et de la biodiversité au Maroc* (Haut Commissariat au Plan, 2006), p. 118
10. S. Biancamaria and Y. Kerr, La Météorologie 70 (2017)
11. T. El Orfi, M. El Ghachi, and S. Lebaut, in *Renforcement de la résilience des hydrosystèmes face aux changements globaux: "de la mesure hydrologique aux modèles de gestion"* (Béni Mellal - Maroc, 2020), pp. 47–52
12. A. Bentayeb and C. Leclerc, *Ressources en Eau du Maroc, Tome III, Domaine atlasique et Sub-atlasique*, Editions du service géologique du Maroc (1977)
13. S. K. McFeeters, International Journal of Remote Sensing **17**, 1425 (1996)
14. H. Xu, International Journal of Remote Sensing **27**, 3025 (2006)
15. K. V. Singh, R. Setia, S. Sahoo, A. Prasad, and B. Pateriya, Geocarto International **30**, 650 (2015)
16. Y. Du, Y. Zhang, F. Ling, Q. Wang, W. Li, and X. Li, Remote Sensing **8**, 354 (2016)
17. F. Menjour, F. Amraoui, and T. Remmal, in *6th International Conference on Cartography and GIS* (Albena, Bulgaria, 2016), pp. 588–597
18. S. Szabó, Z. Gácsi, and B. Balázs, Landscape & Environment **10**, 194 (2016)
19. Q. Guo, R. Pu, J. Li, and J. Cheng, International Journal of Remote Sensing **38**, 5430 (2017)
20. B. Sallak, Les risques hydrologiques d'inondations et la problématique d'aménagement des territoires de piémont: Cas du dir de Taghzirt à ZaouitEchiekh, Province de Béni Mellal, Maroc, Thèse de doctorat, Université de Lorraine, 2019

Wastewater reuse in agriculture sector: resources management and adaptation in the context of climate change: case study of the Beni Mellal-Khenifra region, Morocco

Elhousna Faouzi*, *Abdelkrim* Arioua, *Ismail* Karaoui, *Kamal* Ait Ouhamchich, and *Driss* Elhamdouni

Earth sciences department, Sultan Moulay Slimane University, Faculty of Sciences and Techniques Beni Mellal, Morocco

Abstract. Recent climate change effects worsened water scarcity in Morocco and forced the country to seek alternative water resources such as domestic and industrial wastewater. In this context, we assessed the treatment efficiency of wastewater treatment plants (WWTP) of the BeniMellal-Khenifra region based on physicochemical and biological parameters. Vegetation cover evolution near WWTPs was also analysed using satellite images (Landsat TM and OLI). Six WWTP were evaluated based on treated water quality and a survey of nearby farmers and residents. Results showed treated wastewater is in line with Moroccan standards, and can be reused for irrigation and aquaculture without harmful effects. The survey pointed to the Boujaad WWTP as a model in the region. The vegetation cover evolution before and after WWTP existence showed an important improvement of cultivated lands. As a conclusion, wastewater reuse will allow the BeniMellal-Khenifra region to secure agricultural irrigation to safeguard freshwater quantities and quality despite climate change.

1 Introduction:

Rapid urbanization and climate change increase pressure on the availability of freshwater resources in Morocco [1]. Water scarcity is particularly acute in semi-arid and arid zones, because of the growth resulting from the accelerated demographic and economic development in many Moroccan cities [2]. Limits to water resources availability is not only caused by limited quantities, but also degraded quality resulting from contaminants generated by domestic, industrial and agriculture activities [3]. Therefore, it is necessary to consider unconventional water resources such as wastewater to meet high water demand by different sectors, and to recover the significant wastewater discharged directly into nature without any treatment[4].Wastewater treatment becomes the primary concern as it is associated to the issues of population health and environment degradation. Wastewater can

* Corresponding author: elhousna.faouzi@usms.ac.ma

be used for sectors with quality requirements below human consumption standards, such as agriculture, aquaculture and other social activities[5].

Remotely sensed data has been applied worldwide in land cover classification and monitoring on a regional or global scale [6]. It is particularly important when field data is unavailable, or hard to retrieve [7]. Remote sensing techniques provide reliable information on monitor wastewater treatment plants contribution in vegetation cover development. Landsat TM, OLI images were used for their potential application in identifying the land cover classification and the hydrographic system[8]. A selected time-series of twelve satellite images over the period from 2008 to 2018 were used to map the vegetation cover in the study area (near WWTPs).Those images were obtained under clear sky conditions to reduce induced effect by atmospheric disturbances. Support vector machine (SVM) classification was used to estimate the vegetation cover evolution in areas close to WWTPs.This result will present the importance of wastewater reuse in the agriculture sector, and to highlight the contribution of remote sensing applications in evaluating vegetation cover.

The BeniMellal-Khenifra region has 13 wastewater treatment plants, most of them in poor condition or under construction[9]. The poor condition of WWTPs leads to malfunction, that in turn creates environmental issues, as the treated water is either reused in agriculture despite its high pollution load, or discharged directly into nature to pollute groundwater and surface water. Municipalities in the BeniMellal-Khenifra region cities are using natural lagoon ssystem, due to its efficient process, low economic cost, and adapted to semi-arid arid climate conditions. This lagoons system is largely appreciated in its efficiency in small towns, and recommended for hot climate conditions [10]. lagoon system technique is considerable efficient in eliminating pathogenic germs and ensure a better reuse of wastewater in agriculture sector [11].

2 Materials and methods

2.1 Study area

The BeniMellal-Khenifraregion is located in the center of central part in Morocco. It is 240 km far away from Moroccan capital Rabat [12]. This region is characterised by four large natural landscape: the mountain ranges, piedmont, the Khouribga phosphates plateau, and the Tadla Plain [13]. This region is formed byBeniMellal, Azilal, Fkih Ben Salah, Khenifra, and Khouribga provinces (Figure1) [14]. Climate in the study area is arid to semi-arid. The average annual rainfall is around 100 mm in the arid area and the 600 in humid area. The average temperature is 18°C (40°C in summer and -3, 5°C in winter)[15].

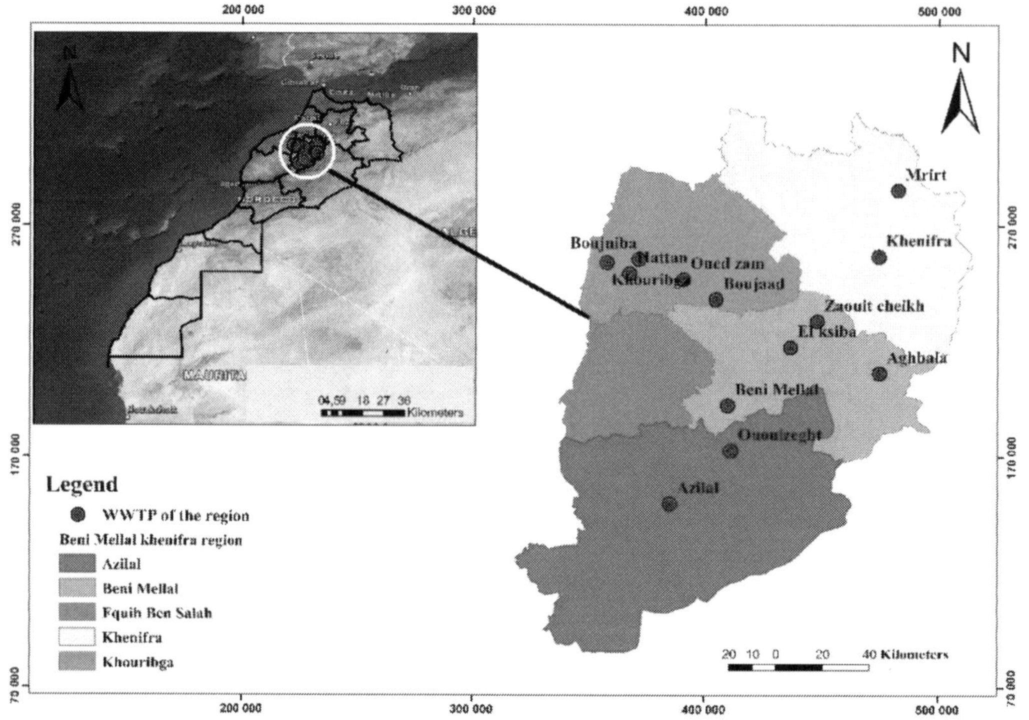

Fig. 1: Geographic location of WWTP in the BeniMellal-khenifra region.

2.2 Methods

The physico-chemical and microbiological composition of water is essential to identify possible reuse of treated wastewater[16]. For this purpose, raw and treated wastewater quality at six WWTPs in the study area was analysed to reveal associated shortcoming and anomalies at each treatment plant. Wastewater samples were taken upstream and downstream of each treatment plant during field visits. Raw and treated urban wastewater samples were collected from the inlet and outlet points for each WWTP during two periods, high quantity of wastewater arriving into the station (morning and evening) and low one (noon hours). Farmers and residents close WWTPs were surveyed. Satellite images were downloaded from the Internet. Applying the Supervised classification SVM, this method aims to find the optimal result with limited information provided by a small set of samples[17].

2.2.2. Water quality parameters

The following physical and chemical parameters have been measured at field(electrical conductivity, temperature, salinity, and pH), while the rest (Biochemical oxygen demand, chemical oxygen demand, total suspended solids, nitrates, nitrites, ammonium, ortho-phosphates, and sulphate) were analysed at Hydraulic Basin Agency laboratory following the AFNOR Moroccan standards [18]. The Moroccan standards for NH_4^+, NO_2^- and PO_4^{3-}, are not set for irrigation purpose, as these parameters are generally absorbed during the treatment process [19].

The sampling procedure is done under favourable conditions, in accordance with Moroccan standard procedure, while adding fixing reagents, and taking all the necessary precautions during the transport of the samples (storage in ice).

https://doi.org/10.1051/e3sconf/202018302005

Electrical conductivity and salinity are good indicators for substances assessment [20]. Both parameters are strongly linked, and are originally coming from a natural or anthropogenic source. Controlling the salinity of irrigation water is of major importance, whether for natural waters or for wastewater [21].
Water salinity is calculated based on electrical conductivity and in situ temperature, by applying the following formula [22].

$$\sum_{i=0}^{5}\left\{\left[ai+\left(\frac{t-15}{1+k(t-15)}\right)bi\right]Rt^{\left(\frac{i}{2}\right)}\right\} [22] \quad (1)$$

With: $Rt = \frac{Cs,t,p}{42.914\, rt}$ and $rt = \sum_{i=0}^{4} Cit^{i}$
S: Salinity (g/L);
Cs, t, p: Electrical conductivity (ms/cm);
T: Temperature in °C; a_i, b_i, c_i and k are dimensionless observations

2.2.3. *Survey*

The survey is aiming to identify the impact of wastewater treated on near population health, social economy, types of adopted cultures, and the WWTP impact on their crop yield. However, a survey targeting individual farmers was conducted in all the exposed and unexposed areas near WWTPs. This survey is dedicated to collectinformation on agricultural crops production trends during last few years (after the existence of WWTP),the impact of wastewater reuse in agriculture sector,environment, public health,wastewater benefits on irrigation, quantity of fertilizer used, types of cultures irrigated by wastewater, and the farmer's suggestion for the current management of their wastewater.

Based on a preliminary survey, the treated and untreated wastewater are both used for irrigation in thestudy area. Most of farmer's in this study area using partly treated water for plots irrigation without a fertilizer because this water contains a significant amount of nutrients.

The collected survey information/responsewerebased on a simple statistical analysis, summarized in figure 3.

2.2.4. *Crop yield and WWTP performance evaluation using satellite imagery*

Satellite images (Landsat TM, OLI) were used to evaluate the evolution of vegetation, and to determinate WWTP performance at upstream and downstream of all treatment plants. Landsat images were obtained from United States Geological Survey (USGS) Earth Explorer portal (https://earthexplorer.usgs.gov). A subset of the images was selected based on a certain number of factors: availability and the atmospheric conditions during the images acquisition were used. Only images captured during the summer time were used, as the climate is the drier so there is no much evaporated water in the atmosphere. The LS5 images are characterized by seven spectral bands width a spatial resolution of 30 m for Bands 1 to 5 and 7. The spatial resolution for Band 6 (thermal infrared) is 120 m, andLS8 images consist of multispectral and thermal bands, and a panchromatic band with higher spatial resolution (15m)[23].The images were georeferenced using the World Geodetic System 1984 datum (WGS-84) and the Universal Transverse Mercator (UTM) projection (Table1).Although the captured images were processed by applying all necessary operations: radiometric and atmospheric corrections, using FLASH the module. Supervised classification (SVM) was applied to detect the vegetation cover evolution before and after WWTP existenc ein the study area. SVM algorithm is a new general learning algorithm based on statistical theory.

Table 1. Landsat image's metadata.

Metadata	Image (2008)	Image (2018)
Sensor	TM	OLI
Output format	GEOTIFF	GEOTIFF
Day/Night	Day	Day
Cloud Cover	0	0
Projection	UTM	UTM
Datum	WGS84	WGS84

3 Results and Discussion

According to water intended for irrigation quality standards in Morocco [24], we noted that the parameters (pH, temperature, BOD(Biological Oxygen Demand), conductivity) of treated wastewater are comply with standards. Indeed, the pH values of all the WWTP in the region spanned between 6.5 and 8.5 as a limit value for irrigation water quality, The Moroccan standard has set an interval of 6.5 and 8.5 for pH as permissible values [16]. The study results observed that the pH of the treated wastewater was suitable for irrigation, explained by a neutral pH around 7.5.

The electrical conductivity (EC) and salinity values were within the limit values for treated wastewater quality intended for irrigation[24]. The limit value is 3 $mS.cm^{-1}$ for conductivity and 7.68 g/L for salinity, explains slight to moderate restriction use of treated wastewater.

It is imperative that farmers first ensure the physical quality of their soil and the proper drainage. Poor drainage could trigger accumulation of salts in the root zone and limit the availability of water in the soil for crops. The risks of soil salinization resulting from irrigation with such water present a criterion detrimental to its quality and can be considered harmful for sensitive crops, particularly at the study area where evaporation is high.

Temperature values are also lower than the Moroccan irrigation standard fixed on 35 °C. The high conductivity (salinity indicator) concentration was caused by high evaporation at semi-arid climate. However, the values recorded after treatment were accepted for Moroccan standard controlling irrigation water, which imposes a maximum of 3000 µS / Cm at 25 °C.

The concentrations of chemical oxygen demand(COD)in the Azilal, Zaouit Echcheikh and Ouaouizeght WWTPs are above the standard set here at 250 mgO_2/L[25]. The treated wastewater from the other three remaining WWTPs was intended to be reused in irrigation. In addition, the concentration of the BOD_5 wastewater taken from the WWTP is not suitable for irrigation. Indeed, the limit of 120 mgO_2 /L set by the Moroccan standard is not respected here [25].

The total suspended solid (TSS) content is 90.9 mg/L at El Ksiba WWTP outlet, which was considered suitable for irrigation use. Indeed, the quality standards referenced for wastewater intended into irrigation indicated that the limit values were around 2000 mg/L for gravity irrigation and 100 mg/L for spray irrigation[24]. The other WWTP with a high

https://doi.org/10.1051/e3sconf/202018302005

TSS load reaching around 250 mg/L were greater than the maximum allowable value. In fact, using a treated wastewater with its high load need to take precaution to avoid clogging of the soil porosity and low permeability.

The NO_3^- concentrations ranging between 1.11 mg/L and 4.09 mg/L at the outlet, hence, it was considered suitable for irrigation purpose. Nitric nitrogen, which is frequently used in agricultural fertilizers, is often a pollutant of groundwater therefore must be closely monitored. In our study area, where farmers used a low quantity of pesticide, the nitrate concentration was far from permissible value of 30 mg/L.

The acceptable sulphate values for treated wastewater intended for irrigations was fixed at 250 mg/L, meaning that all our collected samples were suitable and can be reused for irrigation.

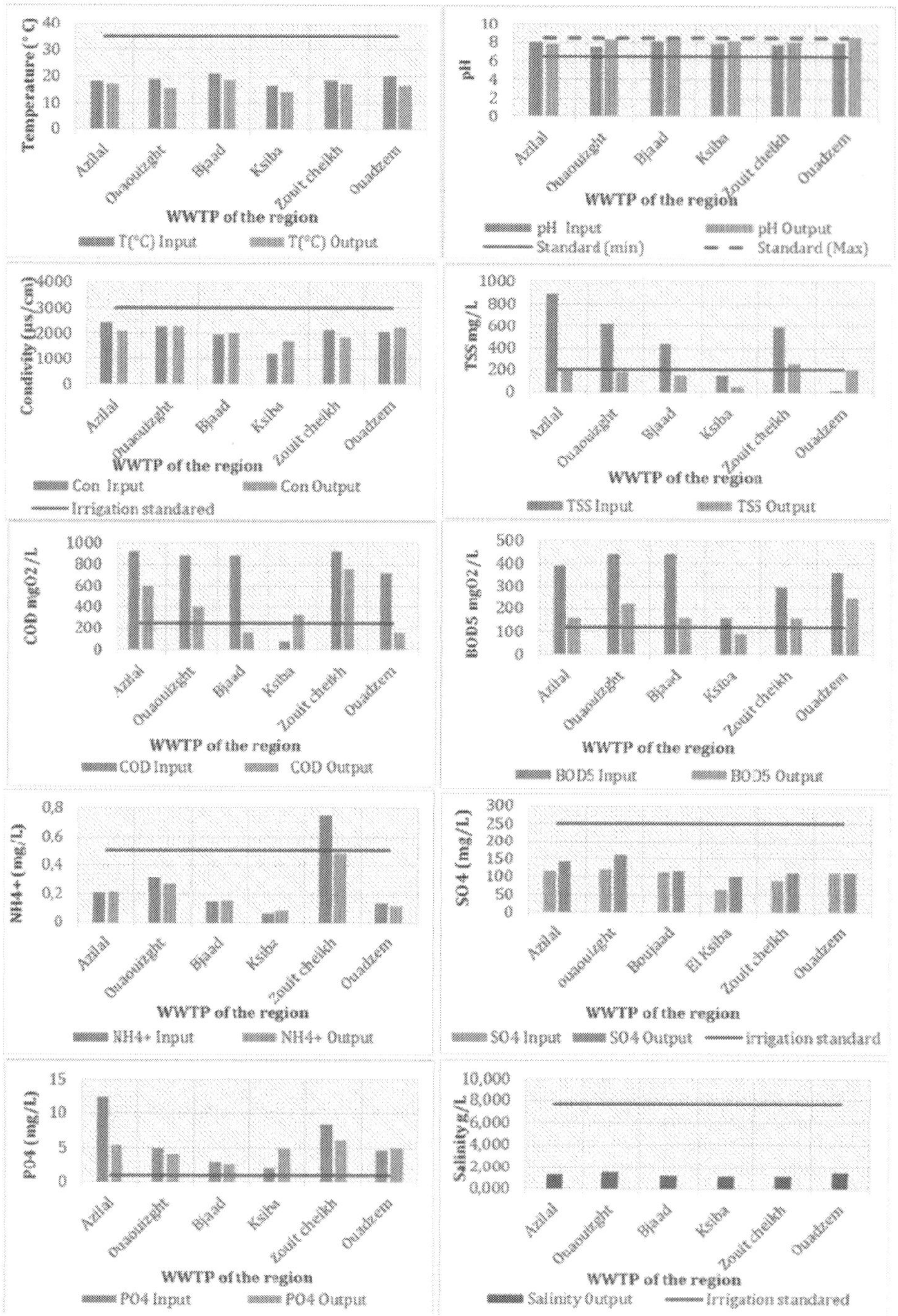

Fig. 2: Physico-chemical and organic analyses results for treated wastewater.

3.1 Survey results analysis

Survey results indicate that agricultural practices in BeniMellal-Khenifra region are linked to perennial culture (arboriculture) and annual culture (cereal and market gardening).

The reuse of treated wastewater in the study area is very varied: the majority of cities are not interested in the reuse of wastewater while others consider it as a reliable and beneficial alternative to irrigation and, at the same time, for agriculture.

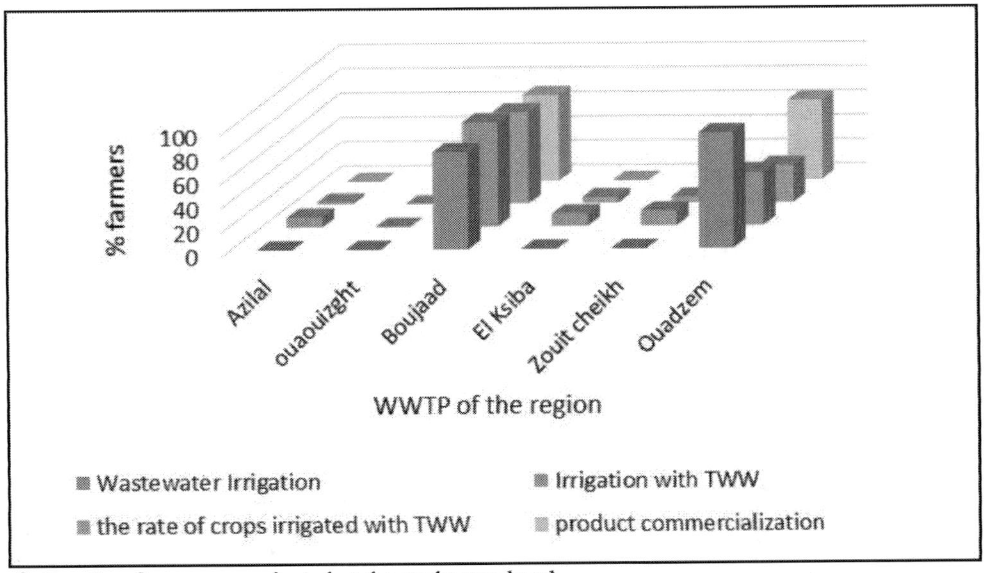

Fig. 3. Results of surveys conducted at the study area level

The information acquired during surveys, field visits, and by comparing analysed samples with Moroccan standards for wastewater intended into irrigation, showed that all WWTPs in the area has the ability to reuse their treated wastewater for irrigation without major negative impacts on the environment.

The classification of WWTPs was done based on surveys and their efficiency history. Results were used to classify the Boujaad wastewater treatment plant as a model treatment plant, because of its performance and experience of reusing treated wastewater in the irrigation of all crops and green spaces in the center of the city. The reuse of wastewater from the Boujaad WWTP has several positive impacts on different aspects, namely: saving water and fertilizers, improving crop yields, improving farming techniques, protecting consumer health and the environment.

3.2 Satellite images analysis:

Satellite images were used to infer the contribution of wastewater treatment plants on vegetation cover development at the study area, we used satellite images to identify and characterize the dynamics of the vegetation cover trend observed in the field. A diachronic analysis of images acquired in 2008 (Landsat TM) and in 2018 (Landsat OLI). The classification results are presented in (figure 4) showed the vegetation cover, and land cover variation:

 https://doi.org/10.1051/e3sconf/202018302005

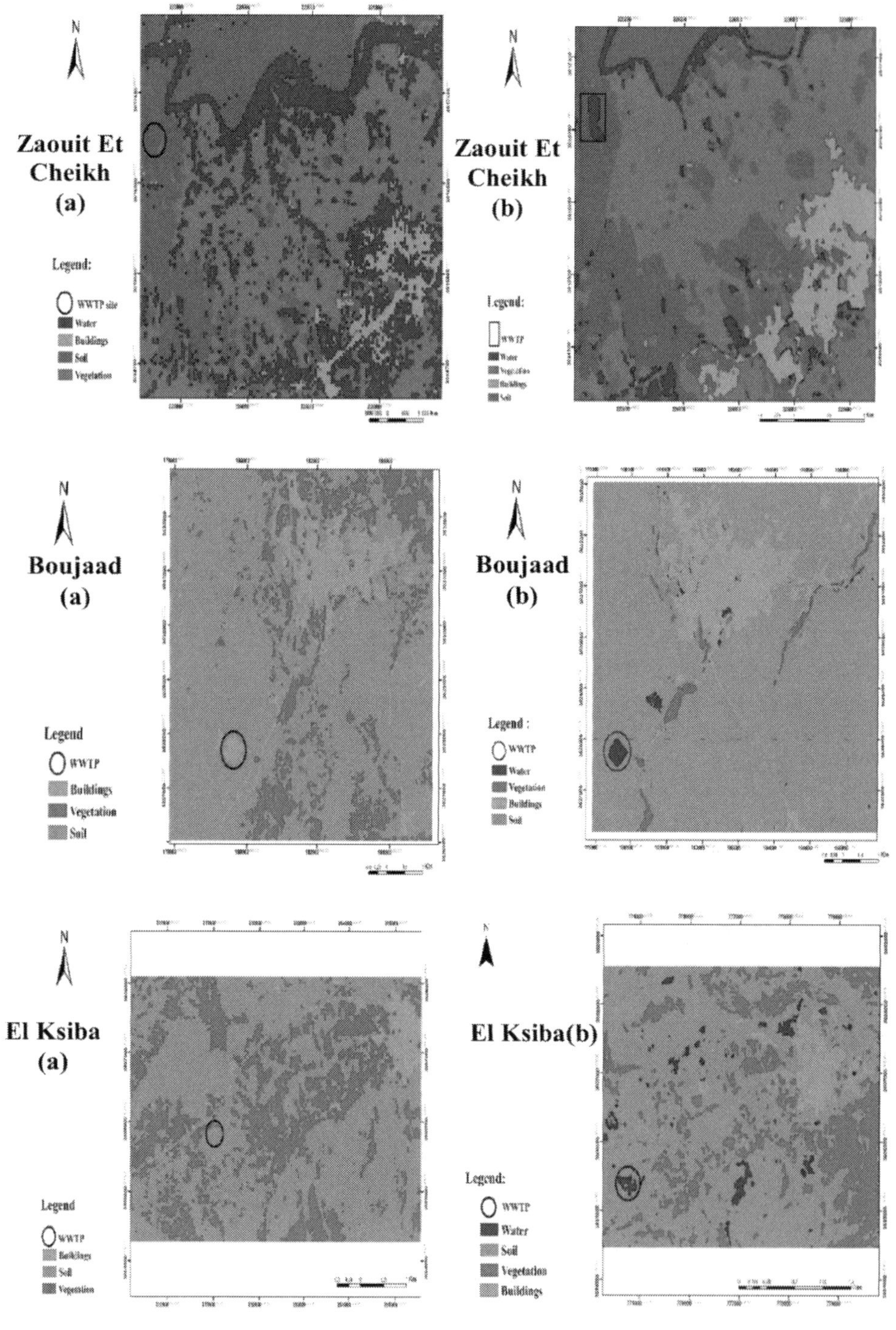

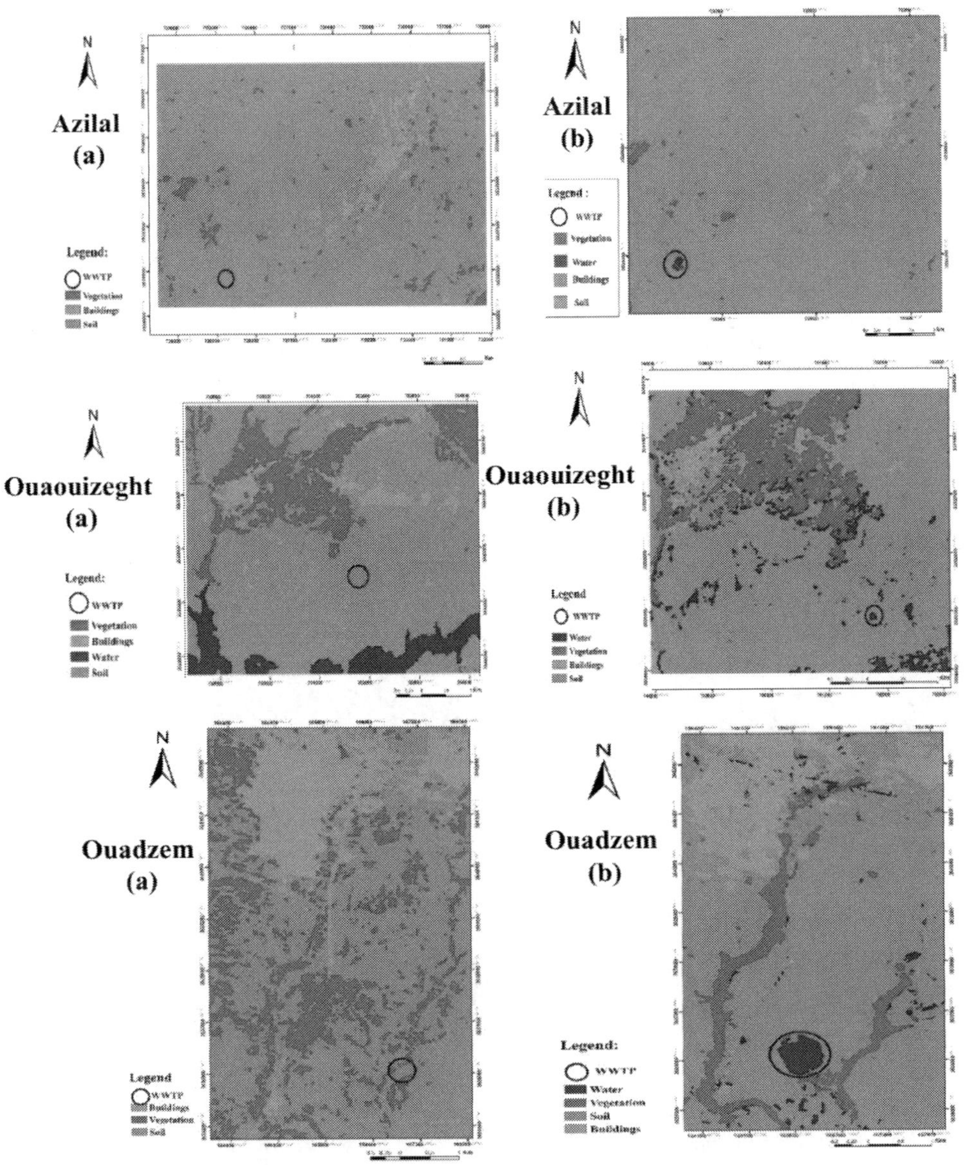

Fig. 4. Land use maps before (a) and after (b) construction of the WWTPs studied

The diachronic study showed that vegetation cover around the WWTPs areas was very important in 2008 (before the existence of WWTPs), possibly because of the diversity of land uses (agricultural and riparian areas). 2008 was a particularly wet year (annual precipitation of 729.14 mm), which may explain the density of vegetation found on the satellite images. The images acquired in 2018 (after the construction of WWTPs), show that dense vegetation formations are located mainly in mountain areas, at the edges of watercourses and in plots near WWTPs where irrigation made use of wastewater (ZaouitEchcheik, Boujaad, OuedZem).Sparse crops were found in the irrigated areas by wells. The rest of the region was devoid of vegetation for most of the year due to the low rainfall in 2018 (522.96 mm)[26].The comparison between 2008 and 2018was done having in mind the exceptional precipitation recorded during 2008. Taking into consideration that 2018 has a low quantity of annual precipitation, it was very clear that WWTP were improving the vegetation cover near all studied treatment plants, especially for cultivated lands, and irrigated perimeters.

4 Conclusions

The studyhighlightedthe importance of WWTP in BeniMellal-Khenifra region (Azilal, Ouaouizeght, ZaouitEchcheikh, El Ksiba, Boujaad and OuedZem) for the agriculture areas, and assessed the physicochemical quality of the treated water. All parameters were in compliance with both national and international regulations on wastewater reuse. Indeed, the performance of the various WWTPs in the study area was appropriate. The BoujaadWWTP was identified as the bestin the study area because it adopts an approach that includes the collection, purification, recovery of treated wastewater, reuse of this water for irrigation and watering of green spaces. Low resolution satellite images were used to evaluate the impacts of WWTP on vegetation. They failed to detect clearly the evidence of the contribution of WWTPs on the vegetation cover and on the hydrographic system, possibly because of their low resolution. On the other hand, field visits and surveys with farmers led to the conclusion that the use of wastewater in the area of irrigation .generated the creation of several cultivated plots. Products cultivated in these plots were marketed locally and regionally without any restrictions.

The authors thank the head laboratory manager at OumEr-Rbia Hydraulic Basin Agency MrAbdelkhalek EL AMRANI IDRISSI, for his technical assistance. Greater appreciation goes to all those responsible personnel and supervisors for WWTP in the BeniMellal-Khenifra region, for their help and contributions. We thank Mr Mohamed Baadi, the university driver for his time devoted to all our missions.

References

1. J. Schilling, K. P. Freier, E. Hertig, et J. Scheffran, *Climate change, vulnerability and adaptation in North Africa with focus on Morocco* , Agric. Ecosyst. Environ. vol. **156**, p. 12-26, août (2012), doi: 10.1016/j.agee.2012.04.021.
2. H. J. Oulidi, A. Fadil, et N. E. Semane,N.E.Eds,*Geospatial Technology: Application in Water Resources Management*. Springer International Publishing, (2020).
3. J. El Addouli, A. Chahloui, A. Berrahou, A. Chafi, A. Ennabili, andL.Karrouch, *Influence de rejets d'eaux usées sur les qualités physicochimique et bactériologique d'eaux utilisées en irrigation*, vol. **4**, no. 56. (2009).
4. I. Karaoui, A. Arioua, A. E. A. Idrissi, and D. Elhamdouni, *Elucidating the mechanisms of water quality deterioration in Oued El Abid River -Morocco-*, *J. Mater. Environ. Sci.*, vol. **8**, no. 9, pp. 3353–3359, (2017).
5. M. A. Hanjra, J. Blackwell, G. Carr, F. Zhang, and T. M. Jackson, *Wastewater irrigation and environmental health: Implications for water governance and public policy*, *Int. J. Hyg. Environ. Health*, vol. **215**, no. 3, pp. 255–269, Apr. (2012), doi:10.1016/j.ijheh.2011.10.003.
6. N. Xu, *Detecting Coastline Change with All Available Landsat Data over 1986–2015: A Case Study for the State of Texas, USA*, *Atmosphere*, vol. **9**, no. 3, p. 107, Mar. (2018), doi: 10.3390/atmos9030107.
7. C. Pipitone, A. Maltese, G. Dardanelli, M. L. Brutto, and G. L. Loggia, *Monitoring Water Surface and Level of a Reservoir Using Different Remote Sensing Approaches and Comparison with Dam Displacements Evaluated via GNSS*, *Remote Sens.*, vol. **10**, no. 2, p. 71, Jan. (2018), doi: 10.3390/rs10010071.

8. L. Sun and K. Schulz, *The Improvement of Land Cover Classification by Thermal Remote Sensing, Remote Sens.* vol. **7**, no. 7, pp. 8368–8390, Jun. (2015), doi: 10.3390/rs70708368.
9. Office National de l'Eau Potable, « *Etude d'impact sur l'environnement, Maroc* », p. 126, (2006).
10. C. R. Steinmann, S. Weinhart, et A. Melzer, *A combined system of lagoon and constructed wetland for an effective wastewater treatment*, Water Res., vol. **37**, no 9, p. 2035‑2042, (2003), doi: 10.1016/S0043-1354(02)00441-4.
11. H. E. Maynard, S. K. Ouki, and S. C. Williams, *Tertiary lagoons: a review of removal mecnisms and performance*, Water Res., vol. **33**, no 1, p. 1‑13, janv. (1999), doi: 10.1016/S0043-1354(98)00198-5.
12. S. Msaad, N. Abbadi, M. Mbarki, S. Rabi, N. Belkhouya, and A. Gamouh, *A Study related to the management of medical and pharmaceutical wastes in BeniMellal-Khenifra region: BeniMellal city as a case of study*, E3S Web Conf., vol. **150**, p. 02019, (2020), doi: 10.1051/e3sconf/202015002019.
13. K. Ait Ouhamchich, I. Karaoui, A. Arioua, A. Kasmi, D. Elhamdouni, F.E. Elfiraoui, Z. Arioua, F. Nazi, and N. Nabih,*Climate Change Trend Observations in Morocco: Case Study of BeniMellal-Khenifra and Darâa-Tafilalt*Regions, J. Geosci. Environ. Prot., vol. **06**, no 07, Art. no 07, juill.(2018), doi: 10.4236/gep.2018.67003.
14. A. Barakat, A. Hilali, M. E. Baghdadi, and F. Touhami, *Landfill site selection with GIS-based multi-criteria evaluation technique. A case study in BeniMellal-Khouribga Region, Morocco*, Environ. Earth Sci., vol. **76**, no 12, p. 413, juin (2017), doi: 10.1007/s12665-017-6757-8.
15. Direction Générale des CollectivitésLocales, *monographie de la région de Béni Mellalkhénifra, Maroc*. (2015).
16. Y. Asfers, H. Taouil, A. Amine, S. Elanz, and S. I. Ahmed, *Study of Physico-Chemical Quality of the Surface Waters of the Estuary of the Western Basin of the OuedOumErrabia-Morocco*, IOSR J. Appl. Chem., vol. **10**, no 05, p. 41‑49, mai(2017), doi: 10.9790/5736-1005024149.
17. A. Yang, S. Jiang, and H. Deng, *A P2P Network Traffic Classification Method Using SVM*, in 2008 The 9th International Conference for Young Computer Scientists, Hunan, China, Nov. (2008), pp. 398–403, doi: 10.1109/ICYCS.2008.247.
18. J. Rodier, B. Legube, et N. Merlet, *Analyse de l'eau 9 e édition*. Dunod, Paris, (2009).
19. N. Legros, *La réutilisation des eaux usées traitées en irrigation comme incubateur d'un processus de pérennisation et de bonne gouvernance des infrastructures d'assainissement : cas pratique de la station d'épuration de Tidili au Maroc*, p. 93, (2017).
20. R. Lal and B. A. Stewart, *Soil Water and Agronomic Productivity*, 1st Edition. CRC Press, (2012).
21. Z. Muyen, G. A. Moore, et R. J. Wrigley, *Soil salinity and sodicity effects of wastewater irrigation in South East Australia*, Agric. Water Manag., vol. **99**, no 1, p. 33‑41, Nov.(2011), doi: 10.1016/j.agwat.2011.07.021.
22. M. E. Mokhtar, M. Chibout, M. Kili, B. EL Mansouri, J. Chao, S.M. EL Kanti, A. Ntarmouchant, and A. Benslimane, *Évaluation de l'intrusion saline dans la nappe de Foum El Oued, province de Laâyoune, Maroc*, Bull. L'Institut Sci. Rabat Sect. Sci. Terre, vol. **40**, p. 53‑69, (2018).

23. Arrété N° 1276-01 conjoint du ministre de l'équipement et du ministre chargé de l'aménagement du territoire, de l'urbanisme et de l'habitat et de l'environnement, *les normes de qualité des eaux destinées à l'irrigation au Maroc*, (oct. 17, 2002).
24. M. Aboelnour and B. A. Engel, *Application of Remote Sensing Techniques and Geographic Information Systems to Analyze Land Surface Temperature in Response to Land Use/Land Cover Change in Greater Cairo Region, Egypt*, J. Geogr. Inf. Syst., vol. **10**, no. 1, Art. no. 1, Jan. (2018), doi: 10.4236/jgis.2018.101003.
25. C. Benmoussa, M. El Yazghi, et S. Mezouar, *Arrêté n° 1607-06 du 29 Joumada II 1427 portant fixation des valeurs limites spécifiques de rejet applicables aux déversements d'eaux usées des agglomérations urbaines*, (juill. 25, 2006).
26. R. Mosteo, M.P. Ormad, P. Goñi, J. Rodríguez-Chueca, A. García, and A. Clavel, *Identification of pathogen bacteria and protozoa in treated urban wastewaters discharged in the Ebro River (Spain): water reuse possibilities*, Water Sci. Technol., vol. **68**, no 3, p. 575-583, août (2013), doi: 10.2166/wst.2013.201.

A decision support model to improve water resources management in agriculture: evaluation of the drip irrigation efficiency in the Ait Ben Yacoub region, East of Morocco

Ihssan Elouadi[1,2]*, *Driss* Ouazar[1], and *Lahcen* El Youssfi[3]

[1]Mohammadia School of Engineering, Mohamed V University, Rabat, Morocco; elouadihssan@gmail.com
[2]High National School of Mines of Rabat, Morocco 3Agronomy and Veterinary Medicine, Institute Hassan II, Rabat, Morocco
[3]Research team of environment and natural resources management, B2DRN Laboratory, Khenifra Higher School of Technology, Sultan Moulay Slimane University, Morocco

Abstract. Globally, climate change is projected to exacerbate water scarcity and increase the recurrence and intensity of droughts. These circumstances call for methodologies that can support the design of sustainable water management policies. Improved irrigation efficiency has been cited as an important way to adapt to climate change. This paper illustrates the potential of hydro-economic modeling for integrating the multiple dimensions of water resources, becoming a valuable tool in the empowerment of sustainable water management policies. The modelling framework is used also to analyze the impacts of climate change-induced drought on water uses in the Ait Ben Yacoub region (the East of Morocco). The evaluation of the conversion process from flurrow irrigation systems into drip irrigation by using the model showed that this policy has some positive effects on the agricultural gross margin and contributes to slightly moderating the impact of climate change on farmer's incomes. Moreover, aoptimal efficiency of drip irrigation will inevitably result in a decrease in the volume of infiltrated water and therefore a very likely drawdown of the groundwater level. This result demonstrates the limitation of this policy and the necessity to accompany this irrigation system by measures of improvement of water supply. It concerns rainwater storage basins that should be covered by the polyethylene geo-membrane as example of promising measures.

Context and objectives

Along the last decades, recent research on climate change has approached the assessment of impacts, vulnerability and adaptation under biophysical component focusing on one specific dimension of climate change, such as the agronomic dimension [1,8], or the hydrological dimension [9,10]. However, the recognition of water management and climate change as multidimensional and multi-scalar concerns [11, 12] evidence the need to integrate biophysical, hydraulic and social economic component.

In line with this, varied types of integrated modelling frameworks have been developed to address the different scales (from the crop to the river basin) and the different dimensions of climate change, water and agriculture (hydrological, agronomic, and socio-economic). However, these frameworks have not always represented the socio-economic dimension of water use in sufficient detail and in some cases, they have undervalued the role of human response to climate impacts.

Trying to better represent economic issues, hydro economic modelling has been extensively used along the last decades as a prominent tool for guiding and implementing water policy decisions [13, 17].

Along this line, this paper presents a novel application of a hydro- economic modelling framework that is used to assess climate change impacts and adaptation in the Ait ben Yacoub region The novelty of the approach presented here lies in the capability of this integrated framework to take into consideration agronomic, economic and hydrologic processes that take place at different scales.

Using this integrated approach, this paper evaluates the impacts of a severe climate change scenario and simulates the effect of Drip Irrigation System as a Climate Change Adaptation Measure on Ait Ben Yacoub region.

*Corresponding author: elouadihssan@gmail.com

https://doi.org/10.1051/e3sconf/202018302006

Water, agriculture and climate change in the "Ait Ben Yacoub" region

In Morocco, a semi-arid Mediterranean country, water resources become increasingly scarce. This scarcity will become acute in the coming years due to: a reduction in water supply (as a result of climate change), and increase in water demand, accentuated by the population growth and the requirements of economic development. Agriculture is regarded as a strategic sector for the country's socioeconomic development. However, this sector faces challenges many; water management among others.

Therefore, Morocco will have to deal with the additional challenges of climate change that will require the adaptation of economic activities dependent on water resources, such as irrigated agriculture, to the new climatic conditions. Dealing with climate change will require a shift in water management and farming decisions towards more sustainable agricultural production and more efficient water allocation, distribution and use.

The study area named "Ait Ben Yacoub " is located in the east of Morocco and characterized by a semi-arid climate where water is a limiting factor, thus constituting an unfavorable condition for agriculture. Major sources of Irrigation water in this region is thewater from AinLaarais with a flow rate of up to 245 l / s and groundwater exploration through wells and boreholes dug by farmers. The study area has 17 watersheds.

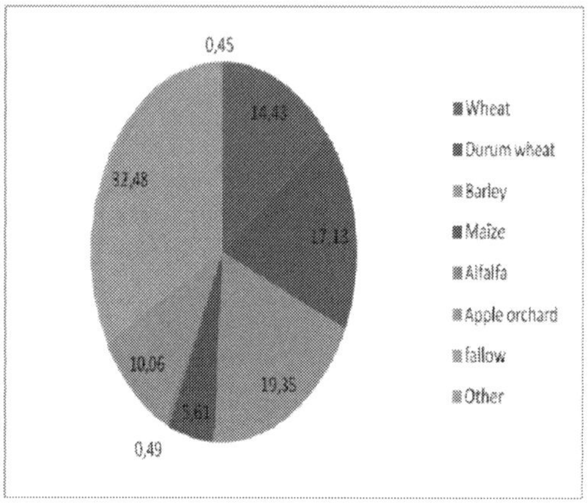

Fig.1:Ratios of crops types within the agricultural lands

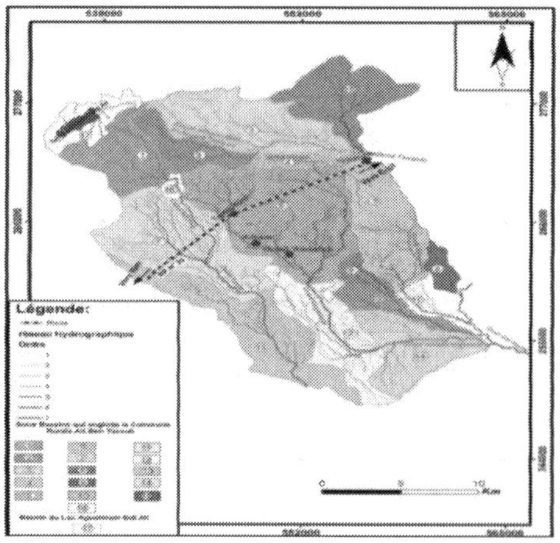

Fig.**2**: watersheds of Ait Ben Yacoub Region in Ait Ben Yacoub region

Methodology: a hydro-economic modelling framework.

To respond to the questions and challenges of climate change, we propose a hydro-economic modeling framework. The developedmodel integrates: economic, agronomic and hydraulic data. It can simulate the behavior of farmers faced with climate change and can takeinto account the competition for resources.

The proposed approach is based on nonlinear optimization techniques. It's a hydrological and economic model that uses water resources so as to maximize the profit at the level of the basin while taking into account a set of constraints which are divided into hydrological, agricultural and resource availability constraints. As a major feature of this model of this model is a detailed disaggregation by spatial units (hydrological units, cropping areas, and grazing land), by agricultural production systems (irrigated and rainfed crops), and by farm sizes.

The Positive Mathematical Programming Method (PMP) approach, as suggested by Howitt [18], has been used to calibrate the models. The fundamental idea of PMP is to reconcile the new optimization of the mathematical programming model with the economic and social reality as evidenced in the database set. Essentially, PMP added calibration constraints so that the base solution is close to the database set. For reasons of computational simplicity and lacking strong arguments for other type of functions, a quadratic cost function is often employed (exceptions: [19, 20]). The general version of this variable cost function to be specified is:

Cv = quadratic cost
d = (N×1) vector of parameters associated with the linear term
Q = (N×N) symmetric, positive (semi-) definite matrix of parameters associated with the quadratic term.
However, the objective function of non-linear programming problem is composed of a quadratic cost calibrated using observed area of crop.

$$\text{Max } Z = px - dx - \frac{1}{2} Q x^2 \quad (2)$$

Subject to

$$Ax \leq b \ [\lambda]$$
$$x \geq 0$$

Where
Z = objective function value
p = (N×1) vector of product prices
x = (N×1) vector of production activity levels
A = (M×N) matrix of coefficients in resource constraints b = (M×1) vector of available resource quantities
λ = (M×1) vector of dual variables associated with the resource constraints.

Results and discussions

1. Simulation 1: climate change

The model simulates the assessment of climate change impact on irrigated agriculture. This scenario provides insight on the behaviour of farmers in conditions of water scarcity. The impact of a reduction of 50% of water allocation, groundwater recharge and rainfall have been simulated at the basin level.

The results of this research show that climate change mayseverely impact irrigation systems in the Ait ben Yacoub region (figure 3). It will reduce considerably the availability of water resources, and willalso reduce also crop yields (between 2% and 10 %) and increase irrigation water requirements. The resulting changes of water stress, crop yields and crop competitiveness will drive changes of crop land allocations. The cultivated area will decrease significantly by 45 % under drought conditions.

Model Results also demonstrate also that climate change will produce significant reductions of the region's income from 6.97 to 6.08 million USD. This reduction in the income concerns all types of farms but the small farms will be more vulnerable to climate change.

The model results highlight the remarkable impact of surface water management on the overexploitation of the groundwater and the risk of its excessivedepletion. A management policy of surface water based on administrative pricing, pumping cost, and water supply and marginal cost is proven inadequate for a sustainable resource management since it underestimates the overall water scarcity at the basin level (Fig. 1). The detailed results of this simulation are presented in the article [14].

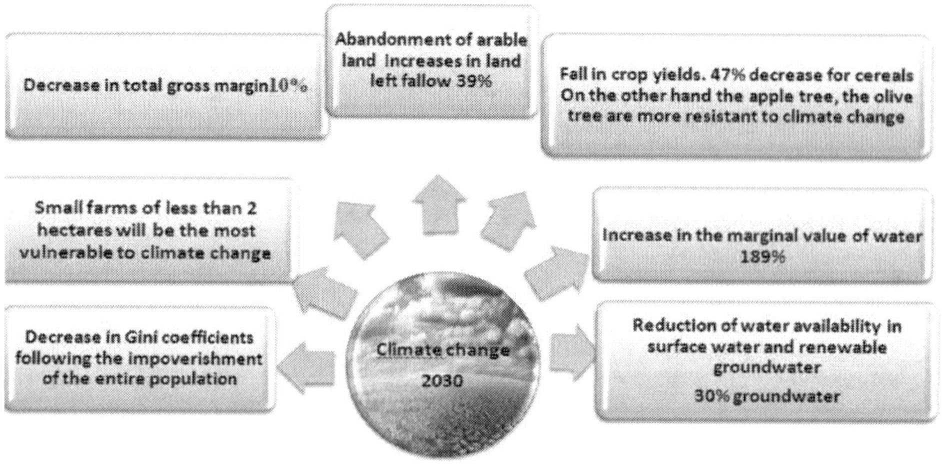

Fig.3: Model results simulation 1 - climate change impact

2. Simulation 2: adaptation measures (drip irrigation)

Climate change has introduced a number of pressing issues concerning water availability and predictability around the world. Old irrigation practices such as furrow irrigation and imprecise sprinkler irrigation are outdated and inefficient. Several researches have confirmed that the adoption of improved irrigation will be important for agricultural adaptation to climate change.

The model evaluates the effect of Drip Irrigation System Implementation as a Climate Change Adaptation Measure on Ait Ben Yacoub region. The evaluation of the conversion process from furrow irrigation systems into drip irrigation by using the model showed that, this adaptation measure will help to slightly reduce the negative effects of the drought in terms of profit. It has some positive effects on the agricultural gross margin and contributes to slightly moderate the impact of climate change on farmer's incomes.

The farmer will tend to gradually abandon cereal crops and alfalfa to grow crops with higher water value such as arboriculture. This change in rotation is not very spectacular for this simulation because it is slowed down by an acute water stress recorded on the scale of underground aquifers. Moreover, a better efficiency of drip irrigation will inevitably result in a decrease in the volume of infiltrated water and therefore a very likely drawdown of the groundwater level.

This result demonstrates the limitation of this policy and the necessity to support this irrigation system by measures of improvement in water supply. It concerns rainwater storage basins that should be covered by the polyethylene geo-membrane as example of measures. The success of any agricultural adaptation strategy in the face of climate change is conditioned by the introduction of new crops resistant to new climatic hazards.

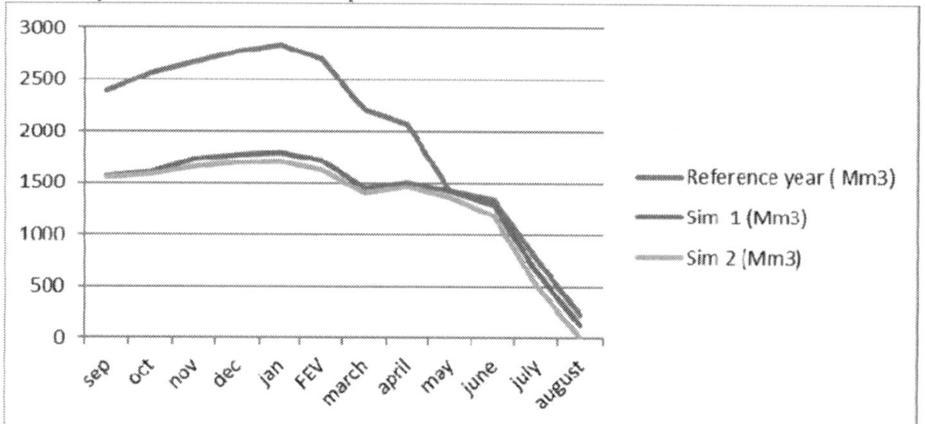

ig.4 : Evolution of the volume of water stored in "itzer-injel" aquifers in million m³ (reference Year and climate change scenario)

Conclusion

Improved irrigation efficiency has been cited as an important way to adapt to climate change [1- 6]. This study tried to contribute to a list of studies that have focused on the role of climate in adoption of irrigation technology.
This study revealed that the policy of shifting to drip irrigation system has some positive effects on the agricultural gross margin and contributes in mitigating the impact of climate change on farmer's incomes. Moreover, an optimal efficiency of drip irrigation will inevitably result in a decrease in the volume of infiltrated water and therefore a very likely drawdown of the groundwater level. This result demonstrates the limitation of this policy and the necessity to support this irrigation system by measures of the improvement of water supply.

References

1. I. Burton, *Adaptation to climate change and variability in the context of sustainable development*, In: Climate Change and Development, L. Gómez Echeverri (Ed.). Yale School of Forestry and Environmental Studies and United Nations Development Programme (UNDP), New Haven, CT, and New York, pp. 153-173 (2000)
2. B. Smit, M.W. Skinner, Mit Adapt Strategies Glob Ch **7**, (2012)
3. Kurukulasuriya, P. and S. Rosenthal. 2003. Climate Change and Agriculture: A Review of Impacts and Adaptations. Climate Change Series No. 91 (World Bank, Washington, D.C, 2002, 2003)
4. T. Cavagnaro, L. Jackson, K. Scow, Climate change: Challenges and solutions for California agricultural landscapes. White Paper CEC-500- 2005-189-SF. California Climate Change Center. www.energy.ca.gov/2005publications/CEC-500- 2005-189/CEC-500-2005-189-SF.PDF. Accessed June 14, 2016, (2006)
5. Jackson, L., F. Santos-Martin, A. Hollander, W. Horwath, R. Howitt, J. Kramer, A. O'Geen, B. Orlove, J. Six, S. Sokolow, D. Sumner, T. Tomich, and S. Wheeler., Potential for adaptation to climate change in an agricultural landscape in the Central Valley of California. Final Paper CEC-500-2009-044-F. California Climate Change Center. Available at www.energy.ca.gov/2009publications/CEC-500- 2009-044/CEC-500-2009-044-F.PDF. Accessed June 14, 2016, (2009)
6. B. Joyce, V. Mehta, D. Purkey, L. Dale, M. Hanemann. Climate change impacts on water supply and agricultural water management in California's Western San Joaquin Valley, and potential adaptation strategies. Final Paper CEC- 500-2009-051-F. California Climate Change Center. Available at www.energy.ca.gov/2009publications/ CEC-500-2009-051/CEC-500-2009-051-F.PDF. (Accessed June 14, 2016), (2009)
7. M. Moriondo, M. Bindi, W. Zbigniew,W. Kundzewicz, M. Szwed, A. Chorynski, P. Matczak, M. Radziejewski,D. McEvoy, A.Wreford, Adapt. Strateg. Glob. Chang. **15** (2010)
8. D. Ventrella, M. Charfeddine, M. Moriondo, M. Rinaldi, M. Bindi, Reg. Environ. Chang. **12**, (2012)
9. B.A. .Joyce, V.K. Mehta, D.R. Purkey, L.L. Dale, M. Hanemann, Clim. Chang. **109** (Suppl. 1), (2011)
10. S. Rochdane, B. Reichert, M. Messouli, A. Babqiqi, M.Y. Khebiza, Water **4**, (2012)
11. T.E. Downing, WIREs Clim. Chang. 3, (2012)
12. H. Meinke, S.M. Howden, P.C. Struik, R. Nelson, D. Rodriguez, S.C. Chapman, Curr. Opin. Environ.Sustain. **1**, (2009)
13. I. El ouadi, D. Ouazar, R. Doukkali , L. El Youssfi, International Journal of Education and Research, **2** (2014)
14. I. El ouadi, D. Ouazar, R.Doukkali, D. Hasnaoui, International Journal of Applied Engineering Research **12**, (2018)
15. I. Blanco-Gutiérrez, C. Varela-Ortega, D.R. Purkey, J. Environ. Manag. **128**, (2013)
16. Brouwer, R., Hofkes, M., Ecol. Econ. 66, (2008)
17. I. Heinz, M. M. Pulido-Velazquez, J.R. Lund, J. Andreu, Water Resour. Manag. **21**, (2007)
18. R.E. Howitt, American Journal of Agricultural Economics, **77** (1995)
19. Q. Paris Q., and R.E. Howitt, American Journal of Agricultural Economics, **80** (1998)
20. Q. Paris Q., and R.E. Howitt, The Multi-Output and Multi-Input Symmetric Positive Equilibrium Problem. In: Heckelei, T., H.P. Witzke, and W. Henrichsmeyer (Eds.): Agricultural Sector Modelling and Policy Information Systems. Proceedings of the 65th EAAE Seminar, March 29-31, 2000 at Bonn University, VaukVerlag Kiel, (2001)

Impact of drought on water quality in the Youssef Ben Tachafine dam (Souss-Massa region, Morocco)

*Fatima*Boudarfa[1]*, *Elmahjoub*Gallouli[1],*Mohamed*El hafa[1], and *Hassan*Oulad Ali[1]

[1] Biodiversity and Ecosystem Functioning Team, Ibn Zohr University, Faculty of Sciences, Department of Biology, Agadir, Morocco.

Abstract. The Massa catchment area has been experiencing an acute rainfall deficit for several years, which has had direct effects on the filling level of the Youssef Ben Tachafine dam. The objective of this work is to study the impact of the drought on the physicochemical quality and the biological balance of this aquatic ecosystem. A monitoring and bi-monthly analyses of the physicochemical and biological parameters (temperature, dissolved oxygen, pH, conductivity, transparency, manganese, iron, nitrate, ammonium, silica, phosphorus, and chlorophyll-*a*) were carried out from January to December 2019. The considerable drop in the level of the reservoir has had a major impact on the degradation of the water quality of the lake and its trophic status. The analysis of the results showed that the trophic state of the Youssef Ben Tachafine dam is between mesotrophic and hypereutrophic.

Introduction

Climate change is now a reality in Morocco, as everywhere else in the world, Morocco is very vulnerable to climate variability. Climate projections established by the National Meteorological Directorate predict an increase in average summer temperatures of around 2 to 6 ° C and a decline of 20% on average rainfall by the end of the century[1]. This global warming has a direct impact on rainfall, on which the filling rate of dams, groundwater recharge and surface water depend. The hydrological regime of all Moroccan watersheds is characterized by a very high interannual variability marked by the alternation of wet and dry sequences, interspersed with years of high hydraulicity or severe drought[2].

The Youssef Ben Tachafine dam is subject to harsh climatic conditions due to the succession of acute drought cycles. Work carried out on the quality of Moroccan lakes under climatic conditions similar to those of the YBT dam[3, 4], has shown that drought has caused significant changes in the general quality of the aquatic ecosystem.

In this context, we conducted a study during a cycle on the YBT dam, from January to December 2019, which focuses on the analysis of hydrological variations, physico-

*Corresponding author: f.boudarfa@gmail.com

chemical and biological parameters. The main goal of this study is to understand the effects of drought on the general quality of the dam and its trophic status.

1 Materials and methods

1.1 General characteristics of the site in study

From the geographical point of view, the Youssef Ben Tachafine reservoir is part of the 940 Km^2 Chtouka plain, which is the natural extension of the Souss plain towards the South-West, It is limited to the south by the Oued Massa on which the dam was built, 90 km south of Agadir, to the north by the Agadir-Biougra road, to the east by the outcrops of the primary formations of the Anti-Atlas and to the west by the Atlantic Ocean (Fig. 1).

From the physiographic point of view, it should be noted that the rugged topography, the wide variety of rocks, the semi-arid climate, the proximity of the desert and the strong winds make this area a large region exposed to erosion and desertification. Demography and agricultural transformations aggravate the situation by depleting natural resources such as natural vegetation and water.

Among the objectives assigned to the Youssef Ben Tachafine dam, is cite the reinforcement of the drinking water supply to the towns of Tiznit and Sidi Ifni as well as to the surrounding centres, in addition the development of irrigation over an area of 19,500 hectares, representing 21.3% of the total agricultural perimeter of the Souss-Massa.

Water inflows in 2019 to the catchment area of the YBT dam are almost nil, which means that this year can be considered very dry, in comparison to previous years. The reservoir reserve has been reduced by more than 50% from $78Mm^3$ in January to less than 39Mm3 in December 2019, as well as the filling rate (from 26% to 13%), this decrease in level is the result of the average volume regulated for the irrigation of the Chtouka plain 275×10^3 m^3/month and the evaporated volume ranging from 258×10^3 m^3 recorded in December to 899×10^3 m^3 in June, plus the average volume withdrawn from drinking water, which is 627×10^3 m^3/month [5].

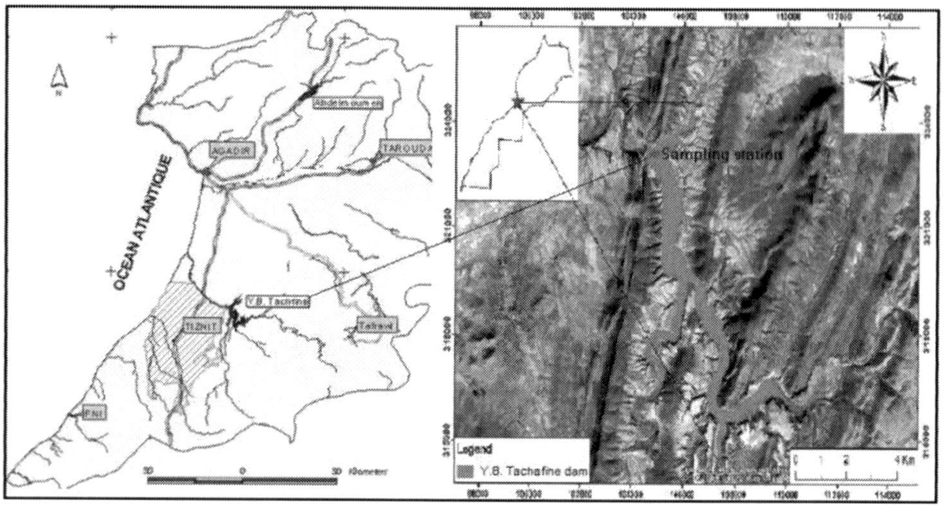

Fig. 1. Schematic representation of the Youssef Ben Tachfine dam.

1.2 Sampling Points and frequency

The sampling station (fig1) is located in the deepest part of the dam [5], the water samples were taken at different levels of the column (surface, 2.5m, 5m, 10m, 15m, and 20m) using a horizontal two-litre WILDKO bottle.These samples were taken between January and December 2019, with a biweekly step and always between 12:00 and 14:00.

1.3 Parameters measured and methods used

1.3.1 In situ measurements

The temperature, conductivity and dissolved oxygen of the water were measured in situ using a multi-parameter instrument (HI 98196 brand HANNA). Transparency was measured in metres using a Secchi disc immersed in the sampling station, and undesirable substances (iron, manganese and hydrogen sulphide) were determined by the colorimetric method using analytical test kits.

1.3.2 Analyses carried out in the laboratory

The water samples were brought to the laboratory and immediately filtered through a 0.45 µm membrane filter. In addition, nutrients (total phosphorus (Pt), ammonium (NH4+), and nitrates (NO_3^-)) were analyzed using an absorption spectrophotometer according to the water analysis protocol[6, 7].For the determination of chlorophyll "a" the water samples were filtered through Whatman GF/C filters (0.45 µm), the pigments were extracted in 30 ml of hot ethanol and the determination was carried out by measuring the absorbance of the extract with a spectrophotometer at wavelengths 665 nm and 750 nm before and after acidification according to the Lorenzen method[8].

Suspended matter is measured by differential weighing of the Whatman 0.45 µm filters before and after filtration of the sample and drying in an oven at 105°C [7].

1.3.3 Trophic status

The trophic status was assessed according to the model proposed by the O.C.D.E[9]. This model is based on annual average concentrations of chlorophyll "a", total phosphorus (Pt)and water transparency (Secchi disc).

Table 1summarizes the various parameters measured and methods used

Table 1.Parameters measured and methods used

Sampling depth in metres	Surface- 1 - 2.5 - 5 - 10 - 15 - and the bottom
Water and plankton sampling	Horizontal Wildco type bottle of 2 L capacity and plankton net of 20µm mesh size.
Temperature (°C), conductivity (µs/cm), pH and O_2-dissolved (mg/l)	Multiparameter apparatus (Hanna brand)
Transparency in meters	Secchi disk
Suspended solids (TSS) in mg/l	Filtration on a 0.45 µm filter then weighed after passing through the oven and the desiccators.
Iron, manganese and hydrogen sulphide in mg/l	Colorimetry (Analysis Kits)
NO_3^-, NH_4^+, Pt, and Si in mg/l	Spectrophotometry
Chl "a" in µg/l	Ethanol extraction

2 Results

2.1 Temperature

Large variations in the dam water temperatures are recorded from one season to the next. During the winter period, the water column is slightly homothermal. The temperature begins to rise at the end of March (19.13 C°) to reach its maximum in August (27.32 C°). A direct stratification settles by creating a thermocline 10 m deep from early June. This thermal cycle classifies the waters of the Youssef Ben Tachafine dam in the category of warm monomictic lakes (Fig.2).

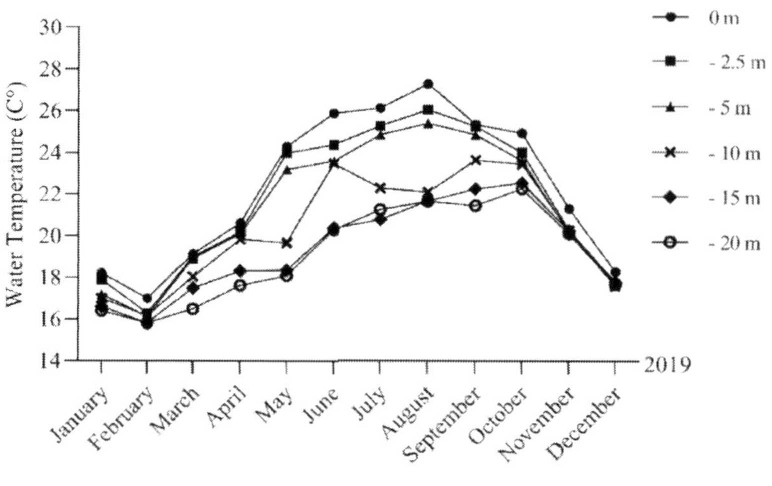

Fig.2. Spatial and temporal changes in water temperature in Youssef Ben Tachafine dam.

2.2 Hydrogen potential (pH)

The pH of the water in the YBT dam remains generally alkaline, fluctuating between 7.51 and 9.13, with the highest values being recorded during spring and summer, particularly in the surface layers, reflecting the intensity of photosynthetic activity in the euphotic layers of the lake. However, a slight acidity of the deep water was observed from April onwards, in parallel with the installation of the thermal stratification (Fig. 3).

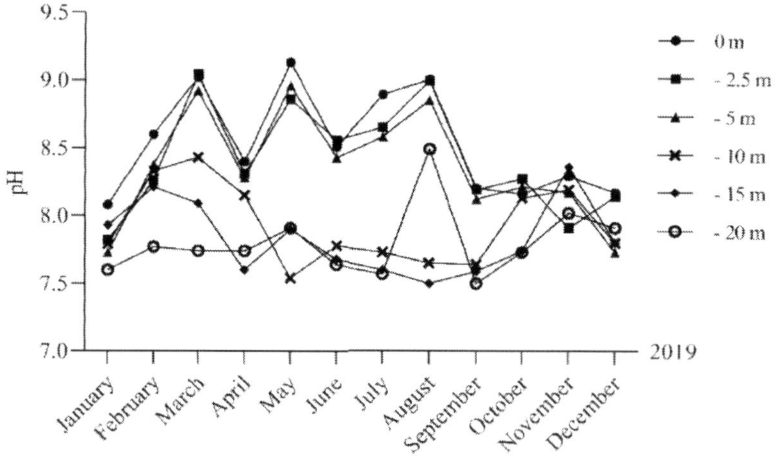

Fig. 3. Spatial and temporal changes in water pH in Youssef Ben Tachafine dam.

2.3 Conductivity

The degree of mineralization is identical over the entire water column, the maximum is recorded in December (600 µs/cm). The seasonal evolution of the conductivity of the water in the YBT reservoir is well marked over time. This being justified by the drop in the water level in the lake, which leads to a concentration of ionisable elements in the water (Fig. 4).

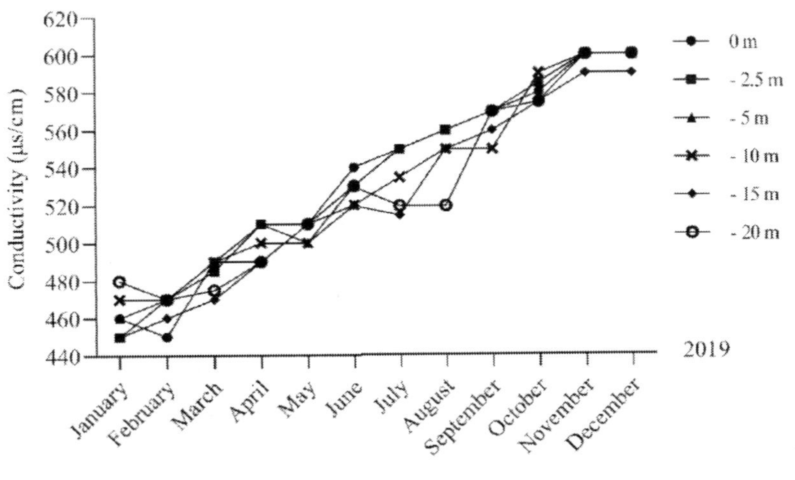

Fig. 4. Temporal changes in conductivity in Youssef Ben Tachafine dam.

2.4 Dissolved oxygen

The water in the reservoir is relatively oxygenated in the surface layers, however the deep layers quickly lose their oxygen reserves and tend towards total anoxia, in parallel with the beginning of the installation of thermal stratification which takes place from early spring to autumn (Fig. 5). The drought conditions led to a marked oxygen deficit. Due to the reduction in the height of the water column, the resuspension of organic matter and detritus promotes biodegradation throughout the water column.

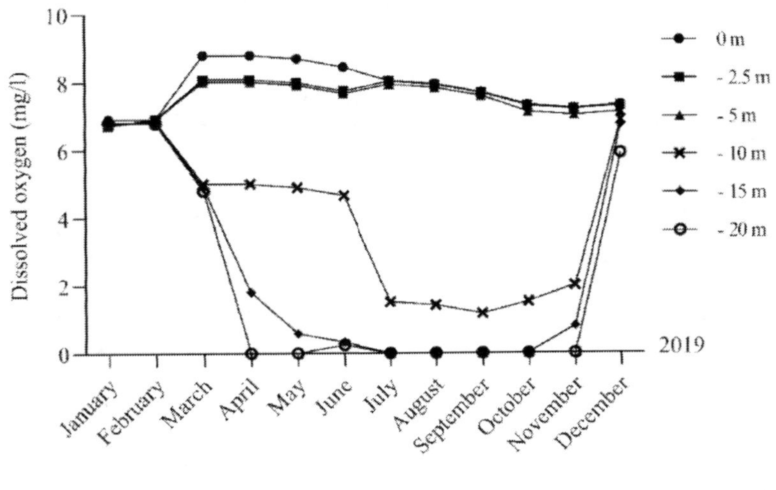

Fig. 5. Spatial and temporal changes in dissolved oxygen content in Youssef Ben Tachafine dam.

2.5 Manganese, iron and hydrogen sulphide

Deep water anoxia generates the formation of hydrogen sulphide and other undesirable chemicals such as $Mn^{2+}$and Fe^{2+}. High values are recorded in summer and autumn at the bottomof the reservoir (Mn^{2+}= 0.9 mg/l, Fe^{2+} = 0.6 mg/l and H_2S = 0.25 mg/l)due to the release of sediment under anaerobic conditions (Fig. 6).

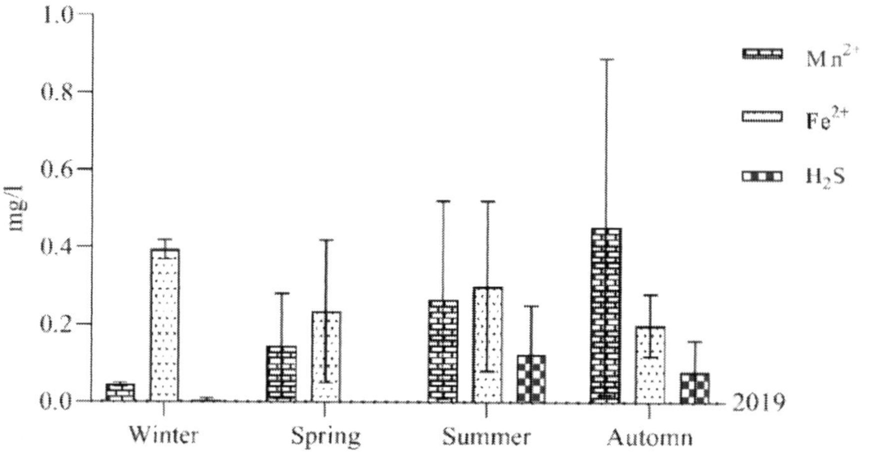

Fig. 6. Seasonal variation of manganese, iron and hydrogen sulphide at Youssef Ben Tachafine dam.

2.6 Chlorophyll "a", TSS and water transparency

The phytoplankton biomass represented by chlorophyll "a" levels during the study period shows two peaks, the first in early spring (9.9 µg/l) and the second in early fall (10.5 µg/l).

The water transparency varies between a minimum of 0.30 m recorded at the beginning of January and a maximum of 1.6 m at the beginning of May, with an average transparency of about 1 m. These low depths of secchi disc disappearance can be explained by the increase in algal biomass in the euphotic zone in spring and fall (Fig. 7) and by the increase in suspended matter during the water mixing period (Fig. 8).

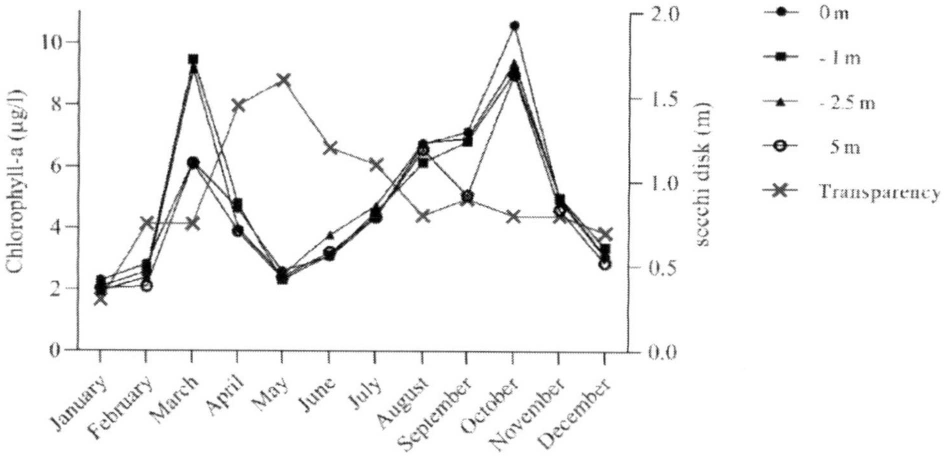

Fig. 7. Temporal changes in the chlorophyll a and the secchi disc depth in Y.B. Tachafine dam.

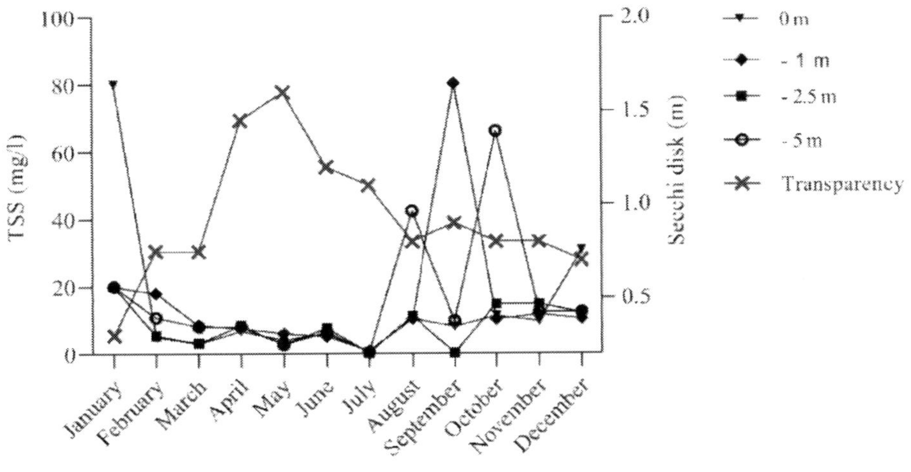

Fig. 8. Temporal changes in total suspended solids and the secchi disc depth in Y. B. Tachafine dam.

2.7 Nutriments

2.7.1 Nitrate and Ammonium

Nitrate levels recorded in the water column of the Youssef Ben Tachafine dam are high in winter (2.82 mg/l), while concentrations decrease to zero in summer in the surface layers (Fig. 9). These anoxic conditions at the bottom of the dam facilitate the reduction of nitrates by denitrification [10]

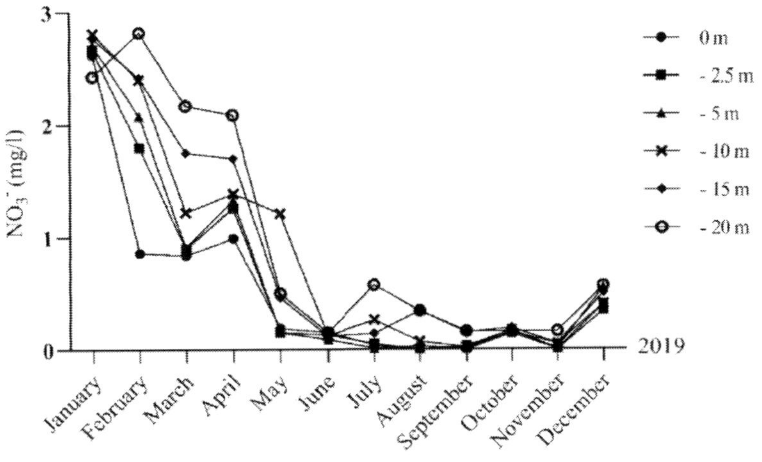

Fig. 9. Temporal changes of nitrates at Youssef Ben Tachafine dam.

Ammonium values vary according to the seasons, the highest concentrations are recorded in hypolimnium in summer ([NH4+] =1.58 mg/l) and autumn ([NH4+] = 1.46 mg/l) (Fig. 10).This is probably the result of the degradation of nitrogenous organic matter.

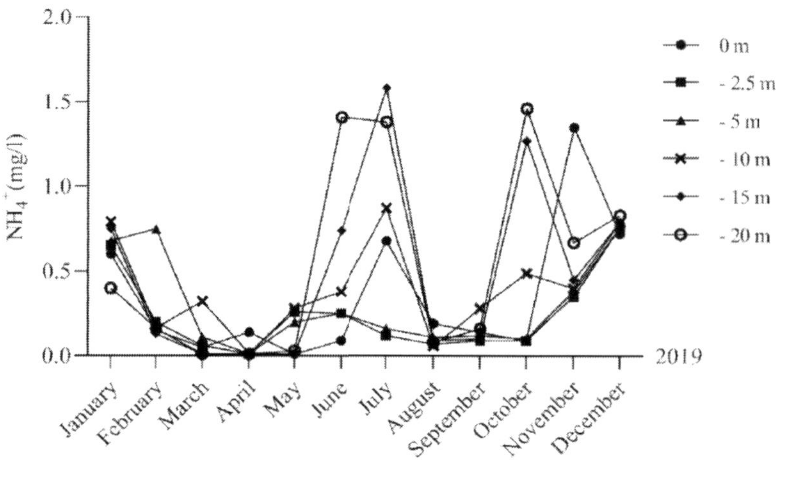

Fig. 10. Temporal changes of ammonium at Youssef Ben Tachafine dam.

2.7.2 Total Phosphorus

Total phosphorus levels show significant fluctuations in the water column, ranging from 0.00 mg/l in summer to 0.34 mg/l in winter, with a slight increase noted in autumn. From April onwards, total phosphorus levels are distributed along an increasing gradient from the surface to the bottom of the lake. This is due to the release of phosphorus from anoxic sediments during summer stratification. [11], this phenomenon has already been observed in other Moroccan reservoirs, such as El Kansera, Al Massira, and Mansour Eddahbi,[3, 4, 12](Fig. 11)

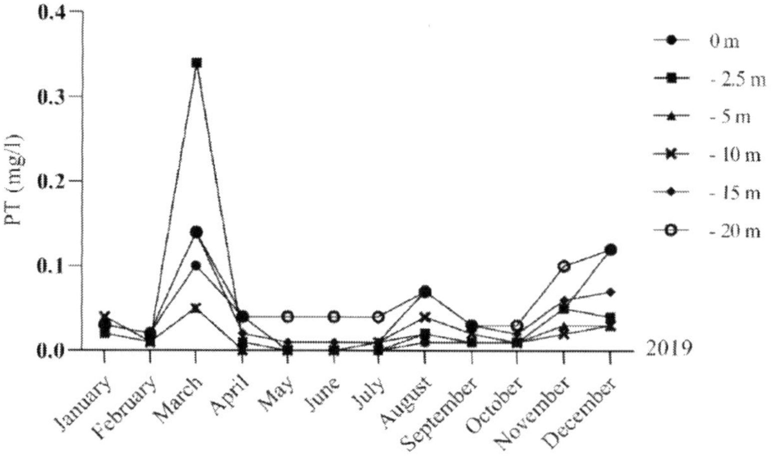

Fig. 11. Temporal changes in total phosphorus in Youssef Ben Tachafine dam.

3 Discussion

The hot and dry climatic conditions of the Massa catchment area have led to a considerable reduction in the volume of water and a significant drop in the filling rate of the YBT dam, leading to a concentration of ionisable elements and consequently a strong mineralisation of the water in this dam. Indeed, the installation of the thermocline between 10 m and 12 m depth during the stratification period reduced the volume ratio of anoxic layers compared to oxygenated layers by intensifying the exchanges between the sediments and the water column.

Under these conditions of total and early deep water anoxia, sediment plays an important role in water quality[13, 14]. The acceleration of the eutrophication phenomenon causes the degradation of the organoleptic quality of the waters of this reservoir by the appearance of hydrogen sulphide, manganese and iron with all their negative consequences on the various uses of the water including in particular the production of drinking water[15-17]. Concerning nutrients, there is a high loss of nitrogenand high denitrification during the stratification period. The anoxic conditions at the bottom of the lake facilitate the reduction of nitrates[18]and the increase of ammonium by denitrification [10].The same phenomenon has been underlined by several authors at the levels of other Moroccan dams[19-21].These ammonium values are higher than those found during a study carried out on the same site more than twenty years ago [22]. Nitrates are rapidly consumed at the surface and decrease throughout the water column, while ammonium ions and total phosphorus are consumed at the surface and released by sediments on the lake bottom. The depletion of phosphorus and nitrates in the surface layers will probably be the limiting factors for phytoplankton production in the Youssef Ben Tachafine dam. The recorded algal biomass is maintained during the study period, despite the absence of external inputs, appears to be the result of internal enrichment of the lake with nutrients from sediment release during the period of thermal stratification and deep-sea anoxia.

In addition, the algal biomass and the re-suspension of the sediments lead to a significant reduction in the transparency of the water in the reservoir. The transparency parameter appears to be a good descriptor of water quality and representative of the trophic level of this reservoir.

Comparing our results with those of Ait Salah[22], especially nitrogen and phosphorus, the two main indicators of eutrophication, it can be seen that the YBT impoundment has evolved towards a eutrophic state. According to the trophic classification proposed by the O.C.D.E. model [9], The trophic status of the YBT reservoir lies between mesotrophy and hyper-eutrophy.

Conclusion

In the light of these results, we can conclude that the drought conditions caused by global warming and hydrological changes over the last two decades have had a negative impact on the water quality of the YBT dam reservoir and its trophic status. One of the first consequences is the early installation of thermal stratification and total anoxia of the deep layers with the presence of manganese throughout the study period. In addition, the reduction in the volume of water in the reservoir and the high water temperature favoured the strong mineralisation of the water and the enrichment of the water column with nutrients from sedimentation and decomposition, with all their negative consequences on the functioning of this precious ecosystem. The evolution of the trophic status of this reservoir from mesotrophy to hyperotrophy is the direct consequence of all these changes.

References

1. I. Motib, M. Batchi, F. Fatah, Conservation Des Ressources Naturelles Pour Une Sécurité Alimentaire Durable Au Maroc, ESJ**16**, 240 (2020)
2. E. Ouharba, Z. Triqui, R. Moussadek, Impact of Climate change on the Bouregreg Watershed Vegetation and Forest of Morocco, IJASRE**5**, (2019)
3. M. Sadani, N. Ouazzani, L. Mandi, J Water Sci **17**, (2004)
4. D.F. Berrada, R. Berrada, A. Benzekri, ANN LIMNOL-INT J LIM, EDP Sciences, 155-166 (1999)
5. A.B.H.S.M, Agence de bassin hydraulique du Sous Massa, (Maroc, 2019)
6. Afnor, In recueil des normes françaises des eaux, méthodes d'essais,622 (1994)
7. J. Rodier, Water analysis: Natural water, seawater.Dunod Paris Edition, (1994)
8. C.J. Lorenzen, Determination of chlorophyll and pheo-pigments: spectrophotometric equations, Limnol Oceanogr**12**,343-346 (1967)
9. O.C.D.E, Eutrophisation des eaux: méthodes de surveillance, d'évaluation et de lutte, Paris,165 (1982)
10. P. Bonin, M. Gilewicz, J. Bertrand, Effects of oxygen on each step of denitrification on Pseudomonas nautica, Can J Microbiol**35**, 1061-1064 (1989)
11. A. Rapin, M. Rabiet, M. Grybos, B. Mourier, A. Fay, T. Kestens, V. Deluchat, Distribution spatiale et mobilité du phosphore sédimentaire dans une retenue hydroélectrique, J Water Sci **30**, 71-76 (2017)
12. S. Kemmou, J.-E. Dafir, M. Wartiti, M. Taoufik, Variations saisonnières et mobilité potentielle du phosphore sédimentaire de la retenue de barrage Al Massira (Maroc), WATER QUAL RES J**41**, 427-436 (2006)
13. E. Gautreau, L. Volatier, G. Nogaro, E. Gouze, F. Mermillod-Blondin, The influence of bioturbation and water column oxygenation on nutrient recycling in reservoir sediments, Hydrobiologia **847**,1027-1040(2020)
14. A.N. Cojean, M.F. Lehmann, E.K. Robertson, B. Thamdrup, J. Zopfi, Controls of H2S, Fe2+, and Mn2+ on Microbial NO3−-Reducing Processes in Sediments of an Eutrophic Lake, FRONT MICROBIOL**11**, 1158 (2020)
15. D. Zogo, H. Soclo, M. Bawa, M. Gbaguidi, Distribution des résidus de fer et de manganèse le long de la colonne d'une retenue d'eau en cours d'eutrophisation: Cas du barrage de l'Okpara à Parakou au Benin (Journée d'Étude du CEBEDEAU, Tribune de l'Eau, 2008)
16. H. Erraji, Les contraintes climatiques et hydrologiques sur le comportement du Fer et du Manganèse dans la retenue du barrage Smir au Maroc, (2016)
17. A. Foutlane, A. Bouloud, K. Ghedda, *Restauration de la qualité des eaux des retenues de barrages*, IAHS Publications-Series of Proceedings and Reports-Intern Assoc Hydrological Sciences**243**, 287-298 (1997)
18. Y. Chen, C. Peng, J. Wang, L. Ye, L. Zhang, Y. Peng, Effect of nitrate recycling ratio on simultaneous biological nutrient removal in a novel anaerobic/anoxic/oxic (A2/O), BAF system, BioresourTechnol**102**,5722-5727(2011)
19. S. Niazi, M. Snoussi, A. Foutlane, Impact of climatic variability on water quality in a semiarid hydrosystem: the Nekkor catchment area (Morocco), SECHERESSE**16**, 183-187 (2005)

20. M. Loudiki, Etude limnologique d'un hydrosystème récemment aménagé dans la région de Marrakech (Maroc): apports du bassin versant, hydrochimie, algues et macrophytes (Thèse, Aix-Marseille Université, 1990)
21. H.A. Hammou, D. Latour, S. Samoudi, K. Mouhri, M. Douma, J. Robin, M. Loudiki, Occurrence of the First Toxic Microcystis Bloom in a Recent Moroccan Reservoir, Water Resour**45**, 409-417 (2018)
22. H. Ait Salah, Organisation et structure des communautés zooplanctoniques de la retenue Youssef Ben Tachfine (sud de Maroc) (Thèse, Univ.Ibn Zohr, Faculté des science Agadir, 1997)

E3S Web of Conferences **183**, 03001 (2020) https://doi.org/10.1051/e3sconf/202018303001
I2CNP 2020

Agro-physiological response of potato to "sustainable" deficit irrigation in the plain of Saïs, Morocco

Omnia El Bergui[1,2], *Aziz* Abouabdillah[1*], *Rachid* Bouabid[1], *Nabil* El Jaouhari[1,3], *Youssef* Brouziyne[4], and *Mohamed* Bourioug[1]

[1]Ecole Nationale d'Agriculture de Meknès, Meknès, Morocco
[2]Institut Agronomique et Vétérinaire Hassan II, Rabat, Morocco
[3]Moulay Ismail University, Faculty of Science, Meknes, Morocco
[4]Mohammed VI Polytechnic University, International Water Research Institute - Benguerir, Morocco

Abstract. Morocco is one of the countries who's facing an insufficient water supply for irrigation, therefore it is seeking for new technologies allowing irrigation management evolution. Deficit irrigation is for this reason one of the recommended approaches to deal with this alarming situation. This study deals with the assessment of Sustainable Deficit Irrigationon potatoes production (Var. Desiree). The experiment was conducted on an experimental plot in open field in Morocco. Three water regimes were applied: T1control (100%), T2 (75%) and T3 (50%) of crop evapotranspiration ETc. This is a complete random block device with four repetitions. Measurements focused on monitoring vegetative, eco-physiological and yield parameters. The results obtained show that: (i)the irrigation regime did not significantly affect the eco-physiological parameters nor the vegetative parameters, except for the height of the plants; (ii) in terms of yield, a deficit of less than 25% of ETc did not lead to a significant difference compared to the control treatment and reached 41,17 tones/ha compared to the 100% ETc with 42,51 t/ha; **(iii)** irrigated treatments at less than 50% of daily ETc revealed extreme yield loss (20,96 Tonnes/ha);(iv) finally, the best agronomic water use efficiency was recorded for irrigated treatments at 75% of Etc.

Introduction

In a climate change and water scarcity context recognized worldwide, Morocco would be ranked among the 45 countries affected by this situation and is likely to reach an extremely high level of water stress by 2040. This situation is challenging and that will impact the Moroccan population and in particular the agricultural industry [1]. The great challenge for the coming decades will then be to increase food production with less water, and thus to improve water use efficiency.

* Corresponding author:aziz.abouabdillah@email.com

Deficit irrigation is one of the most recommended approaches to deal with water scarcity in agriculture. This method is applied by reducing irrigation water inputs without affecting the yield [2]. The crop is exposed to a certain level of water stress either during a given period of time, known as regulated deficit irrigation (RDI), or throughout the entire growing period and is therefore referred to as sustainable deficit irrigation (SDI). Many scientists have observed that the resulting reduction in yield may be insignificant compared to the benefits obtained; notably reduced water consumption.

Potato (*solanum tuberosum*) is one of the most important and widespread crops in the world. It plays a major role in the international and national food system since it is the main non-cereal foodstuff. It has become cultivated in almost all regions of Morocco, occupying an average area of more than 60,000 ha each year (MAPMDREF). Thus, it holds a crucial place among market garden crops in Morocco and plays a very important economic function. The potato has shallow roots and is sensitive to water stress. However, rainfall is insufficient to sustain its production, making irrigation crucial. According to Gultekin et al [3], potato yields were significantly affected depending on the water deficit and the best yield achieved (47.13 Tonnes/ha) was recorded by the treatment receiving 100% of the water needs.Hamida et al [4]have reported that the potato's response to water scarcity indicated that deficit irrigation increased potatowater use efficiency.

In order to verify and test these results; this study was carried out in the region of Fez Meknes, North-West Morocco, addressing thesustainable deficit irrigation of potato. The main objective of this experiment is to assess the agro-physiological response of potatoes to continuous deficit irrigation with different levels of water restriction.

1. Material and method

1.1 Experimental site

The experiment was carried out on an experimental plot of 4000 m^2 (0.4ha) at the National School of Agriculture of Meknes (ENA-Meknès), located in the area of Mejjat-Sidi Slimane Moul El Kifane, Fez-Meknes region, of which the coordinates are 33°50'37. 1 "N latitude and 5°28'30.0 "W longitude. The experimental plot was fallow for more than 4 years before that was planted with onion in 2018 followed by potato in March 2019. The region's weather ranges from Mediterranean to continental with cold winters and hot summers.

The physico-chemical soil analysis carried out in 2019 at the Ferticonseil laboratory of ENA-Meknès, show that the soil texture is loamy-clay with an average of 46% loam, 37% clay and 17% sand. The water content at field capacity (Fc) and permanent wilting point (PWP), estimated by the SPAW software [5]based on the Saxton formula, is 38% and 22% respectively.

In order to determine the water amount to be applied for each irrigation, it was necessary to calculate the Readily Available Water(RAW) equivalent to the Net Maximal Amount of irrigation, which corresponds to the amount of water that a soil can hold in the effective root depth, and which is calculated by using the formula below:

$$RAW = f\,x\,(Fc - PWP)\,x\,Rd\,x\,PHS_{(1)}$$

With
f: Irrigation triggering percentage (5%).
Fcc: Humidity at field capacity(38%)
Hpfp: Humidity at permanent wilting point (22%)

Rd: effective root depth

PHS: **Percentage of humidified soil**corresponding to the percentage of soil moisture estimated at 65% for this installed irrigation system.
All the key parameters of the RAW were calculated at the beginning of the experiment. The factor f is an important indicator for judging the work quality in irrigation management. The value of f chosen for this test is 5%.

The basic meaning of f = 5% is that irrigation is triggered when 5% of the available soil water storage is consumed. The choice of this frequency was based on the study carried out by Lamrahli in 2018 in the same field which compared two values of the f coefficient (5 and 10%) for onion crops and showed that a frequency of 5% gives better results in terms of water absorption and water use efficiency as well as in terms of yield.

1.2 Experimental Protocol

The plant material used for this study is potato (*Solanum tuberosum*) of the 'desirée' variety imported from the Netherlands. Seeding was carried out on March 5, 2019 in the field at a depth of 10 cm, on rows spaced at 80 cm apart and 30 cm tuber spacing.

The experimental design adopted is a complete randomized block design (DBAC), with a single source of heterogeneity with four repetitions and three water regimes.
The three water regimes are as follows:
- Control T1 is a comfort irrigation with 100% crop evapotranspiration ETc throughout the whole cycle;
- Treatment T2 corresponds to sustainable deficit irrigation with an application of 75% Etc during the whole cycle; (i.e. a 25% of water restriction).
- Treatment T3: sustainabledeficit irrigation with 50% ETc applied throughout the cycle; (i.e. 50% of water restriction).

A localized irrigation was provided by a drip system, including drippers each supplying 1.6 l/h, spaced at a distance of 0.4 m from each other.

As this study focused on water supply, the crop was conducted in a non-limiting manner concerning the other factors (tillage, density, fertilization, weeding, etc.).

1.3 Irrigation water needs calculation

The determination of the crop water requirement is based on the water balance method. The daily crop evapotranspiration (ETc) is calculated from reference evapotranspiration data ET0 provided by the meteorological station available at the same experimental site. The formula used for the calculation of the ET0 is the Penman-Monteith equation [6].

The formula for calculating the net water requirement (NWR) of the potato is as follows:

$$NWR = ETc = (ET0 * Kc)_{(2)}$$

The gross daily water requirement applied concerns a percentage of the daily water requirements (DWR) calculated by the crop evapotranspiration formula below:

$$\mathrm{GWR} = ETc/Cu * Ei_{(3)}$$

With

Cu: Uniformity of water distribution coefficient measured at the field and estimated to be 0.92
Ei: irrigation system efficiency estimated at 0.9 for this new irrigation system.

The crop coefficient Kc model used in this study, which depends on the phenological stage of the potato crop, is the one published in the official FAO bulletin 56. This crop coefficient varied between 0.45 and 1.45 depending on the phases of the potato cycle.

1.4 Plant observations and measurements

Measurements were taken every 15 days during the trial period on 36 randomly selected individuals, taking 3 plants for each elementary plot (12 plots).

1.4.1 Growth and development parameters

Plantheight, leaf density and number of stems were regularly monitored during our study as for the development parameters.

1.4.2 Eco-physiological parameters

Measurements included chlorophyll index and leaf temperaturemeasures. It should be noted that leaf temperature measurements were taken using an infra-red thermometer.
The leaf temperature depends on the energy it receives and the energy it eliminates.

1.4.3 Yield Parameters

At harvest time, the upper part of each plant was cut off. Then the wet weight of each plant was determined. All samples were then placed in an oven at 60°C for 48 hours. All yield components (Number of tubers, tuber length and size, tuber weight) were measured per plant and then the marketable tuber yield per hectare.

1.4.4 Biochemical parameters

The extraction and determination of proline was carried out according to the method described by Bates [7].

1.5 Statistical analysis

The data analysis of the studied parameters as well as the calculation of means and the graph conception were carried out with the help of the Excel spreadsheet.

Statistical analyses were performed using SPSS software (version 20). The data were first tested for normality and variance uniformity in order to evaluate the statistical claims. The data were then analyzed using a one-way analysis of variance (ANOVA) followed by the Student-Newman-Keuls (SNK) multiple comparison test.

2 Results and discussion

2.1 The impact of water stress on the growth parameters

The evolution of the number of leaves of potato plants and the average number of stems per plant at the end of the cycle were not affected by the water restriction. Indeed, there was no significant difference between the different irrigation regimes. Our results are in opposition to those from the study conducted in the northern side of Kenya by Kiptoo et al [8]who found that the number of stems per plant was significantly affected by irrigation doses and therefore the highest number of stems/plant was recorded for the treatment receiving 100% Etc. However, data analysis of plant height showed a significant effect of irrigation amount (fig.1) The average plants height generally varies between 43.91 and 56.12cm. It should be mentioned that the highest heights are found in treatments that received high irrigation water quantity, namely treatment T1 (dose = 100% ETc) and T2 (dose = 75%) with mean values of 55.77cm and 53.33cm respectively.

It can be noted therefore that the SDI has led to a decrease in the height of the plants. Indeed, a decrease of 6% and 24% was observed respectively for T2 and T3 treatments compared to the control treatment. These results are similar to those of Arafa et al[9], who found that plant height increased significantly with increasing levels of irrigation applied during the potato growing season.

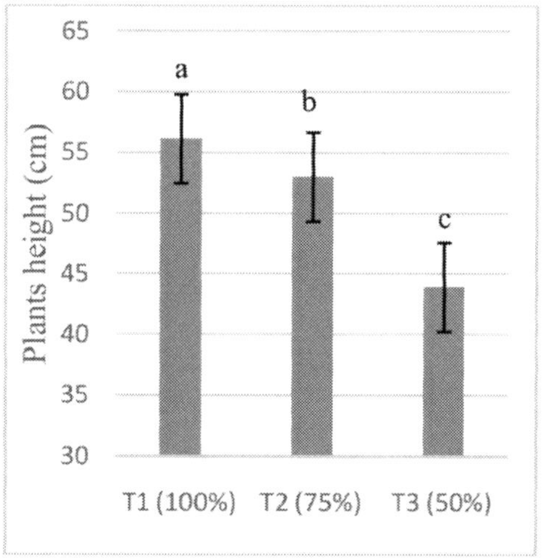

Fig. 1. Average values of potato plant height under different irrigation regimes. The error bars represent the standard deviations

2.2 The impact of water stress on eco-physiological parameters

The chlorophyll content of the potato plants was monitored from day 40 after sowing for all treatments.

According to the graph (fig2, A), it can be noted that the highest values were observed in the plants of the T1 (100%ETc) and T2 (75%ETc) treatments, and the lowest values were recorded by the plants of the T3 treatments (50%ETc). However, analysis of variance did not show a significant effect of theamount of the applied irrigation water on chlorophyll content despite the reduction observed among the three treatments. Thus, the water stress imposed on the plants does not influence their chlorophyll content.

In the same direction, and according to a study conducted by Zin El-Abedin [10], it was shown that the chlorophyll content values of treatments that underwent water deficiency were similar to the values of the treatments irrigated with 100% ETc, which is consistent with the results of Yactayo[11]and our results.

On the other hand, Montagu and Woo[12] demonstrated that water deficiency can destroy chlorophyll and prevent its production.

Concerning leaf temperature, the question often arises as to the possible effects on the metabolism during water stress. For a certain decrease in transpiration due to stomatal closure, the increase in leaf temperature would depend strongly on environmental factors [13].

In the current study, leaf temperatures varied with the intensity of water stress applied to the plants for the different treatments. In fact, the highest leaf temperature value is the one measured in the plants of the T3 treatment. However, the lowest temperatures are recorded for treatments that have experienced no or less water stress, i.e. T1 and T2 (fig2, B).

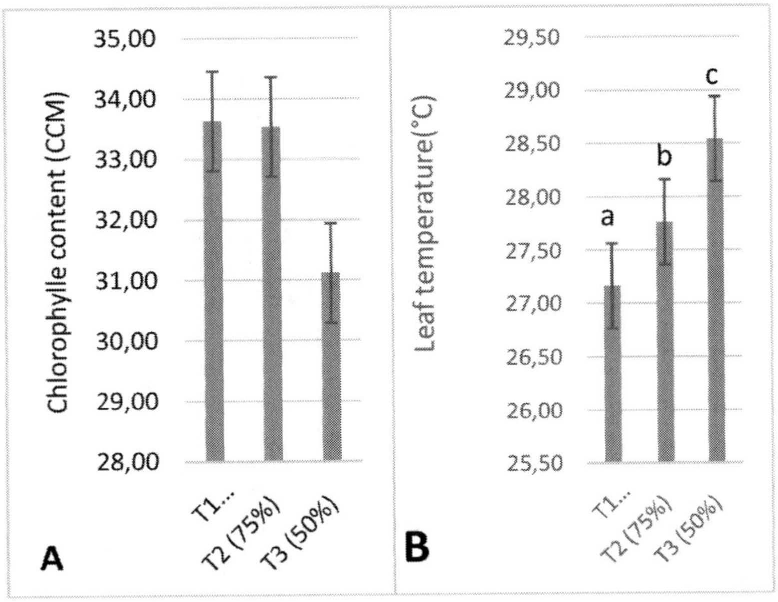

Fig. 2.(A)Values of the relative total chlorophyll content index (TCC) of potato plants in various irrigation regimes. (B): Leaf temperatures °C from different irrigation regimes. The error bars represent the standard deviations.

2.3 The impact of water stress on biochemical parameters

Theproline content of the samples issued from different irrigation regimes is shown in figure 3, which also represents the comparison between these treatments. It can be seen from the graphic that the highest proline concentration was found in the most stressed plants, i.e. 50% of the ETc in T3 treatment, whereas slightly stressed plants (75% of the ETc) and non-stressed plants (100% of the ETc) showed minimum proline levels.

These results are confirmed by the analysis of variance. Indeed, the ANOVA test revealed a significant effect of irrigation dose on proline content. The three treatments are in fact significantly different from one another.

According to Hare [14], free proline will accumulate in various plants in response to a wide range of environmental stresses, so maintaining high water levels prevents the accumulation of osmotic adjustments, such as proline and total soluble sugars [15].Delauney[16]added that it is generally accepted that proline plays a cardinal role as an osmoregulatory agent in plants subjected to hyperosmotic stresses, mainly drought and soil salinity. Indeed, the accumulation of this acid can be part of a general adaptation to poor environmental conditions.

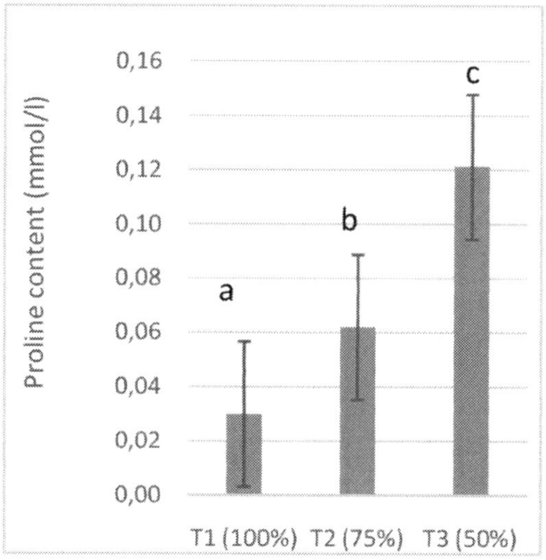

Fig. 3. Different proline concentrations under different irrigation regimes. The same letters indicate values that do not represent a significant difference by the SNK test. The error bars represent the standard deviations

2.4 The impact of water stress on yield parameters

Descriptive data analysis of the average tubers number per plant at harvest showed that the lowest number of tubers per plant was 6.83 tubers per plant recorded among the plants of the T3 treatment (dose=50%) and the highest number of tubers per plant was 8.75 tubers per plant among the plants of the T1 treatment (dose=100%). It should be noted that the means of the tubers number are very close despite the water restriction.
Indeed, the analysis of variance showed no significant effect of the irrigation doses on the tubers number per plant (Fig 4. A).

Our results are in harmony with those of Gultekin, who, according to the results of the Tukey test after two years, showed that the number of tubers per plant was not statistically significant at the 1% significance level. The results obtained in this study were significantly similar to previous studies.

In the same way as the average number of tubers, the average weight of tubers per plant under the different treatments is shown in Figure 4 (B) below. Descriptive analysis of the mean tuber weight data at harvest showed a highly significant effect of irrigation doses on this parameter. In fact, the highest tuber weights were those recorded in the second treatment (dose = 75% ETc) with an average of 1.15 kg per plant and T1 (dose = 100%) with an average of 1.12 kg per plant and the lowest weight was the one recorded in the third treatment (dose = 50% ETc) with an average of 0.62 kg per plant.

Concerning tuber size, it appears from the graph (Fig. 4 C) that T2 and T1 treatments resulted in the largest sizes, 58.40 and 57.31 mm respectively. However, the treatment that recorded the lowest potato size was the T3 with a value of 49.96 cm. Analysis of variance showed that there is a significant effect of the water dose on the average tuber size at harvest.

The highest marketable tuber yield was recorded by the T1 treatment (dose =100% ETc) which is 42.51 Tonnes/ha, and the lowest yield was that recorded by the T3 treatment (dose =50%) which is 20.96 Tonnes/ha.

Analysis of variance showed a highly significant effect of irrigation rate on final potato yield. Thus, the multi-mean comparisons test SNK carried out using the irrigation dose allowed to determine the statistically different means, namely, group (a) is composed of the treatment T1(100%), T2 (dose=75%), and group (b) is composed of the last treatment (fig. 4 D).

Treatment T2 (dose=75%) gave almost equal yield to control T1 (dose=100%), but a 50% restriction of ETc resulted in a 49% yield loss.

Indeed, according to Onder[17], water stress 50% ETc considerably affects potato yield and its parameters. A water deficiency of more than 33% of irrigation needs cannot be suggested. At the beginning of potato season, when the growing period is too rainy, a 33% irrigation deficit may be recommended. However, when the season is not too rainy, full irrigation may be advisable.

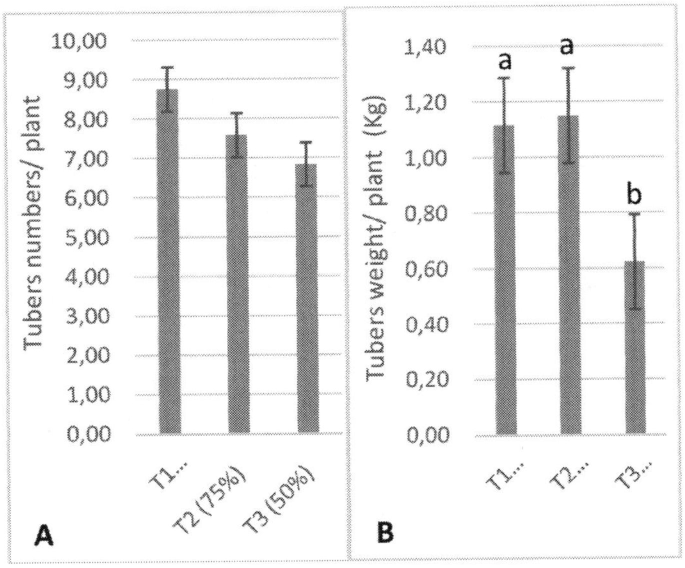

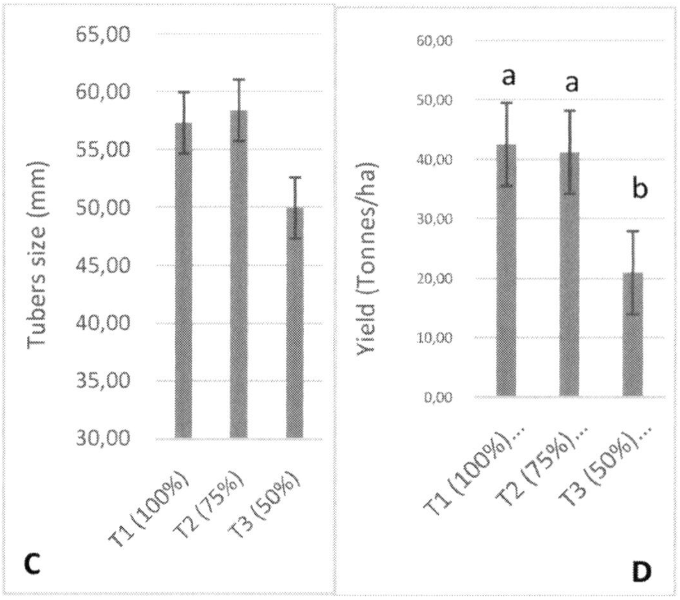

Fig. 4.(A): Different tuber numbers per plant values under different irrigation regimes. (B): Different average tuber weight values per plant under different irrigation regimes. (C): Average tuber size in millimeters at harvest for the different irrigation regimes (D): Estimated yield in Tonnes per hectare for the different irrigation regimes. The same letters indicate values that do not represent a significant difference by the SNK test. The error bars represent the standard deviations.

Finally, the impact of the different irrigation regimes on the agronomic water use efficiency is illustrated in Figure 5, where agronomic efficiency (Kg/m^3) = Yield / Water quantity used [18].

According to the graph, the best agronomic water use efficiency was recorded for the T2 treatment (75%). For the T3 treatment, the water savings achieved do not justify the significant yield losses.

Conclusion

This study assessed the effect of sustainable deficit irrigation with two water regimes 50% Etc and 75% Etc on the potato crop. Water restrictions did not affect vegetative parameters, such as the leaves and stem numbers, except for the plant's height, which decreased with increasing water stress. For the eco-physiological parameters; the chlorophyll content did not show a significant difference for all irrigation doses, whereas leaf temperature was affected. Indeed, as water stress increases, the leaf temperature rises as well. The application of 25% water restriction compared to full irrigation (100% Etc) did not affect yield parameters such as weight, size and number of tubers per plant. However, a severe stress with 50% Etc can result in a considerable yield loss due to a reduction in weight and diameter of the tuber. The best quantitative yields were recorded in treatments T1 (100% Etc) and T2 (75% Etc) with 42.51 Tonnes/ha and 41.17 Tonnes/ha respectively. On the other hand, the plants that underwent extreme stress had a lower yield of 20.96 Tonnes/ha.
The effect of applying a water restriction of about 25% compared to comfort irrigation (100% Etc) resulted in water savings of about 77 mm while maintaining a high yield. In fact, the good water use efficiency was recorded in the plants irrigated at 75% of the crop water requirements.

This study demonstrated that, given the region's conditions, it is possible to optimize the use of irrigation water with less impact on yields; this opens up horizons for greater efficiency in agricultural systems, particularly under potential water shortage scenarios.

Acknowledgements

Thiswork has been carried out under the auspices of the MCRDV research project "COMPETITIVE MECHANISM OF DEVELOPMENT RESEARCH AND VULGARISATION" and we gratefully acknowledge the financial support provided by the Ministry of Agriculture, Fisheries, rural development, water and forest. We are also grateful to the National School of Agriculture of Meknes (ENAM) for their support and cooperation during the project which made this work possible.

References

1. FAO,*Initiative régionale pour faire face à la pénurie d'eau dans la région du Proche Orient et Afrique du Nord Evaluation Nationale Maroc.*(2014)
2. M. English, L. James, C. F. Chen. Journal of irrigation and drainage engineering, **116**(3), 413-426. (1990).
3. R.Gültekin, A. Ertek. International Journal of Agriculture, Environment and Food Sciences. **2**(3) : 93-98. (2018)
4. J. B. Hamida, F. El Mokh, K. Nagaz. *Irrigation déficitaire à l'eau salée: étude du cas de la pomme de terre d'automne irriguée au goutte à goutte de surface et souterrain en milieu aride.* Numéro Spécial - n° 35. (2014).
5. K. E. Saxton, W. J. Rawls. Soil Science Society of America Journal **70**, 1569-1578. (2006).
6. R.G. Allen, L.S.Pereira, D. Raes, M. Smith. *FAO Irrigation and drainage* paper 56. FAO, Rome , **300**(9), D05109. (1998).
7. L. Bates, R. Waldren, I. Teare. Plant Soil 39(1): 205-207. (1973)
8. S. Kiptoo, E.C. Kipkorir, et C.K. Kiptum, Revue africaine de l'éducation, des sciences et de la technologie, **4** (4), 65-77. (2018)
9. M. Arafa, M. Darwisch, N. Elhefnawy. J. Plant Production, Mansoura Univ. **6** (11): 1845 - 1860. (2015).
10. T.K.Zin El-Abedin, M.A.Mattar, H.M. Al-Ghobari, A.A.Alazba. Agronomy, **9**(4): 172. (2019)
11. W. Yactayo, D. A. Ramírez, R. Gutiérrez, V. Mares, A. Posadas, R. Quiroz. Agricultural Water Management, **123**, 65-70. (2013).
12. K.D.Montagu, K.C. Woo.Functional Plant Biology. **26**(2):135-145 (1999)
13. T.C.Hsiao. *Plant responses to water stress.* Annual review of plant physiology. 24(1): 519-570. (1973)
14. P.D. Hare, W.A. Cress. Plant growthregulation, **21**(2) : 79-102. (1997)
15. A. Pirzad, M.R Shakiba, S. Zehtab-Salmasi, S.A. Mohammadi, R. Darvishzadeh, A. Samadi. Journal of Medicinal Plants Research. **5**(12): 2483-2488. (2011)
16. A.J. Delauney., D.P.S.Verma. The plant journal, **4**(2) : 215-223. (1993)
17. S.Onder, M.E.Caliskan., D.Onder, S.Caliskan. Agricultural water management. **73**(1): 73-86. (2005)
18. A. Bouaziz, K. Belabbes.Hommes, Terre et Eaux, **32** (124) :4. (2002)

Making rainfed crops adapted to potential climate change impacts: Modeling sustainable options

Youssef Brouziyne[1*], *Abdelghani* Chehbouni[1,2], *Aziz* Abouabdillah[3], *Jamal* Hallam[4], *Fouad* Moudden[5], *Ali* El Bilali[6], and *Lahcen* Benaabidate[7]

[1] Mohammed VI Polytechnic University, International Water Research Institute, Ben Guerir- Morocco
[2] Institut de recherche pour le développement (IRD), unité mixte de recherche (UMR) Centre d'études spatiales de la biosphère (Cesbio),Toulouse, France
[3] National School of Agriculture, Meknes, Morocco
[4] Soil and Water Management, Soil Health, National Institute for Agricultural Research-Agadir, Morocco
[5] Department of Geology, Laboratory of Geo-engineering and environment, Research Group "Water Sciences & Environment Engineering", Faculty of Sciences, Moulay Ismail University, Meknes, Morocco
[6] University Hassan II of Casablanca, Faculty of Sciences and Techniques of Moahammedia, Morocco
[7] Laboratory of Functional Ecology and Environmental Engineering, Faculty of Sciences and Technology, University of Sidi Mohamed Ben Abdellah, Fez, Morocco

Abstract. Rainfed agriculture is becoming increasingly vulnerable to climate change. This situation is expected to worsen under most future climate projections, which might increase the risks linked to food security and economies which depend on it. Providing insights about the potential responses of rainfed crops to climate change will helps on designing future adaptation strategies. In this study, large amount of data and the agro-hydrological model SWAT have been used to investigate future climate change impacts on rainfed wheat and sunflower crops in a semi-arid watershed in Morocco (R'dom watershed). Downscaled CORDEX climate projections were used in generating future plants growth simulation for R'dom watershed in the 2031 to 2050 horizon under two Representative Concentration Pathways (RCPs): 4.5 and 8.5. The main results of climate change scenarios highlighted that R'dom watershed will undergo significant decrease in water resources availability with more impact under the scenario RCP 8.5. Water productivities of both studied crops could be lower by up to -21% in comparison with baseline situation. Different sustainable management strategies have been simulated using SWAT model under climate change context. The adopted approach succeeded in building up sustainable management strategies toward secured food security in the future.

1 Introduction

Globally, the projectedimpacts of climate change on agricultural productivity are an important concern for decision makers, investors and population. Global farming systems are supposed to support both current and future growing demand for food worldwide; thus, increasing productivity, especially for the mid and long term, should be led through the adaptation of agricultural systems to projected climate change [1]. Many development international agencies and organizations believe that there is still an opportunity to preserve ecosystem balances and are claiming for nature based and sustainable strategies to prepare farming activities to future climate impacts [2].

Rainfed agriculture was always playing crucial role in food security and social integrity for many populations around the world. This farming mode is representing around eighty per cent of the global agricultural lands, and is characterized by low yield levels and high vulnerability to climate effects [3]. Rainfed agriculture in Morocco is particularly vulnerable to the uneven spatiotemporal distribution of rainfall, and is expected to be more vulnerable to the projected climate change especially in arid and semi-arid areas; this problematic will inevitably put the national economy at risks [4].

*Correspondingauthor: Youssef.brouziyne@gmail.com

 https://doi.org/10.1051/e3sconf/202018303002

Several crop growth models have proved their efficacy in miming the real crops development processes around the world, and are generally used to investigate crop's response under stressful conditions (temperature stress, water shortage, pests and weeds stress, restricted fertilization.). Most of these models are relying on a series of functions replicating biological and biophysical processes linking plants to their environments [5]. Some of these models are offering further options through their ability to simulating management strategies so as to investigate the effectiveness of cropping practices in tackling stressors impact (such as: climate change) [6].

Soil and Water Assessment Tool (SWAT), the used model in this study, is a semi-distributed model for physical processes (Hydrology, plants growth, solid transport....) that has been globally used in similar studies [7]. This tool helps in assessing the impact of phenomenon (Ex: Climate change), hydrological and conservation strategies, as well as the effect of agriculture practices [8].

The main purpose of this work is assessing rainfed crops' productivity in a semi-arid watershed, R'dom watershed, under two RCP scenarios (4.5 and 8.5) downscaled from a regional circulation model in the near future (2031 to 2050). SWAT model has been used to simulate a set of sustainable practices to build more resilient farming systems in the area. Results will help to understanding potential climate change impacts on strategic rainfed crops in a typical agro-forestry watershed in the Mediterranean area, and contribute to more adapted agricultural systems in any watershed similar to R'dom.

2 Materials & Methods

2.1 Study area

Located in the north eastern of Morocco, and with a total drainage area of the 1993 km^2, R'dom watershed includes two large plains (Saiss and Gharb). Being a part of one of the largest hydrologic watersheds in Morocco (Sebou basin), R'dom watershed is lying under a Semi-arid climate zone and hosts various socio-economic activities (ex: farming and forestry activities) (table 1).

Table 1. Occupation percentages of major land covers in R'dom watershed

Land use categories		Percentage (%)
Residential centre		3.69
Forest (Oak)		7.98
Pasture & Barren		42.03
Wetlands/Lakes		1.71
Agricultural lands	Wheat	12.23
	Corn	9.83
	Olive trees	10.59
	Onion & potatoes	7.21
	Sunflower	4.73

2.2 Model presentation

SWAT is a physical-based modelthat operates through the division of the basin to several sub-basins and then to smaller sub-units called HRUs (Hydrologic Response Units) [9]. This disaggregation is based on soil types, topography, and land use classes in the study watershed. The model is then simulating the physical processes in each single HRU and gives chance to model management strategies at each HRU (and Sub-basin) level. [10].

In this study, water balance of R'dom watershed was studied at the basin outlet (Souk Elhadstreamgauge) where measured flow out data are available. SWAT is relying on EPIC model (Erosion Productivity Impact Calculator) [11] to simulateplantsgrowth process and to estimateyield. In SWATmodel, plant development is a function of: actual evapotranspiration, leaf area development, light interception and its conversion to biomass and yield. Plant growth can be limited by temperature, water, Nitrogen or Phosphorous stress.

https://doi.org/10.1051/e3sconf/202018303002

2.3 Input data collection and preparation

Being a very important step of any modeling work, data collection and preparation for this study was performed with high precision in order to run SWAT with a start conditions reflecting the reality as much as possible.
As per the topographic of the study watershed, a 30 m digital elevation model (DEM) of R'dom basin has been extracted from the Shuttle Radar Topography Mission (SRTM) (Figure 1).

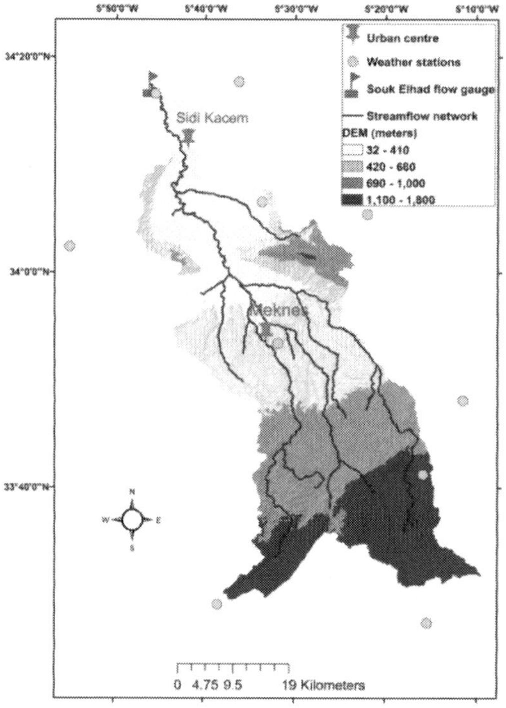

Fig.1. Digital elevation model of R'dom watershed

Soil database of SWAT model was fedby measured soil parameters of the studywatershed based on findings of a study performed by the Moroccan Minister of Agriculture on soil types of the area of Meknes-El Hajeb- Sidi Kacem (Soils study at 1/100,000 scale in the district of Meknes for agricultural development- June 1989). A couple of soil parameters (such as: Saturation hydraulic conductivity and Field capacity...) were estimated using a pedological transfer functions (Soil Water Characteristics software©, Version 6.02.74). Prior to rely on Soil Water Characteristics in estimating the missing parameters, the performances of this software were calibrated against the available soil parameters fromthe study performed by the Moroccan minister of agriculture such as the available water content.
A 30m land cover of R'dom basin was developed based on satellite image processing; after fields' surveys in the main catchment area to have preliminary insight about field situation, two LANDSAT8 image scenes (taken on 2ndApril 2015) were processed using ERDAS Imagine 2014 and ArcGIS 10.2.2. Supervised classification method was used by creating land use classes based on their respective spectral signals.
Observed weather parameters were imported to the SWAT and that included daily temperature (max and min) records and daily rainfall from 10weather stations located inside and around the study watershed (Figure 1). Weather datasetswere from January 2003 to December 2010.
Winter wheat and sunflower were chosen as target crops for this study. The conventional practices used are given in Table 2 [12-14]. Farming practices features were amended according to the real agronomic practices held in R'dom watershed, especially the operations that can have an effect on water cycle such as: season calendar, irrigation management, tillage machinery and frequency...etc (Table 2). In this study, optimal fertilization was supplied to crops so as not be considered a limitation factor to plant growth; only heat and water stress (due to potential climate change impact) were taken into account.

Table 2. Considered farming patterns for winter wheat and sunflower

Farming Practice	Winter wheat	Sunflower
Sowing date	1st November	20th March
Harvest date	1st June	1st July
Tillage techniques	Chisel plow + Disk harrow	Disc plow + Disc harrow
Fertilization N-P-K (kg/Ha)	90-70-80	110-30-40

Default crops database of SWAT is offering a wide generic characteristics of crops, in order to reflect the real crops behaviours against the study region conditions, some cultivar characteristics of the selected crops were amended based on previous studies [14-18](Table 3).

Table 3. Amended plants growth parameters in SWAT database for Wheat and Sunflower crops

		Winter Wheat		Sunflower	
SWAT Parameters		default value	amended value	default value	amended value
T base	Minimum T base (C°)	0	10	6	10
T opt	Optimal T for plant growth(C°)	18	20	25	25
EXT_COEF	Light extinction coefficient	0.65	0.6	0.9	0.7
RDMX	Maximum root depth (m)	1.3	1	2	2.2
CHTMX	Maximum canopy height (m)	0.9	1.8	2.5	2.5

2.4 SWAT simulation

Watershed outlet was manually added to match the geographic location of Souk El Had gauge (Figure 1), the selected station is where flow out process will be monitored to compare measured and predicted flows.
In overall,simulations were from January 2004 to December 2009, the very first two years have been dedicatedto warm up process, the period January 2006 to December 2007 was dedicated to model calibration. Validation was performed over the period January 2008 to December 2009.
SUFFI-2 algorithm [19]of the calibration and uncertainty program SWAT-CUP [20]was used to carry out calibration, validation and uncertainty analysis.
Streamflow and annual crops yields were monitoredduring the calibration and validation processes. Anoptimalgoodness-of-fitbetween observed and predicted water budget and yields components was the ultimate goal. This agreement was examinedby two parameters indicating the statistical indicators and the optimization of the objective functions of this study.
The used statistics tests are:

$$NSE = 1 - \frac{\sum_{i=1}^{n}(Oi - Pi)2}{\sum_{i=1}^{n}(Oi - O_{avg})2} \quad (1)$$

and:

$$R^2 = \left(\frac{\sum_{i=1}^{n}(O_i - O_{avg})(P_i - P_{avg})}{\left[\sum_{i=1}^{n}(O_i - O_{avg})^2 \sum_{i=1}^{n}(P_i - P_{avg})^2\right]^{0.5}}\right)^2 \quad (2)$$

With:
Oi : the recordedcomponent (yield or streamflow) for time period i
Pi: the simulated value for the same period
$\overline{O}$: the mean of recordedcomponent per time period
n: the number of time periods

https://doi.org/10.1051/e3sconf/202018303002

The maximum NSE value is 1.0 and occurs if predicted values are perfectly matching measured ones. The determination coefficient, R^2, provides an idea about how well the correlation between observed and simulated values is [9].

2.5 Future climate data & model scenarios

The used climate change scenarios in this study were generated from the output of the global climate model CNRM CM5 (*Centre National de RecherchesMeteorologique*) based on IPCC Assessment Report (AR5). There are 4 different types of Representative Concentration Pathways (RCPs): RCP 2.6, RCP4.5, RCP6.0 and RCP 8.5 that represent different alternatives of global development (Kim *et al.*, 2013). The CNRM participated in the Coordinated Regional Climate Downscaling Experiment (CORDEX) with different Regional Climate Models (RCMs) [21].
In this study, the downscaled RCM CLMcom-CCLM and two RCPs (4.5 and 8.5) were used. Grid size was 12.5 km and the baseline was from January 1981 to December 2005. The future simulation period for both precipitations and temperature (max and min) was from January 2031 to December 2050. Linear scaling and variance scaling approaches were used during the bias correction in order to show some levels of bias against real measured precipitation and temperature data respectively.

3 Results & discussion

3.1 Future climate conditions

The analysis of the downscaled RCM data showedthat the R'dom watershedwill undergorainfall and mean temperature increase in the period 2031 to 2050 underboth emission scenarios (table 4). In general, a cumulative decrease of annual rainfall by-172.9 mm and -213.6 mm are expected under RCP4.5 and RCP8.5 respectively. Temperature increase is also expected (+1.1°C and 1.7°C respectively for RCP4.5 and RCP8.5). Winter, an important season for cereals, will be the most significantly affected with a major change of +3.4°C in January.

Table 4. Precipitation and temperature changes in climate change scenarios

		Precipitation (mm)			Mean temperature (°C)		
		Baseline	RCP 4.5	RCP 8.5	Baseline	RCP 4.5	RCP 8.5
Winter	December	93.13	-50.2	-48.9	10.8	+0.6	+0.3
	January	58.84	-32.0	-43.8	9.74	+3.4	+3.4
	February	57.69	-26.0	-32.4	11.21	+2.2	+2.5
Spring	march	49.81	-10.4	-23.9	13.44	+0.7	+0.3
	April	39.19	-7.4	-10.7	16.12	+0.0	+0.5
	may	25.91	-21.5	-21.4	19.43	+0.7	+2.0
Summer	June	12.39	-4.1	-8.3	23.64	+1.4	+3.4
	July	1.69	+1.1	+2.2	27.57	+0.0	+2.3
	august	2.69	+2.5	-1.8	27.57	+1.6	+1.3
Autumn	September	19.28	-4.2	-3.9	23.45	+1.2	+1.2
	October	48.42	-4.9	10.4	19.06	+1.1	+1.9
	November	67.83	-15.7	-31.1	14.17	+0.1	+1.0
		476.9	**-172.9**	**-213.6**	**18.2**	**+1.1**	**+1.7**
		Totals			**Averages**		

3.2 SWAT Model calibration

Calibration and validation efforts of SWAT model over R'dom watershed have been achieved as soon as the optimal goodness-of-fit between simulated and observed streamflow was achieved; the goodness-of-fit is when both R2 and NSE values are exceeding 0.5 [8].

Prior to calibration, sensitivity analysis was carried out for both hydrological and crops modules of SWAT. The identified influential parameters of both modules were adjusted using SWAT CUP to reach the best agreement between simulated and recorded process. The most influential parameters for hydrologic component of R'dom are: revealed that CN2 (Initial SCS runoff curve number for moisture condition II), SOL_AWC (Available water capacity of the soil layer), ESCO (Soil evaporation compensation factor), and GWQMN (Threshold depth of water in shallow aquifer for return flow to occur). The most influential parameters for plants growth in R'dom watershed are: HVSTI (Harvest index), WSYF (Lower limit of HVSTI), and BLAI (Maximum potential leaf area index).

After adjusting all the influential parameters for hydrology, R2 and NSE values were respectively 0.85 and 0.68 in calibration, and 0.79 and 0.58 after validation. As per plant growth component, and since only R2 was used, its value was 0.65 in calibration and 0.6 for the validation for wheat and 0.7 in calibration and 0.66 for the validation.

3.3 Future Water balance and crops performances

Running SWAT model in R'dom watershed using future climate variables allowed the investigation of future water resources and crops performances in the study area; table 5 represent both values and changes of future crops water productivities and water yield from baseline situation.

Table 5. Average values of future crops water productivity and water yield, and their changes to baseline

		Wheat CWP *(kg/m³)*	Sunflower CWP *(kg/m³)*	Water yield *(mm)*
RCP4.5	Values average	0.56	0.32	71.17
	Change to Baseline	-18.8%	-15.7%	-20.2%
RCP8.5	Values average	0.54	0.21	65.67
	Change to Baseline	-21.7%	-44.7%	-26.4%

A clear decline of crop water productivities of both crops (wheat and sunflower) and water yield in the R'dom watershed is expected under both climate change scenarios. The extent of change is clearer under the pessimistic scenario for both crops and for the water resources too.

3.4 Adaptation strategies

In order to investigate potential adaptation opportunities of wheat and sunflower to climate change in R'dom watershed, a set of sustainable management strategies were simulated using the calibrated SWAT model over the study area and using the projected climate variables under RCP4.5 and RCP8.5. The selected adaptation strategies were built from the combination of three cropping techniques: No tillage, Early sowing by 10 days and by 20 days from the standard sowing date adopted by local farmers.

In overall, 5 different combinations of adaptations strategies were considered: No tillage (NT), 10 and 20 early sowing days (ESD-10 and ESD-20 respectively), and a combination of no tillage and sowing dates change: NT+ESD-10 and NT+ESD-20. While simulating the adaptation strategies, future total water yield (WYLD) was monitored too as we believe that adapting crops to climate change should not be at the expense of local water resources.

After running SWAT model using the selected adaptation strategies (separately), an analysis of the change to baseline scenario has been carried out; the table 6 represent simulation results.

Table 6. average values and changes from baseline of crops water productivity and water yield after simulation of the adaptation strategies

	RCP4.5			RCP8.5		
	Wheat CWP *(kg/m^3)*	**Sunflower CWP** *(kg/m^3)*	**WYLD** *(mm)*	**Wheat CWP** *(kg/m^3)*	**Sunflower CWP** *(kg/m^3)*	**WYLD** *(mm)*
a. Comparison of indicators in BAU to baseline scenario						
BAU	0.56	0.32	71.17	0.54	0.21	65.67
Change (%)	-18.8%	-15.7%	-20.2%	-21.7%	-44.7%	-26.4%
b. Comparison of indicators in no tillage strategy to BAU scenario						
NT	0.68	0.50	71.7	0.63	0.33	66
Change (%)	+21.4%	+56.3%	+7%	+16.7%	+57.1%	+5%
c. Comparison of indicators in 10 and 20 days early sowing strategies to BAU scenario						
ESD-10	0.64	0.33	72.57	0.62	0.20	66.8
Change (%)	+14.3%	+3.1%	+2%	+14.8%	-4.8%	+1.70%
ESD-20	0.59	0.19	73.9	0.5	0.16	67.9
Change (%)	+5.4%	-40.6%	+3.80%	-7.4%	-23.8%	+3.40%
d. Comparison of indicators in no tillage combined with 10 or 20 days early sowing strategies to BAU scenario						
NT+ESD-10	0.70	0.33	72.21	0.64	0.21	67
Change (%)	+25.0%	+3.1%	+1.50%	+18.5%	0.0%	+2%
NT+ESD-20	0.56	0.19	74.01	0.52	0.16	69.88
Change (%)	0.0%	-40.6%	+4%	-3.7%	-23.8%	+6.40%

Based on these results, both adapting wheat and sunflower crops and preserving water resources in R'dom watershed to future climate change is possible. Most of the simulated adaptation strategies gave good results on at least one variable; and some strategies leaded to inconsistent results across the parameters and the scenarios. Generally, the no tillage (NT), sowing earlier by 10 days (ESD-10), and no tillage and sowing earlier with 10 days (NT+ESD-10) gave the most consistent (positive) change over the three parameters and under the both RCPs.

4 Conclusion

This study attempts to investigate potential opportunities to curtail the vulnerability of two of key rainfed crops in Morocco to future climate change impacts. Results showed that SWAT model succeeded in replicating responses of both hydrology processes and crops development occurring in the study area; both processes will undergo significant effects of climate under both scenarios (RCP4.5 and RCP8.5).
The simulated sustainable adaptation strategies showed promising results as water resources' preservation and increase of crops water productivity is possible according to our SWAT's simulations over R'dom watershed conditions. Among the simulated strategies, no tillage one gave the best compromise between crops productivity, water use efficiency and water resources in the watershed. In the meanwhile, sowing 10 days earlier than usually offers good overall effectiveness to overcome the climate change impacts in the future; combining this strategy with no tillage, gives very promising results as well.

References

1. Y. Brouziyne, A. Abouabdillah, A. Hirich, R. Zaaboul, L. Benaabidate, Agr. Syst.**162**, 154-163, (2018)
2. R. Mrabet, Exp. Agr.**38**, 237-248, (2001)
3. R. Mrabet, R. Moussadek, A. Fadlaoui, E. Van Ranst, Field. Crop. Res.**132**, 84-94, (2012)
4. F. Zahour, M.A. Malik, R. Anser, M. Shehzad, A. Saleem, M. Anser, M. Siddiqui, K. Mubeen, S.H. Raza, *CercetariAgronomice in Moldova,XLVIII,* (2015)
5. P. Steduto, T.C. Hsiao, E. Fereres, D. Raes, In *FAO Irrigation and Drainage Paper*, (FAO Publications: Rome), (2012)

6. H. Briak, R. Mrabet, R.Moussadek, K. Abouamaria, Int. Soil. Water. Conserv. Res.**7**, 176-183,(2019)
7. J.G. Arnold, R. Srinivasan, R.S. Muttiah, J.R. Williams, *J. Am. Water.Resour. As.***34**, 73-89, (1998)
8. D.N. Moriasi, J.G. Arnold, M.W. Van Liew, R.L. Binger, R.D. Harmel, T. Veith, T. ASABE,**50**, 885-900, (2007)
9. J.G. Arnold,;D.N. Moriasi,; P. Gassman,; K.C. Abbaspour,; M.J. White,; R. Srinivasan,; C. Santhi,; R.D. Harmel,; A. Van Griensven,; M.W. Van Liew, ASABE,**55**, 1491-1508, (2012)
10. K.C. Abbaspour, E. Rouholahnejad, S. Vaghefi, R. Srinivasan, H. Yang, B. Kløve, *J. Hydrol.***524**, 733-752, (2015)
11. S.L. Neitsch, J.G. Arnold, J.R. Kiniry, J.R. Williams, K.W. King, *Grassland, Soil and Water Research Laboratory-Agricultural Research Service, (*Temple-Texas, US*)76502,* (2002)
12. A. Ait Houssa, L. Oubaki, K. Reda-Fathmi, S. Drissi, M. Lamghari, M. Benbella, H. Chraibi, *Moroccan Minister of Agriculture*: *Rabat*, (2014).
13. A. Bamouh, In Transfert de Technologie en Agriculture-*PNTTA*, (Agriculture, M.M.o., Rabat), **44**, (1993)
14. A. Si Bennasseur, *Fellah-trade*. (2014)
15. A. Aboudrar, A. Bouaziz, P. Debaek, H. Chekli, *In Proceedings of International Sunflower Conference*, Fargo, USA, (2014)
16. J.S. Zwart, *University of Delft*, Delft, (2010).
17. M. Jlibene, In *La creationgenetique a l'INRA:*Methodologie, acquis et perspectives, (INRA, Ed. Institut National de Recherche Agronomique: Rabat), 57-97, (2009)
18. K.C. Abbaspour, C.A. Johnson, M.T. Van Genuchten, *Vadose. Zone. J.3*, 1340-1352, (2004)
19. K.C. Abbaspour, M. Vejdani, S. Haghighat, In Proc. Intl. Congress on Modelling and Simulation (MODSIM 07),1603-1609, (2007)
20. A. Dosio, Clim. Dynam.**49***,* 493-519, (2017)

E3S Web of Conferences **183**, 03003 (2020) https://doi.org/10.1051/e3sconf/202018303003
I2CNP 2020

Year effect on blooming and harvesting time and fruit weight in two peach varieties grown under Sais valley climatic conditions

Rania Zarzar[1,2], *Abdelali* Blenzar[1], *Adil* Asfers[3], *Abelhadi* Ait Houssa [3], *and Ossama* Kodad[2*]

[1]Département de Biology, Faculté des Sciences. University Moulay Ismail. Meknès, Maroc
[2]Département Arbo-Viti. Ecole Nationale d'Agriculture de Meknès, Maroc
[3]Centre de Formation et de Recherches, Louata, Providence Verte, Maroc

Abstract. Morocco is country concerned by climate warming, phenomenon affecting fruit tree phenology and yield. The present study aims at evaluating blooming and maturity date, fruit weight and yield of two peach varieties grown under Sais Valley conditions during three consecutive years (2017 to 2019). The results showed that, for both varieties, the flowering time was 7 days earlier in 2017 than in 2018. Temperatures were highest in 2017 than in other years of study. For both varieties, the maturity date was earliest in 2017 and 2019, compared to that recorded in 2018. During fruit growth period, the temperatures were higher in 2017 and 2019 compared to those recorded in 2018. The fruit weight was significantly low during the warmest years. From year to year, the fluctuation of flowering and maturity date of the peach in Sais Valley might be related to the variability of temperatures. The warmest years were characterized by early flowering and fruit maturity and low fruit weight.

1 Introduction

Use Peach, *Prunus precica*, is an important commercial fruit species well adapted to Mediterranean climate. In Morocco, during the last decades, new orchards were planted in Sais Valley based on new varieties introduced from different countries mainly from Spain, France and United State. This increase is due to the national market demand for this fruit. However, the studies focusing on the agronomical behaviour of these varieties under climatic conditions of Sais Valley region in Morocco are scarce. The Inter-governmental Panel on Climate Change (IPCC) has provided evidence of accelerated global warming and climate change [1], situation affecting especially Morocco [2,3]. Deciduous fruit trees require specific temperature for breaking dormancy, flowering and fruit maturity. Peach is considered a susceptible to temperature and has a relatively strict chilling requirement [4]. Indeed, several studies have shown that the chilling accumulation variability during dormancy could affect negatively blooming time and fruit set [5,6]. As well the temperature, inconsistencies during fruit growth affect negatively the yield and fruit weight

* Corresponding author: osama.kodad@yahoo.es

[7,8]. It has been also reported that pre-blossom temperatures had a clear effect on the yield [6,9,10].

The irregularity of yield is one of the main problems in fruit production [11]. Thus, in fruit tree, the durability and regularity of orchard production is essential to ensure its economic viability during several decades. The current study focused on the evaluation of climatic conditions effects on flowering, harvest date and fruit weight of two new commercial peach cultivars grown under Sais Valley conditions in Morocco.

2 Materials and methods

2.1 Experimental site and plant material

The present study was carried out at the Agriculture Field Louata of the Providence Verte company in the region of Sefrou, Sais area, situated in north-western Morocco (33.54°N, -4.41°W, 649 m above the sea level). This region is characterised by Mediterranean climate characterized by a hot summer and winters particularly mild and cold. Seventeen-year-old peach trees of *Red Robin* and *Alexandra*, respectively, early and mid-season ripening cultivars were studied. Trees grafted on GF677 (*Prunus persica* x *Prunus dulcis*) and were planted on a loamy-clay soil at 3 x 5 m spacing, drip irrigated and trained according to open-vase system. The nitrogen mixed with potassium sulphate and P were applied by ferti-rrigation during the period from Mars to September ate rates of 100, 60 and 120 unites per hectare, respectively. Four trees per cultivar were chosen for observations of the phenological stages and to determine the average fruit weight. The trees were chosen based on trunk diameter to ensure homogeneity among them. The experiment was carried out during three consecutive seasons (2017-2019).

2.2 Climatic data and chilling accumulation

The weather data collected from meteorological stations located near to the field were used to characterize climatic conditions. Temperature data were considered between October 1st and February 28th during the corresponding observation periods (2017, 2018 and 2019). Chilling availability was determined using Crossa-Raynaud method (CH_{CR})[12]. The chilling hours less than 7.2 °C per day (N) were calculated by Crossa-Raynaud model as: $N = [(7.2 - m)/(M - m)] \times 24$, for $m < 7.2$ °C and $N = 0$, for $m \geq 7.2$; where M and m are, respectively, maximal and minimal temperature (°C)[12]. The mean temperatures were calculated using maximum and minimum temperatures.

2.3 Flowering and harvest date

Blooming time was determined during three growing seasons (2017-2019) by identification of the stages of: beginning of flowering (up to 10% open flowers), full bloom (50% open flowers) and end of flowering (less than 5% of flowers to open). Flowering date is considered as the date when 50% of flowers are open indicating period of full flowering (F_{50}). The maturity of the fruit was determined considering the background colour changing from green to light yellow or white-cream of each cultivar.

2.4 Fruit weight

At the maturity, sample of 15 fruits per tree per cultivar were harvested randomly around the canopy and were immediately brought back to the laboratory. The fruits were collected at commercial harvest time, when the colour of the skin was changed. The fruits were weighted using a precision balance.

2.5 Statistical analysis

Statistical analyses were performed by the SAS software (SAS Institute Inc., 1988), using analysis of variance (ANOVA). Significant differences of means were analysed using the Duncan's multiple range test.

3 Results and Discussion

3.1 Temperature and chilling hours availability

The temperatures variation from October to February is considered the main factor affecting the physiological process of floral buds in the mid-season flowering period of species [9,13]. The analysis of the average temperature recorded during this period showed basically similar values for the three years of study (Table 1). The average mean temperatures recorded were13.8 °C, 13.73 °C and 13.45 °C during 2017, 2018 and 2019 respectively. The average mean of minimum temperature was 7.11 °C in 2019 lower than those recorded in 2018 (7.72 °C) and 2017 (8.21 °C). These results showed clearly that the coldest year was 2019 flowed by 2018 and finally by 2017.

Table 1. Average, maximum and minimum daily temperatures (°C) and chilling hours at Louata (Morocco) for three years.

Annual period	**Dialy temperature**	**Mean dialy temperature (°C)**		
		2016-2017	**2017-2018**	**2018-2019**
October	Average	21.70	23.41	18.55
	Maximum	28.02	29.66	23.29
	Minimum	15.38	17.16	13.82
	Chilling hours[a]	-	-	9.38
November	Average	13.85	16.40	13.15
	Maximum	18.70	23.03	18.80
	Minimum	9.00	9.77	7.60
	Chilling hours	19	1.71	24.63
December	Average	11.63	9.90	13.64
	Maximum	16.57	15.30	21.29
	Minimum	6.69	4.50	6.00
	Chilling hours	100	177.53	77.32
January	Average	9.53	9.71	10.10
	Maximum	15.91	15.71	16.84
	Minimum	3.16	3.70	3.37
	Chilling hours	245	208.39	204.55
February	Average	12.28	9.21	11.83
	Maximum	17.74	14.98	18.90
	Minimum	6.83	3.45	4.75
	Chilling hours	46	242.30	107.87

[a]The chilling hours were calculated using the method described by Crossa-Reynaud

When comparing the average mean temperature from October to the end December (autumn), the results showed that the mean value observed in 2018 (16.57 °C) was higher than those observed in 2017 and 2019 with values of 15.73 and 15.11 °C, respectively. However, during the winter (period from the end of December to October), the average of the mean temperature was higher in 2019 and 2017 (10.97 °C) than in 2018 (9.46 °C). These results indicate that the temperatures were warmer during autumn (fall) than during the winter for the three years of study. This variability of temperature during the winter and spring among the three years affects the chilling hour's availability and flowering time.

Concerning the chilling hour availability (CHA), the results showed that the values were varied between 409 and 629.95 CH by the Crossa-Raynaud model (Table 1). The highest value of CHA was recorded in 2018 (629 CH), followed by 2019 (423 CH) and 2017 (409 CH). During the autumn (from October to the end December), the chilling hours accumulated during 2017, 2018 and 2019 were119 CH, 179 CH and 111CH, respectively. However, during the winter (from January to February) the chilling hours accumulated during 2017, 2018 and 2019 were 290 CH, 450 CH, and 311 CH, respectively. Thus, the chilling accumulation under the Sais Valley conditions was probably realised effectively during the winter than fall.

3.2 hilling hours availability and Flowering date

The results of full bloom date of the peach cultivars studied during the three years (2017–2019) are given in Table 2. The flowering date of *Red Robin* variety was later than that of *Alexandra* cultivar during the three years of study (Table 2). For both varieties, the earliest full bloom date was registered in 2017, followed by that of 2019 and finally that observed in 2018.

Table 2. Harvest date observed during warm (2017) and normal (2018-2019) springs for peach cultivars early-maturing RedRobin and mid-season maturing Alexandra in Valley Sais, Morocco.

Year	Flowering date		Harvest date		FDP (days)	
	Red Robin	Alexandra	Red Robin	Alexandra	Red Robin	Alexandra
2017	04-march	01-march	24-May	28-June	81	119
2018	13-march	09-march	10-June	18-July	89	131
2019	09-march	03-march	27-May	11-July	79	130

These findings could be due to the low chilling recorded in 2017 and 2019 (409 and 423 CH, respectively) leading to a full bloom-advancing for both cultivars (Fig. 1). The latest flowering date observed in 2018 might be due to the highest chilling hours (629 CH) recorded in this year. Correlations analysis showed good relationship between flowering date and the accumulated chill hours for both cultivars among the three years (Fig. 1).

The high temperature observed in 2017 and 2019 during the winter was translated by an early flowering dates in comparison with the results obtained in 2018. These results are in accordance with those reported by El Yaacoubi *et al.* [13], which claimed that in warmer conditions during winter, the fruit tree species flowered earlier than in cooler conditions.

It has been reported that insufficient winter chill may delay or prevent flowering, lead to staggered bloom, and cause various forms of anomalous growth [6,14,15]. However, in the present study, the lower chill hours recorded in winter during the three years of study did not delay the full bloom date of both varieties. Chilling requirement for Red Robin and Alexandra varieties is unknown, but it seems that Red Robin has low chilling requirements than Alexandra because of the flowering date. Cultivars with low chilling requirements always bloom earlier, whereas those with high chilling requirements bloom later [16]. Regardless the chilling requirements of these varieties, the results indicate that under Sais Valley climatic conditions these varieties are not affected by the low chill winter accumulation.

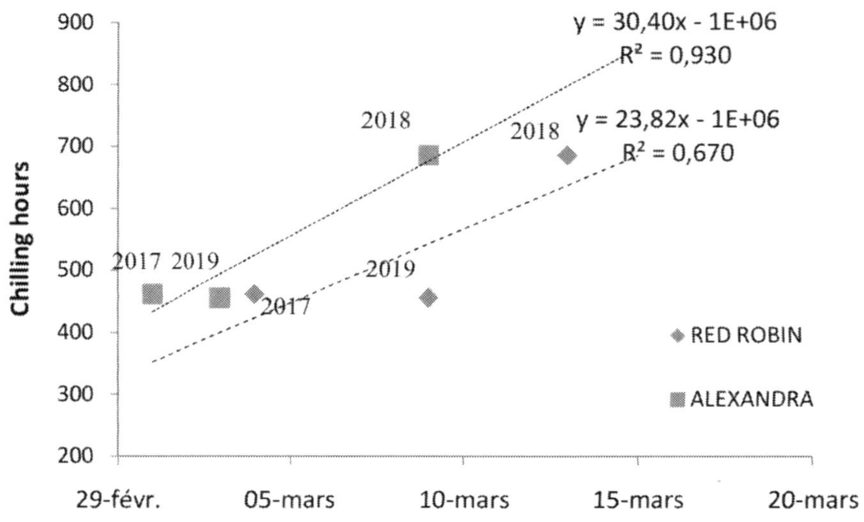

Fig. 1. Relationship between date of full bloom and chilling accumulations expressed as chilling hours (CH) Crossa-Raynaud, for two peach cultivars *Red Robin* and *Alexandra*, observed in valley Sais, during three years (2017-2019).

3.3 Harvest date and chilling hours

The results of harvest date of Red Robin and Alexandra cultivars, during the three years (2017–2019) are given in Table 2. During the three years of study, the fruits of Red Robin reached the maturity stage and harvested earlier than those of Alexandra (Table 2).

The fruit development period (expressed as Julian days from full bloom to the harvest) varied among years between 79 and 89 days for Red Robin; and between 119 and 131 days for Alexandra (Table 2). Shorter fruit development (FDP) day was registered in 2019 and 2017 for Red Robin, and in 2017 for Alexandra. These results could be attributed to delayed bloom that occurred that year as a result of lack of chilling and the resulting warmer temperatures recorded during early fruit growth, as reported in others peach varieties [7].

In the present study, a high positive correlation was found between pre-blossom chill hours accumulation and FDP for Red Robin ($r^2=0,96$) and for Alexandra ($r^2=0,66$). This finding indicates that lack of winter chill affects clearly the FDP of the earliest varieties (Red Robin) and at less level of the mid-season varieties (Alexandra). Temperatures during the early stages of fruit development have been shown to affect the FDP in peach[7]. The monitoring of mean temperature from Full bloom to harvest days showed that during early fruit development the temperatures were higher in 2017 and 2019 in comparison to those recorded in 2018 (Fig. 2). The present results show clearly that the effect of the lack winter chill and temperature during early fruit growth depends on the variety.

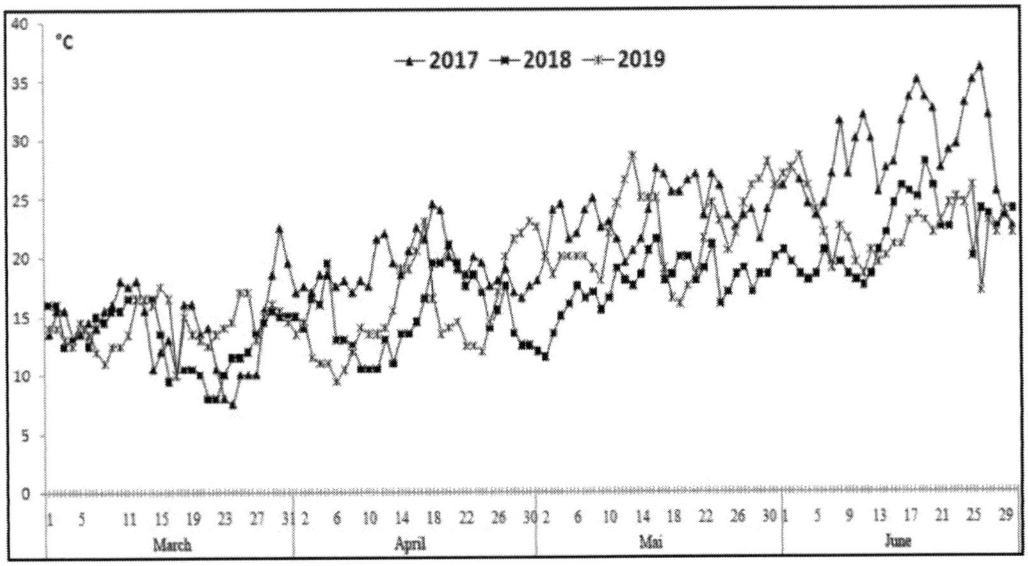

Fig. 2. Monthly mean temperature at valley Sais from March to July (2017 to 2019).

3.4 Fruit weight and temperatures

The statistical analysis revealed significant effect of variety and year on the fruit weight (Table 3). Average fruit weight varied from 142.17 to 160.68 g for Red Robin and from 166.87 to 183.52 g for Alexandra cultivars (Table 3). The lowest value of fruit weight was obtained in 2017 for both cultivars, followed by those obtained in 2018 and finally 2019 (Table 3). This could be due to the negative effect of high temperature during earlier stages of fruit development recorded in 2017 (Fig 2). Similar results were reported in other peach varieties during warmer years [17]. The early cultivar Red Robin and the mid-season Alexandra cultivar showed a slight increase of fruit weight during 2018, with high chill accumulation and delayed fruit ripening (Table 2). In the present study, during warmer year (2017) the fruit development period was short (Table 2) and fruit weight (highly correlated with fruit size) was low (Table 3). Several studies pointed out that during warmer conditions during the spring (corresponding to earlier stage of fruit development) the fruit growth rates was high (short fruit development period) and the trees probably were not able to ensure quickly enough supply resources to support the potential fruit growth rates that apparently accompanied the higher temperatures [17].

Table 3. Effect of variety and year on the fruit weight at harvest during 2017–2019 on peach cultivars in valley Sais, Morocco.

Cultivar	Fruit weight (g)		
	2017	**2018**	**2019**
RedRobin	142.17 c	151.79 b	160.68 a
Alexandra	166.86 b	173.09 b	183.52 a

Average with the same letters in the same column are not significantly different at (p=0,05)

4 Conclusion

The results revealed that the high temperatures affected flowering and harvesting date, and fruit weight of "Red Robin" and "Alexandra" peach cultivars under valley Sais conditions in Morocco. The full bloom and harvesting date were earlier during warmest year, and the fruit weight was reduced. In the context of climate change more attention is needed when planting new orchards in order to reduce anomalies resulting from warmer climatic conditions expected to be more frequent in Morocco in the near future. Selecting precocious and low chilling cultivars may help producers in the warm production region. Additional studies should be carried out in the future, including other varieties, to establish fruit growth models that could allow producers to predict the ripening period of the fruit according to the evolution of temperatures during the fruit growth period.

References

1. IPCC, Summary for policy makers. In: Solomon, S., Qin, D., Manning, M., Chen, Z., Marquis, M., Averyt, K.B., Tignor, M., Miller, H.L. (Eds.), *Climate Change 2007: The Physical Science Basis. Contribution of Working Group I to the Fourth Assess- ment Report of the Inter-governmental Panel on Climate Change.* Cambridge University Press, Cambridge, New York (2007)
2. J. Shelling, K. P. Freier, E. Hertig, J.Scheffran,.Agric. Ecos. Env. 156 (2012) 12– 26 (2012)
3. M. Benassi. Options Méditerr: A SéminMéditerr.80:83–86 (2008)
4. L.R.Wang G.R. Zhu, W.C. Fang,.Peach Genetic Resource in China. China Agriculture Press, Beijing, China(2012)
5. W.R. Okie, J.Bluckburn,.Hortic. Sci. 46(7):1056-1062 (2011).
6. M. Ghrab, M.BenMimoun, M.M. Masmoudi, N. Ben Mechlia, Int. J. Environ. Stud. 71: 3–14 (2014)
7. T. Wert, J.G. Williamson,J.X. Chaparro, E.P. Miller, R.E. Rouse. HortScience. 44(3) : 666-670 (2009)
8. P.C. Anderson, W.B. Sherman,*New low chill peach and nectarine cultivars from the University of Florida*. Proc. Fla. State Hort. Soc. 107: 331-333 (1994)
9. A. El Yaacoubi, G. Malagi, A. Oukabli, M. Hafidi, , J-M. Legave,. Sci. Hortic. 180, 243–253 (2014).
10. J. Rodrigo, and M. Herrero,.Sci. Hortic., 92: 125-135(2002)
11. J. Tromp,. Sci. Hort. 30, 109–116 (1986)
12. P. Crossa-Raynaud, Ann. Serv. Bot. Agron. Tunisie 28, 1–22 (1955)
13. A. El Yaacoubi, A. Oukabli, M Hafidi, I. Farrera, T. Ainane,S. I. Cherkaoui , J-M Legave,. Sci Horti. 249 : 59–64 (2019)
14. J.L. Petri, L.G. Berenhauser, ActaHortic. 662, 53–60 (2004)
15. H. Yu, E. Luedeling, J.Xu, Proc. Natl. Acad. Sci. U. S. A. 107, 22151–22156 (2010)
16. R. Scorza, W.R. Okie. *Acta Horticulturae*, 290: 177–231 (1990)
17. Y.L. Grossman, T.M. Dejong, Tree Physiology, 14, 329–345 (1994).

Year and genotype effects on the fruit physico-chemical traits of sweet cherry cultivated under Southern Mediterranean climate in Morocco

Mina El Baji [1,2], *Zineb* Nohaissi[2] , *Said* En-Nahli [2], *Hafida* Hanine [2], and *Ossama* Kodad [1*]

[1] Laboratory of Bioprocess and Bio-Interfaces, Faculty of Science and Technology, Beni Mellal, Morocco
[2] Departement d'Arboriculture Viticulture. Ecole Nationale d'Agriculture de Meknès. Maroc

Abstract. The aim of the present work was the evaluation of genotype and year effects on some physico-chemical parameters of five sweet cherries grown in Middle Atlas of Morocco. The analysis of variance showed significant effects of the genotype, year and their interaction on all fruit physico-chemical traits focused. '*Coeur de pigeon*' and '*Burlat*' produced the heaviest fruit, whereas 'Van' and 'Coeur de pigeon' fruits were sweeter than others varieties Differences among the three years were highly significant (at $p<0.001$) for all parameters highlighting the annual influence on the fruit quality parameters. Year effect was significant on the physico-chemical fruit traits confirming the climatic conditions effect on fruit quality. Fruits harvested in 2016 and 2017 showed the highest values in terms of physico-chemical traits studied.

1 Introduction

Sweet cherry (*Prunus avium* L.) is deciduous fruit tree belonging to the genus *Prunus*, Rosaceae family, and is regarded one of the most important fruit crops within temperate area of the world [1]. In Morocco, sweet cherry was introduced previous to 1920 by the French protectorate and it was propagated by grafting and also by the seeds in middle Atlas Mountains regions of Morocco. The sweet cherry occupies acreage of around 4000 ha, with an annual production higher than 14,100 tons. The plantation area of this species is limited to medium and high altitude (idle and high mountain atlas) to satisfy the high-chill requirements required to break down dormancy of the introduced varieties. The most popular sweet cherry cultivars grown in Morocco are '*Bigarreau Van*' and '*Bigarreau Burlat*', with 'Napoleon' as pollinizer [2,3]. However, there are other cultivars grown at a small scale such as '*Cerisette*' and '*Coeur de pigeon*' [2].

During the last decade, the production of sweet cherry has increased significantly and new orchards are planted in the middle atlas area because of favourable climatic conditions

* Corresponding author: osama.kodad@yahoo.es

and availability of water for irrigation. Cool nights and template days during the fruit growing period generate suitable conditions for growing high-quality cherries in this area. For the new orchards, in spite of the lack of winter chill observed these last year's [3], the growers continue planting '*Bigarreau Van*' and '*Bigarreau Burlat*'. Other producers have decided to recuperate and plant some older cultivars such as '*Coeur de Pigeon*' and '*Cerisette*' introduced during the protectorate period [2]. However, few data about the chill requirement and fruit quality of these varieties are available for the growers for a bitter management of the orchards. In addition, the accelerated global warming and climate change affecting Morocco [4], including the middle atlas region, could affect negatively the agronomical behavior of these varieties.

Cherry fruit quality has been referred especially to the fruit size and firmness, while these parameters are only a part of quality definition [5]. In fact, fruit size is a determining factor in the choice of the cultivar [6]. Fruit firmness is also an essential attribute of quality (Esti et al., 2002), being considered a key aspect for marketing of cherries [7]. These characters depend on the genotypes [8], and influenced by the environment conditions. Finally, the chemical profile of sweet cherries largely influences the quality of the fruits and the preferences of consumers [9]. Titratable acidity (TA) and Total soluble solids (TSS) are important parameters to determine the adequate harvesting period [10].

The aim of this study was to evaluate the effect of genotype and year on the physico-chemical quality of five sweet cherry cultivars grown in the Middle Atlas Mountains during three consecutive years.

2 Material and Methods

2.1 Plant material

Fruits from the five sweet cherry cultivars '*Burlat*', '*Coeur de pigeon*', '*Napoleon*', '*Cerisitte*' and 'Van' were harvested in commercial orchards situated at Imouzzer Kander city in the Middle Atlas region, at the optimum commercial maturity stage based on fruit maturity and colour development during May and June of three consecutive seasons (2015-2017). Thirty-year-old trees were studied. Trees grafted on Sl64 rootstock (*Prunus mahaleb*) and were planted on a loamy-clay soil at 4 x 6 m spacing, drip irrigated and trained according to open vase system. The trees were chosen based on trunk diameter to ensure homogeneity among trees. The experiment was carried out during three consecutive seasons (2015-2017). The weather data collected from meteorological stations located near to the field were used to characterize climatic conditions.

2.2 Physical Analyses

At the maturity, 15 fruits per trees of both cultivars were harvested randomly around the canopy and were immediately brought back to the laboratory. The fruits were harvested at commercial stage according to the colour of each variety. The physical parameters studied were: fruit weight (FW), fruit length (FL), fruit width (FWi), nut weight (NW), fruit firmness (Ffir), flesh width (FWe) and the firmness of the flesh fruit (N). The linear parameters were measured using a digital caliper with 0.01 mm sensitivity. Fruit and nut weight was measured using an electronic balance with 0.01 g sensitivity. The firmness of the fruit was evaluated by maximum compression force in the equatorial region of the fruit using an electronic firmness tester (AGROSTA®100 Field digital firmness tester).

2.3 Chemical Analyses

These included total soluble solids (TSS, °Brix), titratable acidity (TA, g malic acid/l), pH and maturation index (MI). TSS was determined with a digital refractometer. TA was determined by potentiometric titration with 0.1 N NaOH up to pH 8.1, using 10 mL of juice in 50 mL distilled water(IFU, 1996).The pH value was measured using a digital pH-meter. MI was calculated as the ratio between TSS and TA (Fellers 1991).

2.4 Statistical Analyses

Statistical analyses were performed by the SAS software (SAS Institute Inc., 1988), using analysis of variance (ANOVA). Significant differences of means were analyzed using the Duncan's multiple range test.

3 Results and Discussion

3.1 Genotype and year effects on physical fruit traits

The statistical analysis revealed the significant effects of the genotype and the year on the fruit and nut physical traits evaluated (Table 1). The fruit weight varied between 6.5 g for '*Burlat*' and 4.32g for '*Cerisette*' (Table 2).

Table1. Analysis of variance for physical fruit and nut traits of 5 sweet cherry cultivars

		Mean square					
Source of variation	df	FW (g)	FL (mm)	FWi (mm)	FFir (N)	FWe (mm)	NW (g)
Genotype	4	80.61*	315.8*	251.12*	107.3*	65.8*	0.07*
Year	2	58.06*	86.6*	146.26*	32.28*	63.2*	0.46*
Genotype x year	8	26.67*	42.3*	65.77*	15.94*	24.5*	0.04*
Error	361	0.43	1.18	1.28	0.62	0.55	0.001

* Significant at $P < 0.001$

Over the three years of study the '*Burlat*' and '*Coeur de Pigeon*' showed the heaviest fruit (Table 2), whereas '*Cerisitte*' had the smallest value for fruit weight (Table 2). The varieties '*Burlat*' and '*Van*' cultivars are considered the important varieties planted in the new commercial orchards planted in Morocco. The fruit weight of 'Van' (5.81g) was significant different and lower than that of '*Burlat*' (6.5g) (Table 2). Similar results were reported by Pérez-Sánchez et al. [10] for these varieties grown under Spanish climatic conditions. '*Cerisitte*' and '*Coeur de pigeons*' are traditional French cultivars introduced into the Moroccan system production to offer a wide range of choices for growers. The fruit weight of *Coeur de pigeon* is similar to that recorded in '*Burlat*', whereas '*Cerisitte*' produce small fruit (Table 2). In sweet cherry, the fruit weight is transmitted quantitatively [8], and its expression depends primarily on the genotype [11].

The mean value of fruit length varied between 20.81 mm for '*Burlat*' and 15.59 mm for '*Cerisitte*' (Table 2). Fruit length and width of '*Burlat*' and '*Coeur de Pigeon*' were higher than those of others studied varieties (Table 2). During the three years of study, the fruits of 'Van' ware longest and largest than those of '*Burlat*'. These results are in accordance with those reported in the literature [10,12]. The fruit flesh width of '*Burlat*', '*Coeur de Pigeon*' and '*Van*' was similar (Table 2). (Table 2). For te fresh market, consumers prefer sweet cherries with large flesh [10]. 'Napoleon' cultivar was planted by the growers because of the intermediate flowering date that overlap the blooming period of '*Burlat*' and '*Van*' and is considered by the growers a good pollen donor in the orchards. However, the fruit of 'Napoleon' is not appreciated by the consumers because of the size and yellow color of the skin.

Table 2. Mean value of physical fruit traits of five sweet cherry grown under Middle Atlas Mountains.

Variety	Year	FW	FL	FWi	FFir	FWe	P N
Burlat	2015	6.77b*	20.76b	23.21b	5.81b	8.46a	0.21a
	2016	8.61a	23.56a	25.33a	6.33a	8.68a	0.43b
	2017	4.15c	18.12c	18.55c	3.82c	5.28b	0.25a
	Mean	6.51	20.81	22.36	5.32	7.47	0.29
Cerisitte	2015	5.11a	16.92a	21.72a	5.84a	6.98a	0.33c
	2016	3.78c	15.91b	19.11b	4.77c	4.35c	0.38a
	2017	4.08b	14.86c	19.44b	4.95b	5.94b	0.36b
	Mean	4.32	15.89	20.09	5.19	5.76	0.36
Cœur de pigeon	2015	6.58a	21.43a	22.85a	7.51a	7.95a	0.25b
	2016	6.54a	20.27b	22.16b	5.38b	7.87a	0.32a
	2017	6.23a	20.62b	21.70c	5.43b	6.98b	0.25b
	Mean	6.45	20.77	22.24	6.11	7.60	0.27
Napoleon	2015	4.70a	20.66a	18.55a	4.80a	6.56a	0.26c
	2016	4.59b	19.54c	17.72c	4.77a	6.01b	0.34a
	2017	4.34c	20.13b	18.16b	4.29b	5.33c	0.30b
	Mean	4.55	20.11	18.15	4.62	5.96	0.30
Van	2015	5.41c	19.58b	21.86b	7.24c	7.55b	0.22b
	2016	6.93a	21.01a	23.42a	8.14a	8.27a	0.38a
	2017	5.05b	18.94c	20.76c	7.69b	7.01c	0.23b
	Mean	5.80	19.84	22.01	7.69	7.61	0.28

*Different and same small letters indicate significant ($P<0.05$) and no significant differences ($P>0.05$), respectively, between the means according to Ducans's test.

Fruit flesh firmness is the most important trait to evaluate sweet cherry quality for the fresh market. 'Van' fruits had high firmness (7.69N), flowed in decreasing level by '*Coeur de Pigeon*' (6.1 N), '*Burlat*' (5. 31N), '*Cerisette*' (5.13 N) and finally '*Napoleon*' (4.62 N) (Table 2). Interesting the firmness of 'Coeur de Pigeon' fruit found to be similar to that of '*Burlat*' fruit. The firmness of '*Burlat*' fruits was lower than that of '*Van*' fruits. Similar results have been already reported under other climatic conditions [11]. In sweet cherry, the fruit firmness is related to the ripening period; higher fruit firmness was recorded in the late ripening cultivars [13].

Significant differences were obtained among the cultivars for stone weight (Table 1). '*Cerisette*' showed highest nut weight (0.35g), whereas '*Coeur de Pigeon*' (0.27g) and '*Van*' (0.28g) produced fruits with low stone weight (Table 2). The analysis of variance showed significant differences among varieties for flesh width (Table 2). The mean value of flesh weight varied 5.75 mm for '*Cerisette*' and 7.61 mm for '*Van*' (Table 2). The varieties '*Van*', '*Burlat*' and '*Coeur de Pigeon*' showed largest flesh width in comparison to that of '*Naopoleon*' and '*Ceristette*'.

The year effect was also significant for all studied traits (Table 1). Similar results have been reported in other cherry cultivars [14]. The mean values of fruit weight, length, and width were higher in 2016 than in 2015 and 2017 (Table 2). However, fruit firmness and fruit flesh width were higher in 2015. In sweet cherry, the final fruit size and weight is related to the cell length than to the numbers of cells [15]. Thus, climatic conditions and supply resources limitations affected more final cherry size during cell enlargement (stage III of fruit development) than cell division (during stage I) [16,17]. The data of rainfall registered in 2016 showed that the amount of precipitation during the May and June (corresponding to the stage III of fruit development) was higher than that registered in 2015 and 2017 during the same period (Table 3). Thus, the high values of fruit weight and dimension and fruit flesh width obtained in 2015 and 2016 might be due to the high rainfall registered during the period of fruit development in comparison to that recorded in 2017. However, the mean fruit firmness was lower during the rainfall year (2017) like that obtained during the drought year (2015). It has been reported that fruit firmness decrease with rainfall increasing [17]. These findings are in accordance with those obtained in the present study for the year 2015 and 2016. However, during 2015 (drought year) the fruit firmness was lower (Table 2). Sekse et al. [18] reported that during warm and sunny years the cherries produce softer fruits than the cold and rainy years. In 2017, the temperatures were higher (2°C of difference) than those registered during 2015 and 2016 (Table 3). The warmer climate and drought stress recorded during fruit development in 2015 might explain the lowest value of fruit firmness obtained in this year.

Table 3. Mean temperature and total precipitation during the development period of the fruit in Imouzer Kander location.

	Temperature (°C)			Rainfall (mm)		
Months	2015	2016	2017	2015	2016	2017
April	20	21	25	58	41	46
Mai	25	25	24	61	79	40
Juin	27	27	30	32	22	14
Mean	24	24	26	50	47	33

The significance of the year × genotype interaction for all the physical parameters (Table 1) indicates that the magnitude of the variation from year to year depends on genotype. Several authors reported that physical fruit trait depends on the genotype [8,10,12]. In fact, the mean values of fruit weight and fruit fresh width were stable and high during the three years of study (Table 2), whereas for '*Burlat*' and '*Van*' these characters varied significantly from year to year (Table 2).

3.2 Genotype and year effects on chemical fruit traits

The analysis of variance showed a significant effect (*at* $P<0.001$) of genotype and year on all chemical parameters (Table 4). The chemical parameters of the studied varieties are given in Table 5. The mean value of TSS varied between 9.85 °Brix for '*Burlat*' and 16.51 °Brix for 'Coeur de Pigeon'. The fruit of '*Van*', '*Coeur de Pigeon*', '*Napoleon*' and

'*Cerisitte*' showed similar and significantly higher values of SSR than that of 'Burlat' fruits (Table 5).

Table 4. Analysis of variance of chemical fruit attribute of five sweet cherry cultivars

		Mean square		
Source of variation	df	TSS	pH	TA
Genotype	4	161.11*	0.56*	81.77*
Year	2	64.79*	0.32*	57.31*
Genotype x Year	8	96.83*	1.80*	55.83*
Error	15	24.04	0.10	11.46

* Significant at P < 0.001

The mean value of TSS of 'Van' was significantly higher than that of '*Burlat*' fruits during the three years of study (Table 5). These results are in accordance to that reported in others cherries [12,19]. However, higher values of TSS have been reported for '*Burlat*' by Pérez-Sánchez et al. [10] under Spanish climatic conditions.

Table 5. Mean value of chemical parameters of five sweet cherry cultivars grown in Middle Atlas Mountains

Variety	Year	TSS (°Brix)	Ph	TA (malic acid/l)
Burlat	2015	9.88b*	3.55b	6.30b
	2016	12.10a	3.09c	8.11a
	2017	7.60c	4.03a	5.35c
	Mean	9.86	3.56	6.59
Cerisitte	2015	14.20b	3.55a	9.77c
	2016	15.40a	2.85b	14.47a
	2017	15.25a	3.56a	10.15b
	Mean	14.95	3.32	10.46
Cœur de Pigeon	2015	12.35c	3.30b	9.25b
	2016	17.01a	3.41a	11.73a
	2017	20.20b	3.02c	11.75a
	Mean	16.52	3.24	10.91
Napoléon	2015	14.95b	3.73a	5.56c
	2016	16.30a	3.41c	9.41a
	2017	14.10c	3.65b	7.70c
	Mean	15.12	3.60	7.56
Van	2015	11.01c	3.43b	7.20c
	2016	19.50a	3.61a	7.87b
	2017	15.60b	3.07c	11.60c
	Mean	15.37	3.37	8.89

* Different and same small letters indicate significant (P<0.05) and no significant differences (P>0.05), respectively, between the means according to Ducans's test.

Concerning the titratable acidity, the mean value fluctuated from 6.58 malic acid/l for '*Burlat*' to 10.9 malic acid/l for '*Coeur de Pigeon*' (Table 5). The values obtained in '*Coeur de Pigeon*' and '*Cerisitte*' were similar, 10.91 and 10.46g malic acid /l, respectively. In the present study, 'Van' fruit showed highest value of titratable acidity than than of 'Burlat'. Similar resultants have been reported for these varieties [12]. The pH varied between 3.24 for '*Coeur de Pigeon*' and 3.59 for '*Napoleon*' cultivar (Table 5). The mean value of the pH of '*Burlat*', '*Van*' and '*Napoleon*' were similar. At commercial harvesting pH of '*Burlat*' was 4.21, whereas that of 'Van' was between 3.6 and 3.82[19]. In the present study, the values are lower than those reported in the literature, probably due to a different date level of ripening at harvest and growing conditions.

The year effect was significant for all the chemical parameters (Table 5). Similar results were reported in others cherries [20, 21]. The mean values of TSS were higher in 2016 as compared to the other two years (Table 6). Cherries produced in 2016 and 2017 contained the highest total soluble solids and titratable acidity (Table 5). However, in 2015 the different varieties produced fruit with lowest TSS and titratable acidity (Table 5). These differences might be attributed to the climatic variations registered during 2015-2017 (Table 3). In 2015, the climatic conditions were extreme during fruit developing (high temperature and drought stress) which probably affected the chemical attribute of fruit.

The analysis of variance showed that year × genotype interaction was significant for all chemical attributes (Table 4) indicating that the magnitude of the variation from year to year depends on genotype. Similar results were reported for sweet cherry [21]. '*Cerisitte*' cultivar produced fruits with stable weight and flesh width mean values during the years of study (Table 5), whereas the others varieties showed high variability of these traits among the years (Table 5).

4 Conclusions

The results obtained in the present work confirm the negative effect of extreme climatic conditions on the expression of fruit traits in sweet cherry. Significant influences of the year and the genotype on the quality of cherry fruits produced under climatic conditions of Morocco were observed. Fruits produced in 2016 and 2017 were heavier and sweeter than those produced in 2015. The high temperature and drought stress (lack of rainfall) might be related to the low values of physico-chemicals traits of the studied varieties obtained in 2015. However, the behavior of the varieties was different from year to year. The present results indicate that additional studies should be carried out in the future, including other varieties, to establish fruit growth models that could allow producers to predict the ripening period of the fruit according to the evolution of temperatures during the fruit growth period.

References

1. H.Demirsoy, L. Demirsoy. *Fruits*,**59,** 219–223 (2004)
2. A.Oukabli, Bull. Trans. Tech, 114 (2004)
3. O. Kodad, S. En-Nahli, H. Hanine, M. El Baji, Agriculture du Maghreb,**97,** 62-63 (2016)
4. M. Benassi. Options Méditerr: A Sémin Méditerr,**80**, 83–86 (2008)
5. S. Predieri, R. Dris, F. Rapparini, Food, Agric. Envi,**2,** 307-309 (2004)
6. A. Oukabli, M. Laghezali, Fruits,**55**(1), 83-92 (2000)
7. E.D. Cittadini, *Sweet cherries from the end of the world: Options and constraints for fruit production systems in South Patagonia, Argentina*. The Netherlands. Wageningen University. 20-29 (2007)
8. P. E.Hansche, V. Beres, R. M. Brooks, P. Am. Soc. Hortic. Sci,**88**, 173–183 (1966)

9. M. Fazzari, L. Fukumoto, G. Mazza, M.A. Livrea, L. Tesoriere, L. Di Marco, J. Agric. Food Chem,**56**, 3561–3568 (2008)
10. R. Pérez-Sánchez, M.Á. Gómez-Sánchez, Mr. Morales-Corts, J. Food Qual,**33**, 490–506 (2010)
11. B.Gonçalves, J. Moutinho-Pereira, A. Santos, A. Paula Silva, E. Bacelar, C. Correia, E. Rosa, Tree Physiology,**26**, 93–104(2006)
12. K. Vursavus, H. Kelebek, S. Selli, J. Food Engin,**74**, 568–575 (2006)
13. V. Usenik, D. Kastelec, F. Štampar, Food Chemistry,**90**, 663-671 (2005)
14. A.Khadivi-Khub, Trees **28**: 669–685(2014)
15. J.W. Olmstead, A.F. Iezzoni, M.D. Whiting. J. Ame. Soc. Hort. Sci, **132,** 5 697-703(2007)
16. V. Blanco, P.J. Blaya-Ros, R. Torres-Sánchez, R. Domingo, Plant*s* . **9**, 94 **2020**
17. J. Blažková, I. Hlušičková, J. Blažek. Hortic. Sci. (Prague),**29,** 92–98 (2002)
18. L. Sekse, M. Meland, T. Reinsnos, S. Vestrheim, Euro. J. Hortic. Sci, **74**(6),268-274 (2009)
19. A.A. Hayaloglu, N.Demir, J. Food Sci,**80**,564-570 (2015)
20. R.Kurlus, J. Fruit. Orn. Plant. Res, **12**, 35-39 (2004)
21. J. Neilsen, D. Neilsen, F. Kappel, T. Forge, Hortscience,**49**, 215–220. (2014)

Partition of rainfall variation effect on crop productivity into amount and distribution effects

Mohammed Jlibene[1*]

Independent and previously-researcher at INRA Maroc, Morocco

Abstract. Rainfall is the most influential factor of rainfed agricultural production in semi-arid Mediterranean and Sub-Saharan environments. While effects of rainfall amount and distribution are confounded, their partition will help developing strategies to manage green water in these environments. A series of 39 years of country average cereal yield and rainfall data was used (1981-2019). Cumulated rainfall over months of November, Rainfall of the month of January, February and March of Central location of Marchouch (33.5613° N, 6.6919° W) taken as a representative of agricultural land, were used because of high correlations with yield. Regression of cereal yields on rainfall showed scattered points below a top frontier line representing highest water productivity (yield/rainfall), accounting for effect of amount independent from distribution. Regression of all 39 yields on rainfall accounts for both confounded effects of amount and distribution. Difference indicates effectofdistribution of rainfall. Cumulated rainfall over the 4 months, varied from 64.20 mm to 447 mm with an average of 219 mm. Yields ranged from 3.59 q/ha to 22.88 q/ha with an average of 12.29 q/ha. Top water productivity indicating effect of amount was 3.98 kg/mm. Distribution of rainfall reduced actual yields by 13% to50% depending on season and cereal crop. The model can be adapted to rainfed crops in semi-arid environments.

Introduction

Rainfall influences cereal yields in rainfed Mediterranean arid environments. The total rainfall over the months of September until may correlated positively with cereal yield in Morocco (r=0.70), using cereal yield and Marchouch rainfall data of a series of 39 years, Whenever there is an increase or a decrease in rainfall, cereal yields fallow the same trend, for the last 39 cropping seasons, with 2 exceptions of increased rainfall fallowed by decreased yield and 5 exceptions of decreased rainfall amount fallowed by increased yield. Increased yield fallowing decreased rainfall may have more likely resulted from a favorable distribution within the season.

*Corresponding author: jlibene.mohammed@gmail.com

Water productivity may be used as an indicator to appreciate the effect of total rainfall on yield, including both amount and distribution [1, 2]. The effect of rainfall distribution has not, up to date at least to our knowledge, been isolated. Breaking down those effects will help setting strategies for rainfall management in rainfed crop production systems in drought prone environments.

The objective of this communication is to present a method of partitioning confounded effects on yield into single effects due to amount and to distribution.

Method

Regressing yield on rainfall for a long data series shows scattered points below a frontier line which represents top yields for each level of rainfall amount, or high water productivity (Figure-1). The frontier regression line is expected to show effect of amount of rainfall under optimal distribution, measured by the coefficient of regression.

Regressing yield on rainfall for the whole series is expected to show effects of both amount and distribution confounded.

Difference between the two will shows effect of distribution, independent from amount of rainfall. Since distribution is intended to be a deviation from optimal distribution, its effect is expected to be negative, and effect of the amount of rainfall to be positive. If the two regression lines are parallel, effect of distribution is the difference between the intercept.

The effect of distribution is an average effect. Deviation of individual yield from frontier regression line may involve other factors like sowing date, temperature, diseases or insects.

Moroccan average yield of cereal including bread wheat, durum wheat and barley and rainfall of Marchouch location for the last four decades (1981-2019) have been used in this study [3]. Since country rainfall average data was not available, we used data of the INRA experimental of Marchouch, located in the center of Moroccan agricultural lands (33.5613° N, 6.6919° W). This location happens to correlate with most agricultural locations in Morocco (Tadesse, personal communication., 2020) and therefore was used as a representative of total agricultural lands of Morocco.To improve the correlation between rainfall and yield, the months of November, January, February and March were used because of their best correlations to yield. The correlation was improved to 0.78.

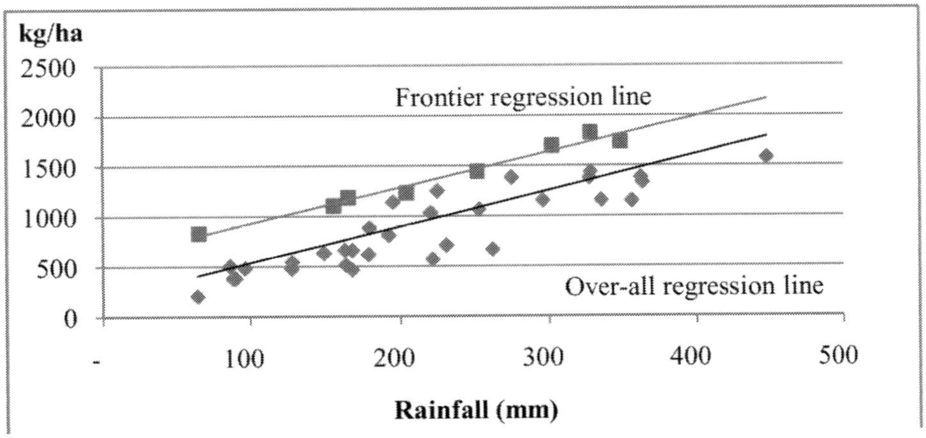

Fig.1: Regression of yield on rainfall. The frontier regression line in top indicates situations of well distributed rainfall. Slope represents amount effect on yield, and difference among intercepts represents distribution effect.

Regression parameters for cereals in general, and for individual species, bread wheat, durum wheat and barley, are in results and discussion section below.

Results and discussion

Considering the total rainfall of the four months during the last four decades, rainfall, yield and water productivity, were highly variable, as expected, 7-folds, 6-folds, and 5-folds respectively. Both variations in total amount of rainfall and its distribution account for 65% yield and water productivity variations.

Table 1: Statistics (minimum, maximum, average and range) of rainfall, cereal yield and cereal water productivity. Data from Morocco, 39 seasons. Rainfall of INRA location of Marchouch representing average rainfall of agricultural lands of Morocco. Yields are country average, official statistics (ONICL, 2020).

Statistics	Rainfall (mm)	Yield (kg/ha)	Water productivity (kg/mm)
Minimum	64.2	359	3.26
Average	219	1229	6.03
Maximum	447	2288	16.73
Range	1 : 7	1 : 6	1 : 5

Effects of amount and distribution by species

According to this model, the effect of amount of rainfall is positive and rainfall dependent, and is measured by the coefficient of regression indicating water productivity, and effect of distribution is negative and independent from rainfall amount.

Table 2: Effects of amount and distribution of rainfall on major cereal crops in Morocco.

Species	Amount	Amount & distribution	Distribution*
Bread Wheat	5.03x + 968; R^2 = 0.96	4.19x + 544; R^2 = 0,56	0.84x - 424
Durum Wheat	4.86x + 930; R^2 = 0.97	4.24x + 436; R^2 = 0.57	-0.62x - 494
Barley	3.54x + 578; R^2 = 0.97	3.6x + 178; R^2 = 0. 65	0.06x - 400
All Cereals	4.62x + 781; R^2 = 0.99	3.98x + 357; R^2 = 0.60	-0.64x - 424

*Non-significant coefficients of regression, considered as estimates of 0.

Bread wheat showed highest water productivity (5.03 kg/mm) followed by durum (4.86 kg/mm) and barley (3.54 kg/mm). Durum wheat was most affected by rainfall distribution, followed by bread wheat and barley.

Table3: Effects of rainfall distribution on cereal crops productivities estimated at minimum, average and maximum rainfall to indicate variation.

	Yield by level of rainfall			Yield loss (%) due to distribution, at		
Species	Minimum 64.2mm	Average 219mm	Maximum 447mm	Minimum 64.2mm	Average 219mm	Maximum 447mm
Bread Wheat	1291	2070	3216	-33%	-21%	-13%
Durum Wheat	1242	1994	3102	-40%	-25%	-16%
Barley	805	1353	2160	-50%	-30%	-19%
All Cereals	1078	1793	2846	-39%	-24%	-15%

The potential of wheat water productivity in semi-arid environments of Morocco was estimated at 22 kg/mm of rainfall consumed during the crop cycle [1]. The cereal productivity country wise is way below this figure, indicating the potential for improvement.

Average yield loss from erratic distribution of rainfall ranged from 13% to 50%, depending on the species and level of rainfall. It is quite high, which warrant search for efficient rainfall water management measures.

Among measures that can be adopted, the fallow system allow to transfer water from season to other, vegetative covering of the soil and mulching which reduces evaporation losses, appropriate cultural practices to improve water productivity, supplemental irrigation… Water productivity was improved by adopting these measures [2] which are expected to mitigate the effect of erratic distribution as well.

The rainfall effect partitioning method can be adapted to other rainfed crops in rainfall dependent and drought prone environments.

References

1. M. Jlibene. Options génétiques d'adaptation du blé tender au changement climatique ; Variétés à résistances multiples : sécheresse, cécidomyie, septoriose, rouilles brune et jaune. Edition INRA. (2009).

2. M. Karrou, M. El Mourid, M. Boutfirass, M. El Gharous, Al Awamia **123-124** (2008)

3. ONICL. Cereal statistics, Office National Interprofessionnel des céréales et Légumineuses (2020). www.ONICL.org.ma

Effect of the concentrate-based finishing on the proximate composition, organoleptic quality and fatty acids profile of the Beni-Guil sheep meat

Kamal Belhaj[1,2,*], *Farid* Mansouri[1], *Marianne* Sindic[2], *Yassine* Taaifi[1], *Mohamed* Boukharta[3], *Hana* Serghini-Caid[1], and *Ahmed* Elamrani[1].

[1]Laboratory for Agricultural Productions Improvement, Biotechnology and Environment (LAPABE), Faculty of Sciences, University Mohammed First, BP-717, 60000 Oujda, Morocco.

[2]Analysis Quality and Risk Unit, Laboratory of Food Quality and Safety (QSPA), Gembloux Agro-Bio Tech, University of Liège, 5030 Gembloux, Belgium.

[3]Institute of Agricultural Industries, High school of Charlemagne, 4500 Huy, Belgium.

Abstract. The aim of this study is to compare two sheep farming practices (lamb farming with or without finishing period on concentrate), on carcass characteristics, fatty acid profile and organoleptic quality of the meat. The study was performed on 24 Beni-Guil lambs. Finished animals (F) had access to creep feeding until weaning and were fed a mixed ration of barley and alfalfa hay during a finishing period of 45 days; however, non-finished animals (NF) remained on the pasture. This comparison was carried out by analyzing the carcass characteristics, proximate composition and fatty acids profile of the meat. Ultimate pH and meat lightness were lower for F-lambs than NF-lambs who didn't go through the finishing period. The concentrate-based finished practice produced carcasses with better fatness state and conformation. The intramuscular fat content is more important in finished-lambs' meats (3.81 vs 1.82) which show more juiciness and a high meat's color lightness. However, meats of NF-lambs presented a high proportion of polyunsaturated fatty acids (PUFA: 20.70 vs 16.82), particularly the PUFA n-3 (3.97 vs 1.17) and consequently a low n-6/n-3 ratio (3.92 vs 12.72). Finally, we recommend, finished lamb meats for the large-scale market and meat of pasture-raised lamb for the niche market.

1 Introduction

In Morocco, the trend in the distribution and consumption of sheep meat has changed considerably in recent years, both in terms of quality and quantity. The consumption and the price of fresh sheep meat depend on certain specific qualities and labels when it is intended for a target or niche market (small scale production of pasture lambs), but also influenced by its cost price and household incomes when it is intended for markets of large distribution (lambs raised under an intensive production system in which heavy carcasses are produced). Generally, in Morocco, traditional production is still dominant and sheep husbandry is performed with local breeds, using traditional methods based on pastures with pastoralism practices. In this system of traditional production, the sheep proliferate according to the natural reproduction period and lambs suckle their mothers until the age of weaning (3 months). As a rule in Morocco, where the lambs are slaughtered at 6–8 months of the animals are fattened ranging from 30 to 45 days before slaughter, to increase the slaughter live weight and to improve the aesthetic quality (commercial quality) of the lamb's carcass, depends on market demand.

Consumers of red meat tend to prefer lamb meat from pasture, considering that such lamb meat is much healthier, tastier, and more natural than meat from concentrate-based production systems. The meat quality is influenced by several factors such as the feeding system, breeding conditions, rearing season, and age at slaughter. Ekiz, Demirel [1], Ricardo, Fernandes [2], and Yalcintan, Ekiz [3]have shown that lamb meats produced under a concentrate-based system generally presents have higher carcass yield, better carcass conformation, and high fatness state. However, other authors indicate that meat from grazed lambs has a darker color and shows more favorable fatty acids profile than meat from the concentrate-fed system[4-6]. BeniGuil-

* Corresponding author: belhaj.kamal90@gmail.com; k.belhaj@ump.ac.ma

PGI Lamb meat is the main sheep meats produced in the eastof Morocco.It is very appreciated by the consumer because of its taste and its nutritional quality[7]. This has been revealed in our previous studies concerning the carcass and meat physicochemical characteristics of Beni-Guil sheep [7, 8]. In this study, we supposed that the fatness state has significant effects on the meat quality and the fatty acids profile of the Beni-Guil sheep meat. For this purpose, we investigated the effect of the concentrate-based finishing period (indoor finishing on concentrate and alfalfa hay (F); raising on pasture without finishing phase (NF)) on the fatty acids profile and organoleptic quality. Except for our previously published works, no others studies have evaluated the variation of meat quality of Beni-Guil sheep reared in Eastern Morocco according to the two common finishing strategies.For this purpose, we investigated the effect of the concentrate-based finishing period (indoor finishing on concentrate and alfalfa hay (F); raising on pasture without finishing phase (NF)) on carcarss traits, organoleptic proprieties and fatty acids profile of Beni-Guil sheep meat.

2 Material and methods

2.1. Animal material and Experimental design

We compare two finishing strategies (n=12 for each group): (1)raising on pasture with indoor finishing period (45 days) on concentrate and alfalfa hay (F); (2) raising on pasture without finishing period (NF). The lambs were weaned at the age of 3 monthsand slaughtered at 7 months of age.
The study was carried out on twenty-four *Longissimus lumborum* (LLM) muscles of twenty-four female lambs of Beni-Guil sheep breed (n = 24). The animals used in this study were born and raised in the Ain-Beni-Mathar region in the south-east of Morocco. The choice of female lambs was made randomly from the same breeders in the two treatments. Before treatment, the studied animals were 5.5 months old, and after the treatments, the lambs were slaughtered at an age of 7 months with a live weight of 34.11kg±3.2 and 25.67kg±2.31, respectively for the finished and unfinished lambs.The finishing period based on concentrate-feed (barley and alfalfa hay: (1 to 1.5 kg/day) for 45 days was the only difference between the animals of the two studied treatments. The slaughter procedure was carried out according to the conventional Islamic method (bleeding, skinning, pollarding, evisceration, slot, post-mortem inspection, refrigeration).

2.2. Objective and subjective carcass measurements

Carcass measurements and observations included fatness state and carcass conformation for prediction of lamb carcass quality, were subjectively evaluated under the supervision of a veterinarian and according to the European system (EUROP) as described byCarrasco, Ripoll [9]. Carrasco, Ripoll [7]. The live weight (LW) and cold carcass weight (CCW) are used to calculate the carcass yields or commercial yield (CY) using the following formula: CY = (CCW/LW) × 100

2.3. Analytical methods

2.3.1. Evaluation of meat quality

The meat ultimate pH (pH_{24}) was measured in *Longissimuslumborum* muscle (LLM), using a pH meter with a penetration probe. The meat color was estimated according to the international center of lighting using the CIELAB system using a Chromameter 400 Konica Minolta CR400(L* = Lightness, a* = Redness, b * = Yellowness) [10]. The meat juiciness was estimated by the application of a mechanical force. The samples were sliced into 1 cm thick steaks, with a diameter of 4 cm^2, placed between 18 pre-weighed Whatman papers. The water holding capacity (WHC) was calculated as the difference in sample weight before and after applying a mechanical force of 2.250 Kg for 5 min following the method described by Grau and Hamm [11]:

$$WHC (\%) = [(IWM - FWM / FWM) * 100]$$

Where IWM: Initial weight of the meat, FWM: Final weight of the meat

2.3.2. Samples preparation

Strip loin cuts were collected from carcasses of Beni-Guil female Lambs 24hours*post-mortem* in LLM. The samples were lyophilized and stored at -20 °C for ulterior analyses.

2.3.2.1. Proximate composition

Dry matter was determined by drying in a stove at 100 °C ± 3 °C according to the Association of Official Analytical Chemists (AOAC)[12]. Intramuscular fat (IMF) was extracted and quantified according to Bligh and Dyer [13]. Ash was measured by incineration in the oven according to AOAC[12]. Total proteins were calculated according to the Kjeldahl method using the conversion index of 6.25[12].

2.3.2.2. Fatty acids profile analysis

Before the lipid profile analysis, the extracted intramuscular fat was methylated to fatty acids methyl esters (FAME) using BF3 at 14% weight in methanol [14]. The Agilent gas chromatography (GC: HP6890 series, Agilent Technologies, USA) coupled with a flame ionization detector (FID) was used to analyze the fatty acids profile, using the Helium as carrier gas (flow rate: 1.7 ml/min). The GC-FID was equipped with a capillary column Omega wax type (length column: 30 m; Diameter: 0.25 mm; film thickness: 0.25 μm). The temperatures of the injector and the detector were set at 150 and 250 °C, and the oven temperature was at 210 °C. The injection mode used was the splitless type by injecting a volume of 1μL. A fatty acid methyl esters standard, containing 37 components (Supelco, Bellefonte, PA, USA), was used to identify the individual peaks. The average percentage of each fatty acid was used to calculate the sum, ratios, and lipid indices of the meat.

2.4. Statistical analyses

Theobtained results weregiven as means±standard deviation of triplicate determinations. The statistical analyses were carried out using the software SPSS version 20while considering the finishing strategy as a source of variation. The means were compared using the *t*-test and the significance was declared at $p<0.05$.

3 Results and discussion

The effects of the finishing period on carcass traits, organoleptic quality of meat, and the fatty acids profile are presented in Tables 1, 2, 3, and 4.

3.1. Carcass characteristics

The carcass yield is an important criterion for the assessment of carcass commercial-quality in rearing sheep destined to meat production. This parameter (dressing percentage) allows the professional to predict the potential commercial transactions of the carcass (profitability)[8].The results indicate that the finished animals (group F) hada higher ($p<0.001$)cold carcass weight (16.37kg vs 11.55 kg) with higher commercial yield percentage than those not finished (group NF) (48.43 vs 45.11; Table 1). A similar result was obtained by several researchers [1, 3,5]. Borton, Loerch [15], and Karim, Porwal [16]reported that the intensive production systems produce animals with a higher dressing percentage compared to extensive systems. They explain this result by the digestive tract which is more developed in animals produced extensively due to the higher coarse food compared to those produced intensively.The finishing period had a significant effect on the carcass' subjective characteristics ($p<0.001$). Moreover, F animals have shown a higher ($p<0.001$) fatness state (3.51 vs 1.61) and conformation score (2.82 vs 1.42; Table 1).Previous studieshavereported that animals from production systems based on concentrate-fed produce carcasses with higher fatness and conformation scores due to (i) the consumption of energy-rich food ration (barley), and also due to (ii)the physical activity or the greater energy expenditure of grazing lambs toobtain food[1, 3, 5, 17].

Table 1. Finishing period effect on the carcass characteristics of Beni-Guil sheep meat

Parameters	Finishing period		*P*-value
	Lambs Group F	Lambs Group NF	
Live weight at slaughter (kg)	34.11±2.63	25.67±3.1	0.000
Commercial yeild (%)	48.43 ±0,03	45.11±0.01	0.002
Fatness score	3.51±0,46	1.61±0.32	0.000
Conformation score	2,82±0,12	1.42±0.13	0.002

Group F = indoor finishing lambs on barley and alfalfa hay; Group NF = lambs raised on pasture without finishing period; Conformation: Excellent=5 ;Very good=4 ;Good=3 ;Fair=2; Poor=1 ;Fatness score: 1=Very low; 2=Low; 3 Average; 4=High; 5= Very high;

3.2. Organoleptic quality

Meat pH is a chemical parameter that determines the final quality of the meat and has direct and indirect effects, practically, on the meat's organoleptic properties[8, 18]. The obtained values varied between 5.7 and 5.8. These recorded values were within the normal ranges for commercial meats [19].The meat from F lambs had a lower($p<0.05$)pH_{24} values than that of NF animals (Table 2).Grazing lambs are associated with low energy forage diets (low energy ration) and have relatively small glycogen reserves compared to intensively reared lambs. Similar results were reported by several authors[3, 5,6]. This difference could be explained by the quantity ofconcentrated food ingested by lambs finished with barley and alfalfa hay which promotes the propionic fermentation in the rumen and consequently, the production of propionic acid (precursor of muscle glycogen).

Meat juiciness is among the main intrinsic factors influencing meat palatability and acceptability[17-18]. The NF lambs show the highest values of WHC (23.96% vs 15.91%; $p>0.001$). This recorded difference could be explained by the quantity of intramuscular fat and by the quantity of water from the food ration ingested by the animals (more grazing = more water in the meat).Priolo, Micol [17]reported that fat content is positively correlated to meat juiciness and that grazing lambs produces carcasses with lower fat content due to the consumption of a low energy diet and intense physical activity. Hence, it could be concluded that the meat of finished animals would be morejuicious than that of unfinished ones.

Meat color or visual appearance is the defining quality of its buying, which is mainly associated with its freshness [20, 21]. Consumers of ruminant meat prefer bright red meat and associate it with high quality and product freshness. The color coordinates (lightness, redness, yellowness, chroma, and hue angle) are presented in Table 2. The lightness values (L*) for the two studied groups, ranging between 37.03 and 34.15 are indicative of bright red color[22]. Meat color can be influenced by numerous factors such as intramuscular fat, ultimate pH, myoglobin content, and its chemical state, physical activity, age, and live weight at slaughter [17, 23-25]. The meat of NF lambs was darker than that of F lambs. Similar results were reported by Dıaz, Velasco [26] and Yalcintan, Ekiz[3] on lamb's meat. No differences were found in the yellowness index (p>0.05). The greater values of the L* index were recorded in the meat of F animals (37.03 vs 34.15). This result can be ascribed to intramuscular fat which increases the meat lightness and the physical activities associated with pasture. Dıaz, Velasco [26], Cartier and Moevi [27] and Yalcintan, Ekiz [3] reported that pasture-fed animals produced darker meat (lower L*) due to a higher myoglobin content and lower intramuscular fat and they explained this difference by physical activity, higher myoglobin content in animals raised on pasture and low energy diet (less intramuscular fat). Moreover, lambs raised on pasture contain a high concentration of oxidative type I muscle fibers, which are rich in myoglobin and contribute to a higher muscular myoglobin content. Continuous physical exercise, associated with extensive rearing systems, induces the transition of glycolytic muscle fiber (fast-twitch fibers) to oxidative muscle fiber (slow-twitch fibers) [5]

Table 2. Finishing period effect on the proximate composition and meat quality characteristics of Beni-Guil sheep meat

Parameters	Finishing period		*P*-value
	Lambs Group F	Lambs Group NF	
Proximate composition			
Dry matter (%)	25,71±1,99	24.51±0.28	0.328
Total protein (%)	21,41±0,56	21.61±0.56	0.561
Intramuscular Fat (%)	3,82±1,72	1.82±0.13	0.002
Ash (%)	1,02±0,06	1.05±0.01	0.118
Organoleptic quality			
Ultimate pH (pH_{24})	5,72±0.2	5.8±0.16	0.034
WHC = (Expressed Juiciness (%))	15,91±1,23	23.96±2.05	0.000
Meat color			
Lightness (L*)	37,03±1,31	34.15±1.24	0.000
Redness (a*)	12,62±0,96	13.33±0.49	0.497
Yellowness (b*)	13,43±0,79	13.24±0.56	0.056
Chromaticity (C*)	18,46±0.75	18.79±0.64	0.278
Hue angle °	46,80±3,14	44.80±1.13	0.035
a*/b*	0,94±0.1	1±0.04	0.040

Group F = indoor finishing lambs on barley and alfalfa hay; Group NF = lambs raised on pasture without finishing period; IMF: Intramuscular Fat; WHC: Water holding Capacity; the scale of a*, b*, L*: a* and b*: 60 to 60, L*: 0 to 100. Scales of chroma and hue: chroma: 0–60, hue: 0–360°.

3.3. Proximate composition

The proximate composition of the LL muscle of the studied groups is given in Table 2. Regarding the intramuscular far (IMF), the results show that F lambs have higher (p<0.01) fat content than NF lambs (3.82 vs 1.82). Similar results were found by Ekiz, Demirel [1] and Priolo, Micol [17]. Intensive production systems produce carcasses with higher IMF content than extensive production systems. The concentrate-fed production system is associated with high energy diets (high nutrition level) and consequently a high intramuscular fat. Also, these results could be attributed to the limited exercise associated with intensive production systems[26].Intramuscular fat also has a positive effect on meat juiciness (increasing the sustained juiciness), tenderness (dilution of muscle fibers), and flavor (Flavor carrying components are hydrophobic and are therefore dissolved in fat) of the meat[5]. As a result, the meat of F animals will be more tender and juicer with better flavor. No significant effect has been shown regardingdry matter, protein content, and ash (p>0.05).Numerous authors support our findings and have concluded that production systems had no effect on dry matter, crude protein, and ash[28, 29].

3.4. Fatty acids profile analysis

Tables 3 and 4 present the effect of concentrate-based finishingon the fatty acids profile of LLM of Beni-Guil sheep meat. The predominant fatty acid was C18:1, followed by C16:0, C18:0, and C18:2n6. Total saturated fatty acids were present at higher proportions than monounsaturated fatty acids, with the lowest proportions corresponding to polyunsaturated fatty acids. In general, the values were similar to others reported for Beni-Guil sheep breed[7, 30].The prevailing individual fatty acids (FA) were oleic acid, followed by palmitic, stearic, and linoleic. The saturated fatty acids (SFA) present the highest proportion of total fatty acids in the meat of NF lambs, while the monounsaturated (MUFA) fatty acids dominate the profile of F lambs meat (42 vs 40 vs 16% respectively for MUFA, SFA, and PUFA).

Table 3.Finishing period effect on the fatty acids profile of Beni-Guil sheep meat

Fatty acids (%)	Finishing period		*P*-value
	Lambs Group F	Lambs Group NF	
C10:0	0.12±0.05	ND	-
C12:0	0.15±0.06	ND	-
C13:0	0.02±0.00	ND	-
C14:0	1.88±0.32	1.62±0.3	0.116
C14:1	0.22±0.02	0.31±0.06	0.451
C15:0	0.4±0.11	0.40±0.03	0.989
C15:1	0.13±0.01	0.25±0.01	0.041
C16:0	20.47±1.15	19.58±0.12	0.010
C16:1n7	0.35±0.04	0.39±0.02	0.027
C16:1n9	1.97±0.35	1.45±0.8	0.050
C17:0	1.5±0.62	0.91±0.02	0.010
C17:1	1.09±0.28	1.01±0.02	0.259
C18:0	15.15±0.89	16.14±0.25	0.001
Cis/trans-C18:1n9	38.26±3.43	34.38±0.83	0.001
Cis/trans-C18:2n6	9.33±2.82	8.54±0.76	0.327
C18:3n6	0.8±0.07	1.99±0.02	0.000
C18:3n3	0.45±0.1	0.64±0.03	0.001
C20:1n9	0.18±0.03	0.27±0.01	0.577
C20:2	0.6±0.2	1.14±0.02	0.000
C20:3n6	0.39±0.1	0.6±0.03	0.003
C20:3n3	0.08±0.02	ND	-
C20:4n6	4.42±1.34	4.45±0.45	0.957
C20:5n3	0.42±0.02	2.16±0.02	0.000
C22:2	0.11±0.09	ND	-
C23:0	0.28±0.1	0.40±0.03	0.001
C24:0	0.76±0.05	2.09±0.3	0.000
C22:6n3	0.21±0.02	1.17±0.2	0.000
C24:1n9	0.25±0.02	ND	-

Group F = indoor finishing lambs on barley and alfalfa hay; Group NF = lambs raised on pasture without finishing period; NS: not significant; ND: not detected

The fatty acid profile of finished lambs was different from the not-finished ones.The concentrate-based finishing period had significant effects (p<0.05) on individuals, sums, ratios, and indices of fatty acids (Tables 3 and 4). Also, our findings show that the finishing period can significantly affect certain saturated fatty acids such asC17:0.High concentration of heptadecanoic acid (C17:0) recorded in the meat of F animals could be attributed to the greater ingestion of concentrate, which promotes the rumen bio-hydrogenationand, consequently, the synthesis of volatile fatty acids, particularly the propionic acid (precursor of odd fatty acids)[31]. F animals'meatcontains the highest proportion of MUFA and the lowest percentage of polyunsaturated fatty acids (PUFA).The difference in MUFA fraction is mainly due to the percentage ofC18:1), which is considered as a beneficial FA [32]. The finished lambshave a higher content of C18:1 (38.26% vs 34.38%; p=0.001). This difference can be explained by the nature of the feeding system characterized by its concentraterichness (barley) in the finishing period, which decreases the stay of feed-in rumen, thus decreasingthe bio-hydrogenation of unsaturated fatty acids[33]. Similar results were reported by Scerra, Caparra [34] in milk from ewesfeeded by concentrate and pasture.Regarding the PUFA fraction, the meat from NF lambs shows the highest percentage (16.82% vs 20.70%), leading them to a more favorable PUFA/SFA ratio (0.42 vs 0.5). This difference mainly concernsPUFAn-3content (1.17% vs 3.97%, respectively for F and NF),and particularly the proportion of C20:5 (0.42% vs 2.16%) respectively for F and NF animals).As a consequence, this highest concentration of PUFA n-3 resulted in a significant reduction of n-6/n-3 ratio (12.76 vs 3.97; p<0.001).This difference can be attributed tothe pasture-fed system, which is richer in PUFA n-3 than the concentrate-fed system; particularly, the linolenic acid amount is high in herbage. Several authors have also suggested that the pasture rearing system produces meat with favorable PUFA/SFA and n-6/n-3 ratio, compared to those reared under concentrate-fed production system[5, 26, 34, 35].In accordance with our results, Scerra, Caparra [34], Biondi, Valvo [36], and Mazzone, Giammarco [37] have also demonstrated that a diet based exclusively on green herbage, compared to a concentrate based feed, leads to a favorable PUFA/SFA and n-6/n-3 ratio.Concerning the lipid health indices ((Thrombogenicity Index = (TI), Atherogenicity Index = (AI)), which should not exceed the value of 1 [31], the not finished animals present the favorable indices inferior to 1. The registered results for the two studied groups are comparable to those reported by Sinanoglou, Batrinou [32] for Greek lambs.

The nutritional recommendations for humans in terms of lipid diet are to reduce total lipid intake,specifically the SFA and rebalance the PUFA ratio of the n-6 and n-3 series by increasing the intake of n-3[38, 39].The recommendation in human nutrition suggests that the n-6/n-3 ratio should be between 2 and 4[38]. So, from the dietary point of view, the F lambs present the disadvantage of a lower concentration of n-3 fatty acids, a thrombogenic index superior to the recommended value, and, consequently, a higher n-6/n-3 ratio in comparison to the NF lambs.

Table 4. Finishing period effect on the sums, ratios, and indices of fatty acids profile of Beni-Guil sheep meat

Sums and ratios of FAs	Finishing period		P-value
	Lambs Group F	Lambs Group NF	
SFA	40.73±2.33	41.26±0.51	0.411
MUFA	42.44±4.02	38.04±0.9	0.001
PUFA	16.82±5.98	20.70±1.00	0.028
UFA	59.26±2.33	58.74±±0.3	0.411
DFA	74.42±1.5	77.88±0.16	0.256
OFA	3.14±1.01	2.67±0.18	0.105
UFA/SFA	1.46±0.14	1.42±0.01	0.321
PUFA/SFA	0.42±0.16	0.50±0.01	0.091
PUFA n-6	14.94±3.34	15.58±0.85	0.663
PUFA n-3	1.17±0.37	3.97±0.55	0.000
n-6/n-3	12.72±2.2	3.92±0.22	0.000
TI	1.16±0.13	0.96±0.02	0.000
AI	0.47±0.05	0.44±0.02	0.083
h/H	2.49±0.28	2.61±0.06	0.162
Nutritive value	2.6±0.12	2.58±0.03	0.450

Group F = indoor finishing lambs on barley and alfalfa hay; Group NF = lambs raised on pasture without finishing period; Lambs group F: With the finishing period; Lambs group NF: Without the finishing period; SFA: Saturated fatty acids; UFA: Unsaturated fatty acids; PUFA: Polyunsaturated fatty acids;DFA: Desirable Fatty Acids (C18:0 + UFA); OFA: Odd Fatty Acids;IT = Thrombogenic Index [C14:0+C16:0+C18:0] / [(0.5*MUFA) +(0.5*∑n−6) +(3*∑n−3) +(n−3/n−6)]; IA = Atherogenic Index [(4*C14:0) + C16:0] / [(PUFA)+(MUFA)], calculated as per Ulbrichtet Southgate (1991) without the inclusion of 18:0 which is considered to be neutral on serum cholesterol;h/H = Hypocholesterolemic (h) /Hypercholesterolemic (H);h /H = (18:1n9c + 18:2ω6 + 20:4ω6 + 18:3ω3 + 20:5ω3 + 22:5ω3 + 22:6ω3) / (14:0+16:0);Nutritive value:(C18 :0+C18 :1) /C16 :0

4 Conclusion

Two farming practices for lambs' fattening were studied, a pasture rearing supplemented with indoor finishing on concentrate (F) during 45 days compared to lambs' fattening exclusively on pasture (NF). The results show that concentrate-based finished lambs have better subjective carcass traits and better meat juiciness. However, non-finished lambs (raised on pastures, without going through the finishing period) show meat with a better fatty acid profile and well-balanced ratios of PUFA/SFA and omega-6/omega-3. Thus, from an organoleptic point of view but especially from a productivity and economic profitability point of view, the meats of finished lambs based on concentrate are recommended for the mass market. However, from a nutritional and dietetic point of view, the meats of lambs exclusively pasture-raised will be recommended for the niche market, where informed consumers look for less fat meats with high nutritional quality, even if, because they are less fat, their organoleptic quality is slightly lower than that of the previous ones and cost more.

Currently, it is recommended to provide consumers with products of consistent quality to enhance their confidence in the product and encourage repurchases. Therefore, it is crucial to investigate the full impact of the two production practices (F, NF) to better understand the finishing strategy effects and its duration on the meat quality characteristics of Beni-Guil sheep meat to ensure that the quality is maintained and that consumer expectations are met. Thus, to enhance the quality of the meat's IMF of finished lambs (F), particularly PUFA n-3, it is essential to improve the composition of their diet such as the incorporation of seeds rich in PUFA n-3.

References

1. B. Ekiz, G. Demirel, A. Yilmaz, M. Ozcan, H. Yalcintan, O. Kocak, and A. Altinel, Small Ruminant Res., **114**, 26-34 (2013)
2. H. Ricardo, A. Fernandes, L. Mendes, M. Oliveira, V. Protes, E. Scatena, R. Roça, N. Athayde, L. Girão, and L. Alves, Small Ruminant Res., **130**, 141-145 (2015)
3. H. Yalcintan, B. Ekiz, O. Kocak, N. Dogan, P.D. Akin, and A. Yilmaz, Arch. Anim. Breed., **60**, 225-233 (2017)
4. B. Aurousseau, D. Bauchart, X. Faure, A. Galot, S. Prache, D. Micol, and A. Priolo, Meat Sci., **76**, 241-252 (2007)
5. Y. Hanekom, Stellenbosch: University of Stellenbosch, Sout Africa, **153** (2010)
6. A.G. D'Alessandro, G. Maiorano, M. Ragni, D. Casamassima, G. Marsico, and G. Martemucci, Small Ruminant Res., **114**, 97-104 (2013)
7. K. Belhaj, F. Mansouri, M.-L. Fauconnier, M. Boukharta, H.S. Caid, M. Sindic, and A. Elamrani, Med. J. Nutrition Metab., **11**, 175–185 (2018)
8. K. Belhaj, F. Mansouri, A. Tikent, A. Ouchatbi, M. Boukharta, C.H. Serghini, and A. Elamrani, Rev. Elev. Med. Vet. Pays Trop., **73**, 21-26 (2020)
9. S. Carrasco, G. Ripoll, A. Sanz, J. Álvarez-Rodríguez, B. Panea, R. Revilla, and M. Joy, Livest. Sci., **121**, 56-63 (2009)
10. H. Calnan, R. Jacob, D. Pethick, and G. Gardner, Meat Sci., **119**, 41-50 (2016)
11. R. Grau and R. Hamm, Naturwissenschaften, **40**, 29-30 (1953)
12. AOAC, Official Methods of Analysis of the Association of Official Analytical Chemists (Association of Official Analytical Chemists, Washington, 1990)
13. E.G. Bligh and W.J. Dyer, Can. J. Biochem. Physiol., **37**, 911-917 (1959)
14. A. Ben moumen, Université Mohamed Premier, Morocco,157 (2015)
15. R. Borton, S. Loerch, K. McClure, and D. Wulf, J. Anim. Sci., 83, 679-685 (2005)
16. S. Karim, K. Porwal, S. Kumar, and V. Singh, Meat Sci., **76**, 395-401 (2007)
17. A. Priolo, D. Micol, J. Agabriel, S. Prache, and E. Dransfield, Meat Sci., **62**, 179-185 (2002)
18. A. Valros, C. Munsterhjelm, E. Puolanne, M. Ruusunen, M. Heinonen, O.A. Peltoniemi, and A.R. Pösö, Acta Vet. Scand., 55, 75 (2013)
19. J.F. Tejeda, R.E. Peña, and A.I. Andrés, Meat Sci., **80**, 1061-1067 (2008)
20. A. Clinquart, B. Leroy, O. Dottreppe, J. Hornick, I. Dufrasne, and L. Istasse, Les facteurs de production qui influencent la qualité de la viande des bovins Blanc Bleu belge, inProceedings of theJournée du Centre d'Excellence du Secteur agricole et son Management,CESAM, 26 Mai 2000,Mons, France(2000)

21. M. Beriain, P. Bas, A. Purroy, and T. Treacher, Effect of animal and nutritional factors and nutrition on lamb meat quality, in Proceeding of the Cahiers Options Méditerranéennes, CIHEAM, 2000,Grignon, France (2000) .
22. C. Salifou, A. Youssao, G. Ahounou, P. Tougan, S. Farougou, G. Mensah, and A. Clinquart, Ann. Med. Vet., **157**, 27-42 (2013)
23. R. Lawrie, Meat Sci., (Permgamon Press: Oxford, NY, 1966).
24. B. Ekiz, A. Yilmaz, M. Ozcan, and O. Kocak, Meat Sci., **90**, 465-471 (2012)
25. J.-b. Liu, J. Guo, F. Wang, Y.-j. Yue, W.-l. Zhang, R.-l. Feng, T.-t. Guo, B.-h. Yang, and X.-p. Sun, Small Ruminant Res., **123**, 251-259 (2015)
26. M. Dıaz, S. Velasco, V. Caneque, S. Lauzurica, F.R. De Huidobro, C. Pérez, J. Gonzalez, and C. Manzanares, Small Ruminant Res., **43**, 257-268 (2002)
27. P. Cartier and I. Moevi, Le point sur la qualité des carcasses et des viandes de gros bovins (Institut d'élevage, Paris, 2007)
28. A. Rowe, F. Macedo, J. Visentainer, N. Souza, and M. Matsushita, Meat Sci., **51**, 283-288 (1999)
29. M. Pompa-Roborzynski and W. Kedzior, Arch Tierz, **49**, 268-274 (2006)
30. K. Belhaj, F. Mansouri, M. Sindic, M.-L. Fauconnier, M. Boukharta, C.-H. Serghini, and A. Elamrani, JABS, 13, 166-171 (2019)
31. G. Oriani, G. Maiorano, F. Filetti, C. Di Cesare, A. Manchisi, and G. Salvatori, Meat Sci., **71**, 557-562 (2005)
32. V.J. Sinanoglou, A. Batrinou, F. Mantis, I. Bizelis, and S. Miniadis-Meimaroglou, Small Ruminant Res., **113**, 1-10 (2013)
33. I. Petrova, V. Banskalieva, and V. Dimov, Small Ruminant Res., **13**, 263-267 (1994)
34. M. Scerra, P. Caparra, F. Foti, V. Galofaro, M. Sinatra, and V. Scerra, Meat Sci., **76**, 390-394 (2007)
35. J. Wood, M. Enser, A. Fisher, G. Nute, F. Whittington, and R. Richardson, Cah. Options Mediterr., Series A, **67**, 133-141 (2003)
36. L. Biondi, M. Valvo, M. Di Gloria, E.S. Tenghi, V. Galofaro, and A. Priolo, Small Ruminant Res., **75**, 17-23 (2008)
37. G. Mazzone, M. Giammarco, G. Vignola, L. Sardi, and L. Lambertini, Meat Sci., **86**, 474-478 (2010)
38. A.P. Simopoulos, Biomed. Pharmacother., **56**, 365-379 (2002)
39. N.L. Howes, A.E.D.A. Bekhit, D.J. Burritt, and A.W. Campbell, Compr. Rev. Food Sci. Food Saf., **14**, 22-36 (2015)

Mulching effect on weed control and faba bean (*Viciafaba* L. Minor) yield in Meknes region, Morocco.

Abdellatif Boutagayout[1,2*], *Laila* Nassiri[1], *El Houssine* Bouiamrine[1]and *Saadia* Belmalha[2]

[1]The Environment and Soil Microbiology Unit, Faculty of Sciences-Moulay Ismail University, B.P.11201 Zitoune, Meknes 50000, Morocco.
[2]Department of Plant and Environment Protection, Ecole Nationale d'Agriculture de Meknès, Route Haj Kaddour, BP S/40 - 50000 Meknès, Morocco.

Abstract. In order to protect human health and the environment and to prevent the occurrence of weed ecotypes, global efforts are being made to reduce the world's high dependency on synthetic herbicides for weed control. In this perspective, a field experiment was conducted during 2018/2019 season at the educational farm of the National School of Agriculture of Meknes (Morocco) to evaluate mulch effect of black plastic, oat straw and sycamore leaves on both faba bean and weed growth at flowering and harvest stage. The experiment was performed according to a randomized complete block design (RCBD) repeated four times. The results showed that the highest weeds infestation was recorded in weedy check (Control). Sycamore leaves, oat straw and black plastic mulches have significantly reduced weeds density and biomass, thus provided efficiency by 84.8, 80.6 and 86.8% respectively. In addition, the oat straw, black plastic, and sycamore leaves have improved faba bean yield by 42, 52.9, and 43% respectively. In conclusion, mulching practice can be a promising alternative of weed management for faba bean crop in Meknes region.

Introduction

Faba bean (*Viciafaba* L.) is one of the most important cultivated species of the fabaceae family [1]. Due to its economic, environmental and agro-ecological performances, faba bean contributes to the sustainability of farming systems. The varieties of *Viciafaba* L. have considerable importance as a low-cost, and protein-rich (20-25% seeds) food, which makes it the main source of protein in Africa [2]. Like all legumes, faba beans have the ability to fix atmospheric nitrogen by a symbiotic relationship with bacteria, particularly the genus of rhizobium [3], thus, increases soil fertility and the productivity of agricultural lands. Unfortunately, this craze is waning over time [4]. Until the late 1970s, Morocco was the world's second largest exporter of food legumes, mainly faba beans. Eventually, Morocco imported about 9 % of annual needs to offset its production [5].

Abiotic and biotic factors are major constraints that limit yield and cause instability of faba bean [2], in addition, faba bean crop is very sensitive to weed competition, thus weed

*Correspondingauthor:abdellatif.boutagayout@edu.uca.ac.ma

management is very important for crop production [6]. For example, the yield of a bean weed infested crop is reduced by up to 80% [7]. Regarding of the harmful effects of herbicides use on humans and environment [8], a reasonable and careful attitude must be envisaged to limit or even stop the use of agrochemical products [9].

In this context, several methods provide information on agroecological practices such as mulching that can lead to sustainable weed management [10]. In organic farming, mulching is very useful as one of the strategies for integrated weed management [11], by providing a barrier to light, heat or moisture exchange [12]. It reduces evaporation, preserves moisture, and regulates structure and temperature [13]. On the other hand, mulching practice has several advantages, such as improving soil structure and quality by increasing infiltration and water retention capacity, and providing a refuge for many insects and earthworms [9]. Moreover, it facilitates root penetration and development, and thus nutrient uptake from deeper soil layers [14], also it can reduce erosion [15].

The present study was conducted in an experiment field to evaluate the effect of organic and inorganic mulch on weeds density and biomass, as well as on faba bean yield components.

1. Material and methods

1.1 Site and Field trial

Faba beans (*Viciafaba* L. var. minor) (local cultivar) were sown in 2018-2019 season in one of the fields of National School of Agriculture of Meknes, Morocco (33° 50' 3.322" N 5° 28' 18.887" W). The study area has a Mediterranean climate; winters are cool, and rainy, and the summers are hot and dry. Laboratory analysis showed that the soil texture is silty clay with 1.5% of organic matter. Soil pH is 8.78, the total nutrients content in the soil are 13.0ppm of phosphorus, and 26.8ppm of potassium. The faba bean seeds were sown using a seed machine attached to a tractor set at a depth of 5-6 cm, a distance of 50 cm between rows, and at a density of 100 Kg/ha. The faba bean crop was grown without any chemical fertilizer input. Irrigation and other agronomic practices were done properly and uniformly to grow a successful crop. The growing season for faba bean is from mid-December to mid-June.

1.2 Treatments and experimental design

Three treatments were tested; black polyethylene (black plastic), oat straw (4t/ha) and plane tree leaves (Sycamore leaves) (4 t/ha) were applied between the rows immediately after faba bean emergence to compare these treatments with weedy check (Untreated). The experiment was laid out in a randomized complete block design, with four replications.

1.3 Data collection

Weeds density and biomass were determined by using square of 0.5m x 0.5m, randomly launched in each plot at the flowering and maturity stage. The dry weight of weeds and crops was determined after oven drying at 105 °C until complete drying and constant weight [16]. The average weed biomass for each treatment was calculated and expressed in g/m^2. Also, weed control efficiency (WCE) was assessed based on percentage reduction in dry weed biomass [4], the WCE efficiency was calculated as following:

$$\text{W.C.E}\ (\%) = \frac{\text{Weed dry matter in weedy check} - \text{weed dry matter in treated plot}}{\text{Weed dry matter in weedy check}} \times 100$$

At maturity stage, the plant height, the number of pods, the number of seeds, the number of branches, the seed yield (tons/ha) and the 100 seed weight in gram were recorded from 10 plants in each net plot.

1.4 Statistical analysis

The results were subjected to analysis of variance (ANOVA) ; using statistical software IBM SPSS Statistics 21 to assess mulching practices effects on the considered variables. The normality of the data was evaluated through the Tukey test. Unless otherwise stated, the significance level was $p \leq 0.05$.

2. Results and discussion

2.1 Weed flora associated with faba bean

Table 1. Common weeds associated with faba bean crop during the experimentation

Species	Common name	Family
*Polygonumavicular*L.	Prostrateknotweed	Polygonaceae
*Glebioniscoronaria*L.	Crown daisy	Asteraceae
*Calendula arvensis*L.	Field marigold	Asteraceae
*Sonchusoleraceus*L.	Common sowthistle	Asteraceae
*Helminthiaechioides*L.	Bristlyoxtongue	Asteraceae
*Fumariaparviflora*Lam.	Fineleaffumitory	Papaveraceae
*Papaver rhoeas*L.	Common poppy	Papaveraceae
Chenopodium album L.	Fat hen	Amaranthaceae
*Galium aparine*L.	Cleavers	Rubiaceae
*Biforatesticulata*L.	MundaneEuropean bishop	Apiaceae
Lamium amplexicaule L.	Henbitdead-nettle	Lamiaceae
Coronillascorpioides(L.) Koch	Yellow crownvetch	Fabaceae
Vicia sativa L.	Common vetch	Fabaceae
*Medicagopolymorpha*L.	Burclover	Fabaceae
*Sinapisarvensis*L.	Wild mustard	Brassicaceae
*Vaccariahispanica*Mill.	Cowsoapwort	Caryophyllaceae
*Cichoriumintybus*L.	Common chicory	Asteraceae
*Diplotaxisassurgens*Thell.	Mursenaper	Brassicaceae
*Bromusstirilis*L.	Barren brome	Poaceae
Phalaris minor Retz.	Littleseedcanarygrass	Poaceae

The total weed species identified in the floristic surveys carried out at the experimental field as well as the common names and botanical families of each species are revealed (Table 1). The total weed flora identified was very diverse; with 20 species distributed over 13 families. The most dominant weed species found in the faba bean field were *Polygonumavicular*, *Glebioniscoronaria*, *Vaccariahispanica*, *Papaver rhoeas* and

Fumariaparviflora etc. broad-leaved weeds (dicotyledonous) are the most dominant, while Grasses weeds (monocots) are represented only by two species (*Bromusstirilis* and *Phalaris minor*).

Finally, the asteraceae family is the most frequent in the study field with 4 species followed by the fabaceae (3 species).

2.2 Weed density

Organic and inorganic mulches provide better management of the weed community compared to the control (Table 2), although the use of plastic mulch showed no significant difference compared to the use of oat straw and plane tree leaves, a slight advantage was observed with plastic mulch. This can be explained by the inhibitory effect of black plastic against weed germination and by exerting physical pressure on weeds [17]. However, another study was carried out in Lithuania by Jodaugienė*et al.* [18] to evaluate the effect of organic mulch (wheat straw, peat, wood chips...) on weeds in common bean crop (*Phaseolus vulgaris* L.), showed that organic mulches were very effective in reducing weed germination and emergence.

For their part, Döring*et al.* [19] found that straw mulch did not affect weed numbers, coverage, and above-ground biomass; this may have been due to the relatively modest quantities of straw applied or to its infestation by weed seeds.

Table 2. Effects of oatstraw, plastic, and plane treeleavemulches on weeddensity at the flowering and maturity stage

Treatemts Stage growth	weedy check	Oatstraw	Black plastic	Plane treeleaves
Flowering stage	468.5a	89.5b	65b	83.5b
Maturity stage	562.5a	92.5b	75.5b	85.5b

In the row, the mean values followed by the samelower case letters are not significantlydifferent.

On the other hand, a study carried out on an organic field faba bean crop [20], revealed that oat straw mulch (4 or 6 t/ha) resulted in effective weed control, especially photosensitive weeds such as *Chenopodium album* L. and Matricaria spp.

2.3 Weed Biomass

Table 3 shows that the effect of mulching on biomass is very important. The different types of mulch used, especially black plastic, increases faba bean biomass. The plastic mulch improves the biomass of faba bean by about 57% compared to the control, while the organic mulches each increase it by about 49%.

Table 3. Effects of oatstraw, black plastic, and plane treeleavesmulches on weed and fababeanbiomass at the flowering and maturity stage

	Treatemts Stage growth	weedy check	Oatstraw	Black plastic	Plane treeleaves
Weedsbiomass (g/m^2)	Flowering stage	453.89a	65.81b	58.33b	60.54b
	Maturity stage	586.40a	142.18b	78.72b	99.93b
Fababeanbiomass (g/m^2)	Flowering stage	75.57a	160,92b	181.23b	150.54ab
	Maturity stage	94.72a	171.99ab	213.87b	183.75b

In the row, the mean values followed by the samelowercaseletters are not significantlydifferent.

Concerning weeds, their biomass varies very significantly compared to the control, and lower extent between the different types of mulch; the maximum value of weed biomass is recorded in the second survey of the control (without mulch) with 586.45 g/m^2. However, the lowest weed biomass is observed in mulch treatments, especially plastic-based mulch, at the flowering stage (58.33 g/m^2). Similarly, a study showed that the use of plastic mulch (black or white), wheat or rapeseed necessarily reduced weed biomass compared to the control, moreover, there was no significant difference between the treatments used [21].

2.4 Weed Control efficiency (WCE)

Weed control effectiveness against weeds through the use of mulch ranges from 75.7% to 87.1% (Table 4); the maximum value is obtained with plastic mulch, in the first survey (87.1%), followed by the application of sycamore leaves (86.7%), in the same period. In general, this control efficiency is greater during the flowering stage than during crop maturity. On average, mulching with black plastic mulch had the highest value (86.8%), followed by plane tree leaf mulch (84.8%) and oat straw (80.6%), respectively. In the same way, Kouelo *et al.* [22] showed that mulches provide 80% of weed control efficiency.

Azadbakht *et al.* [21] found that wheat and rapeseed mulch treatments can reduce weed biomass compared to controls at rates of 83% and 81%, respectively. Similarly, Mehmood *et al.* [23] reported similar results that application of black plastic mulch showed a maximum reduction in weed density, fresh and dry weed biomass and suggested that organic mulches such as dry eucalyptus leaves and wheat or rice straw could be used for effective weed control in cornfields.

Table 4. Weed Control efficiency (WCE) of mulches against weeds in faba bean field

	Weedy check	Oatstraw	Black plastic	Sycamoreleaves
Flower stage (%)	-	85.5	87.1	86.7
Maturity stage (%)	-	75.7	86.6	83
Average (%)	-	80.6	86.8	84.8

2.5Yield and yield components of faba bean

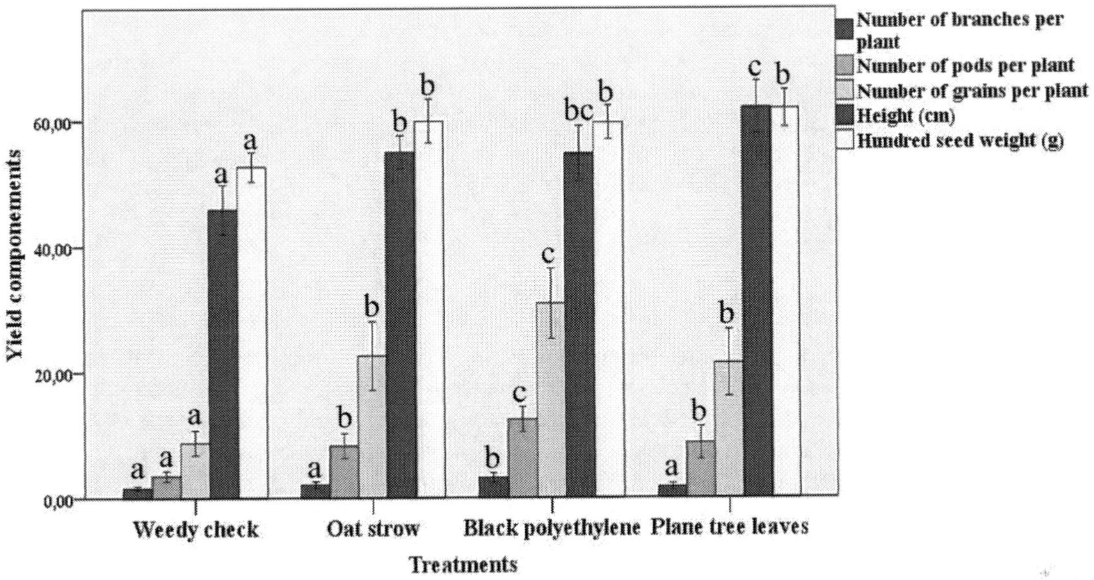

Fig.1. Effect of oatstraw, blackpolyethylene, and plane treeleavesmulches on fababeanyield components

The influence of mulch tested on the yield components of faba bean is illustrated in Figure 1. Average yields are significantly higher than the control. Indeed, the best yield is obtained with plastic mulch. This yield gain is more marked on the number of seeds, pods and branches, while for plant height the plane tree leaves mulch that ensures better faba bean growth (Figure 1). Similarly, the hundred seed weight is also increased compared to weedy check, but remains the same between the different types of mulch. Therefore, an improvement in yield per unit area of approximately 42, 52.9 and 43% is achieved with oat straw, black plastic, and plane tree leaves, respectively (Figure 2).

Tukey's test indicates that there is no significant difference between the values of the yield components obtained by oat straw and plane tree leaf mulches; indeed, the organic mulches provided intermediate yields between the control and black plastic, reflecting the mulches ability to control weeds and prevent their growth and propagation [24].

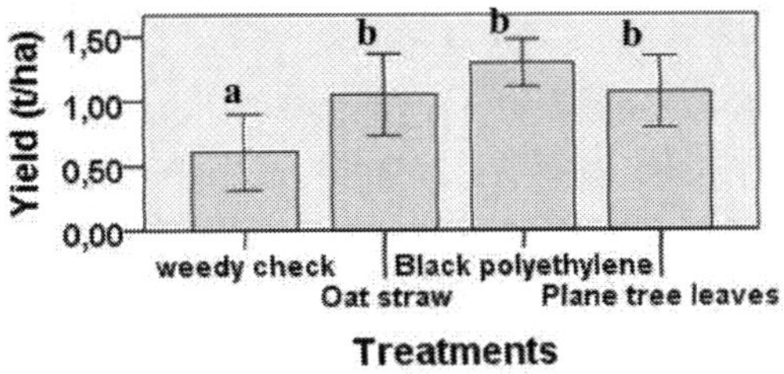

Fig. 2. Fababeanyield (tons per hectare)

Another study showed that mulching significantly improved grain yield, as well as crown diameter, leaf area index, and growth rate of maize ($p<0.0001$) [22].Mesfine*et al.* [24] also obtained an increase in sorghum grain yield.

Mulching can affect yield in many ways, including increasing soil moisture and water use efficiency in arid and semi-arid conditions [24-25], as well as increasing nitrogen, carbon and phosphorus levels [22].

For their part, Duppong *et al.*[26] reported that the height of a medicinal plant was significantly higher when oat straw was applied than in other treatments (flax and wool straw) and even higher than the positive control (manual weeding).

It should also be noted that plant residues not only affect the soil, but can also affect the germination, survival, growth, and competitiveness of weeds and crops. However, the use of mulch can be expensive due to labor costs and other costs associated with mulch preparation. Still, regardless the costs, it can be noted that environmentally friendly methods such as the application of plant residues allows long-term environmental durability[21].

Conclusion

The negative impacts of herbicides on humans, animals, and environment has attracted the attention of researchers to find new environmentally-friendly methods for weed control. From the data presented, we conclude that the mulch practice could be one of the possible strategies to reduce the use of herbicides in the faba bean crop, mainly by reducing density and biomass of weed species. Indeed, the weed control efficiency of straw mulch, black plastic, and plane tree leaves has been around 80.6, 86.8, 84.8% respectively, which allow an improvement in faba bean yield components (height, number of grains, pods, and branches…), therefore they allowed an increase in faba bean yield (tons/ha) by 42, 52.9 and 43% compared to the weedy check.

References

1. H. Labuda, *Flowering and characteristics of useful traits of some faba bean (Viciafaba L. var.major Harz) cultivars and breeding lines*,ActaAgrobot, **65 (4)** (2012)
2. S. G. Kumari& K. M. Makkouk,*Virus diseases of faba bean (Viciafaba L.) in Asia and Africa*, Plant Viruses, **1** (2007)
3. P. Ruisi, D. Giambalvo, G. Di Miceli, A.S. Frenda, S. Saia &G. Amato, *Tillage effects on yield and nitrogen fixation of legumes in Mediterranean conditions*,Agron.**104**, 1459-1466 (2012)
4. B. Hajjaj, R.Mrabet, M. Bouhache, A. Taleb, &A. DOUAIK, Rev. Mar. Sci. Agron. **4**(2016)
5. M. Fouad, N. Mohammed, H. Aladdin, A. Ahmed, Z. Xuxiao, B.Shiying, &Y. Tao, Elsevier, 113-136 (2013)
6. M. Bouhache, &S. Benmansour,AMPP, **6**, 51-62 (2014)
7. M.G. Dawood,*Weed management, folic acid and seaweed extract effects on Faba bean plants and associated weeds under sandy soil conditions*,Int. J. Agric. Eng.,**19**(2018)
8. E. Pannacci, F.Tei, & M. Guiducci,*Mechanical weed control in organic winter wheat*,Ital. J. Agron.,**12**(2017)
9. J. Pousset,Ed.Agridécisions,ISBN291219914X, Paris, 704 (2003)
10. B. Melander, I.A. Rasmussen, & P. Bàrberi,*Integrating physical and cultural methods of weed control-examples from European research*,Weed Sci., **53**, 369-381 (2005)
11. A. Matković, D. Božić, V. Filipović, D.Radanović, S. Vrbničanin, &T. Marković*Mulching as a physical weed control method applicable in medicinal plants cultivations*,Lekovitesirovine, **35** (2015)

12. A. M. El Mahi,*Chemical Weed Control InFaba Bean (ViciaFaba L.) Using Two Foliar Applied Herbicides*,Master thesis, University of Khartoum, 117 (2015)
13. M.A. Kader, M. Senge, Mojid M.A., & K. Ito,*Recent advances in mulching materials and methods for modifying soil environment*, Soil Tillage Res,**168** (2017)
14. P. Ranjan, G.T. Patle, M.Prem, &K.R. Solanke,Curr. Agric. Res. J., **5**(2017)
15. S.G. Telkar, A.K. Singh, K. Kant, S.P.S. Solanki & D. Kumar, Biomolecule Reports Ed. 4 (2017)
16. R. Benniou, K. Zabi, …& H.Belguet, Revue Agriculture Ed.,**1**, 230-239 (2016)
17. D. C. Chang, J.H. Cho, Y.I. Jin, J.S. Im, C.G. Cheon, S.J. Kim, & H.S. Yu, *Mulch and planting depth influence potato canopy development, underground morphology, and tuber yield*,Field Crops Res., **197** (2016)
18. D. Jodaugienė, R. Pupalienė, M. Urbonienė, V. Pranckietis,& I. Pranckietienė, *The impact of different types of organic mulches on weed emergence*,Agron. Res., **4**(2006)
19. T. F. Döring, M. Brandt, J. Heß, M. R. Finckh, & H. Saucke,*Effects of straw mulch on soil nitrate dynamics, weeds, yield and soil erosion in organically grown potatoes*, Field Crops Res., **94** (2005)
20. L. F. P. Massucati&U. Kopke,*Effect of straw mulch residues of previous crop oats on the weed population in direct seeded faba bean in Organic Farming*,inProceedings 26th German Conference on Weed Biology and Weed Control,Braunschweig, Germany, 11-13 March, 2014(2014)
21. A. Azadbakht, M. T. Alebrahim, & A. Ghavidel,*The effect of chemical and non-chemical control methods on weeds in potato (Solanumtuberosum L.) cultivation in ardabil province, Iran*,APPL ECOL ENV RES Ed.,**15**(2017)
22. A. F. Kouelo, P. Houngnandan, A. Azontonde, M. Benmansour, J. Bekou, & T.AkploAgron. Afr., **29**, 65-78 (2017)
23. T. Mehmood, S. Khan, A. Qayyum, A.R. Gurmani, W. Ahmed, M. Liaquat& A.Farid*Evaluation of Organic and Inorganic Mulching as an Integrated Weed Management Strategy in Maize Under Rainfed Conditions*, Planta Daninha, **36**, 13 (2018)
24. T. Mesfine, G. Abebe, & A. R.M. Al-Tawaha, *Effect of reduced tillage and crop residue ground cover on yield and water use efficiency of sorghum (Sorghum bicolor (L.) Moench) under semiarid conditions of Ethiopia*,World J. Agric. Res., **1**(2005)
25. S. Chandra, R. D. Singh, V. K. Bhatnagar& J. K. Bisht, *Effect of mulch and irrigation on tuber size, canopy temperature, water use and yield of potato (Solonumtuberosum)*, indian j Agron, **47**, 443-448 (2002)
26. L. M. Duppong, K. Delate, M.Liebman, R.Horton, F. Romero, G. Kraus & P. K. Chowdbury,CROP SCI, **44**(2004)

Assessment of non-destructive spectroscopy and chemometrics tools for the development of green analytical methods to determine the shelf-life of olive oils

Aimen El Orche[1,*], *Mohamed* Mbarki[1], *Amine* Cheikh[3], and *Mustapha* Bouatia[2].

[1]Laboratory of Chemical Processes and Applied Materials University of Sultan MoulaySlimane, Beni-Mellal, Morocco.

[2]Laboratory of Analytical Chemistry, Faculty of Medicine and Pharmacy, Mohammed V University, Rabat, Morocco

[3]Faculty of Medicine, Abulcasis University, Rabat, Morocco.

Abstract. The development of sustainable and environmentally friendly analytical methods for agri-food products and the modification of reference methods is an essential issue to be treated in green analytical chemistry. The potential application of non-destructive spectroscopic techniques with chemometrics tools to achieve these principles are examined in this work. In this study a new sustainable analytical approach based on the use of fluorescence spectroscopy and multivariate analysis methods of Machine-Learning(Support Vector Machine regression) and chemometrics (Partial Least Square regression) have been developed to control the quality of virgin olive oils in Morocco according to their shelf life. The spectral data of 45 samples were first analyzed by principal component analysis method (PCA), the PCA method shows an important classification of the three groups of olive oil according to their shelf life. The use of the regression methods SVM and PLS shows a high ability to predict the quality of olive oils, this ability is shown by the high value of R-square and the low value of root mean square error of calibration and cross-validation (RMSEC, RMSECV), the validation of these models by cross-validation shows the potential of this sustainable analytical approach in the determination of the quality of virgin olive oils.

1 Introduction

Olive oil is an essential vegetable oil in the Mediterranean countries; nowadays, this nutrient is attracting the interest of many consumers around the world, thanks to its significant nutritional and organoleptic properties and its contribution to the protection of human well-being [1]. These qualities are particularly related to its composition rich in fatty acids, in particular oleic and linoleic acid [2], and its high content of minor compounds with bioactive characteristics, mainly phenolic compounds, tocopherol and chlorophyll. These compounds are also considerably reduced during the storage of olive oil, although new products are created due to the oxidation process. In numerous markets, the storage of olive oil may range from 6 to 24 months, resulting in a deterioration of the quality of olive oil. [1].

Determining the shelf life of olive oil constitutes an important challenge to protect consumers against fraud because many sellers expose in the market oil stored for a long time as much as freshly produced oil. Currently, different methods are used to assess the oxidative deterioration of olive oil. Routine methods include the peroxide value (PV), which determines the quantity of primary oxidation products, and the acid value, which determines the acidity of olive oil and the extinction coefficients at 232 and 270 nm, which measure the formation of primary and secondary oxidation products [2]. Other analytical method based on the use of high performance liquid chromatography HPLC and gas chromatography GC which consists in determining the molecular composition of oils [3,4],

Nevertheless, these techniques are complex to implement and interpret, costly and time-consuming, and use

1 Corresponding author: aimen.elorche@gmail.com

expensive reagents of both organic and inorganic nature, since these reagents have an impact on the environment. For this reason it is necessary to develop innovative and sustainable analytical tools capable of carrying out rapid and reliable quality controls of virgin olive oil in order to determine the shelf life of olive oil and protect consumers against fraud.

The combination of chemometric and spectroscopic techniques to develop methods according to the concept of green chemistry has become an important challenge over the last decades. A wide variety of chemometric methods have been used so far in the field of analytical chemistry and studying food products [5-7], These techniques are characterized by the rapidity and do not involve the use of reagents which leads to the destruction of the ecosystem.

The objective of this report is to show the importance of fluorescence spectroscopy coupled with recognition algorithms in the development of rapid and sustainable analytical methods for the control of virgin olive oil according to its shelf life.

2 Materials and methods

2.1. Sampling

A total of 45 samples of freshly produced virgin olive oil and two groups of expired olive oil stored for a period of time fluctuate between 12 and 24 months. These samples were generally stored in a dark place with a temperature interval of 10°C +1.

The analysis of the samples was performed using fluorescence spectroscopy, the excitation wavelength was fixed at 400 nm and the observation interval between 415 nm and 785 nm.

2.2. Multivariate data analysis

Firstly the spectral data are analyzed by principal component analysis (PCA) in order to reduce the dimensionality and to exploit the dataset [8].

To develop training models able to predict the quality of olive oil the partial least square PLS method and support vector machine SVM have been applied to the spectral data set.

Partial least squares regression, or PLS regression, is a fast, efficient and optimal method for a well-controlled covariance minimization criterion. Its use is recommended when a large number of explanatory variables are used, or when there are strong collinearities between variables [9].

The SVM method is one of the supervised learning algorithms. This method solves the problem of pattern recognition. It aims to find an optimal separator that optimizes the margin between two classes of data, using a limited set of training sequences [10].

3 Results and discussion

3.1. Principal component analysis

A preliminary examination of the spectra was carried out by the PCA applied to the fluorescence spectra. Figure 1 shows a score plot of the first two principal components. From the score plot PC1-PC2 it can be seen that the first component accounts for all of the total variability in the data set (96% of the total variance) and the second principal component accounts for 2%. These two components show a classification of olive oils according to their freshness.

https://doi.org/10.1051/e3sconf/202018304003

The results obtained by the exploratory analysis method PCA show that the spectral fluorescence data provide information on the molecular properties of the different olive oils, allowing their classification according to their shelf life.

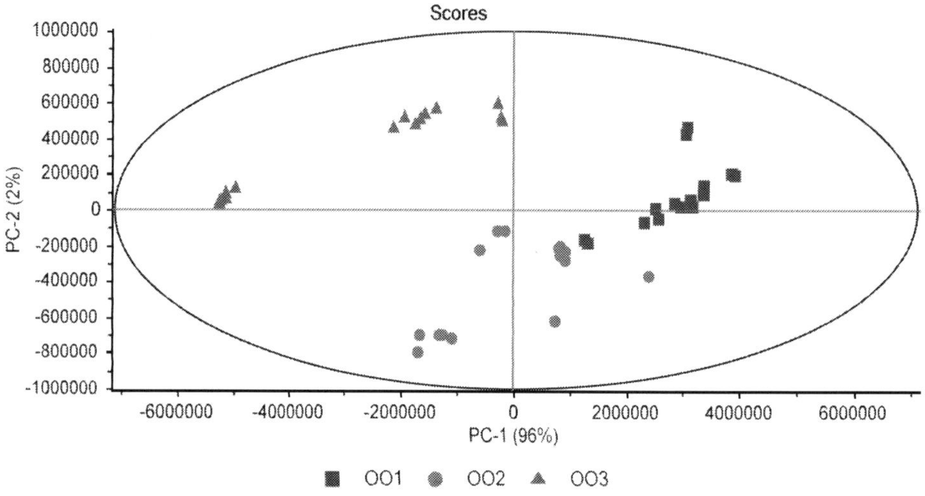

Fig.1: Score plot of the first two PCs obtained by PCA applied to the fluorescence data of olive oils(OO1: olive oil fresh, OO2: olive oil stored 1 year, OO3: olive oil stored 2 years).

For the development of sustainable analytical methods capable of determining the shelf life of olive oils during the storage process, it is necessary to choose statistical methods of multivariate analysis which are part of the regression type learning methods, these methods will be applied directly on the spectral data obtained by fluorescence spectroscopy.

3.2. Partial Least Square regression

In order to develop models capable of predicting the quality of olive oils (shelf-life), the PLS regression has been applied to the spectral data.

The application of PLS regression shows a very high capacity, this capacity was represented by the value of R-square that reach 99% and the low value of RMSEC that represent0.08 using the first four latent variables.

To assess the predictive capacity of the model, the cross-validation method was applied, the results found by the cross-validation are considered performant, this performance was demonstrated by the value of R-square and low value of RMSECV which represents successively 91% and 0.23 as shown in the figure 2.

The application of the PLS method on the spectral data of different olive oil categories shows important and reliable results (figure 2), this reliability is explained by the high value of R-square coefficients which reaches 99% and a low error of 0.08 (RMSEC), this high value shows that there is a strong dependency between the reference value and the value predicted by the model. The predictive evaluation of the model developed by the PLS algorithm using the cross-validation shows that the developed model is able to predict the shelf life of the stored olive oil with high reliability explained by the R-square value and the error value which represent respectively 91% and 0.23%.

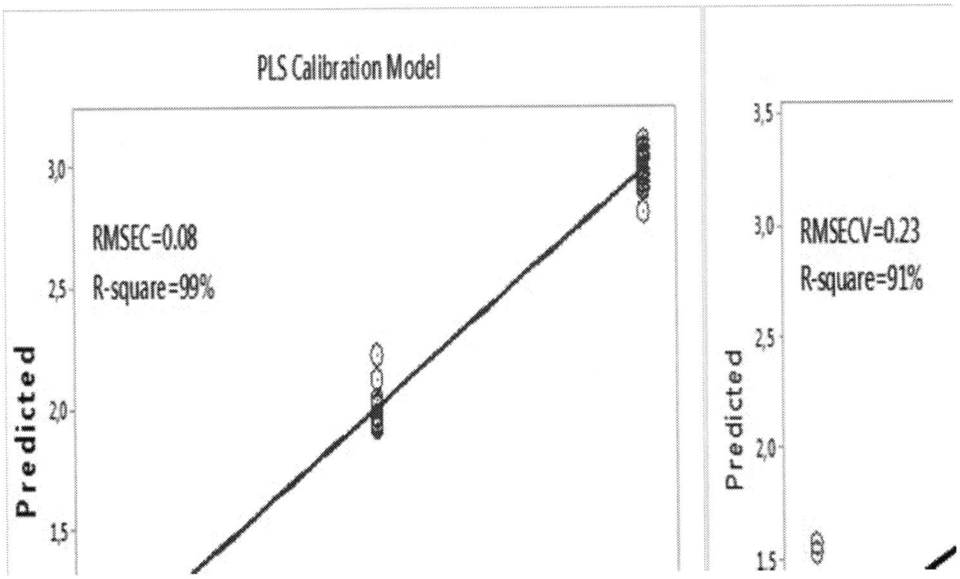

Fig.2: PLS regression results of calibration and cross-validation using 4 latent variables (LV).

3.3. Support Vector Machine regression

The analysis of the spectral results by the SVM regression also shows a very high capacity represented by the high value of R-square 99.99% and the low value of RMSEC 0.0005 as shown in the figure 3.

on the other hand, the application of the SVM regression method on the spectral data of the three categories of olive oil also shows an important capacity of the constructed model, this capacity is represented by the strong linearity between the reference value and the value predicted by the model expressed by the R-square value and the RMSEC error which represent respectively 99% and 0.0005 as shown in the figure.

the application of cross-validation for the evaluation of the constructed SVM model shows the high capability of this model in the determination of the shelf life of olive oils, this capability is shown by the value of R-square and the error which represent 97% and 0.19 respectively.

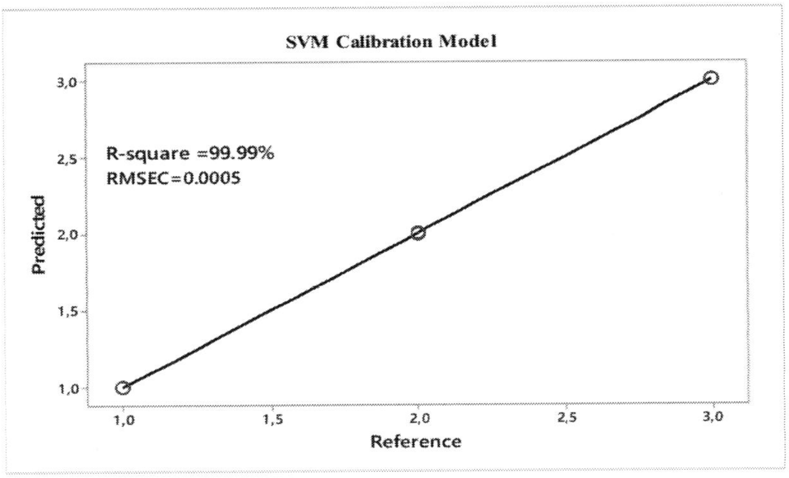

Fig. 3. SVM regression calibration model using linear kernel type algorithm.

According to the results found, the two multivariate statistical methods PLS and SVM give efficient results, these results show that the application of the SVM regression has a higher R-square and lower RMSE compared to the results found by the PLS regression.

4 Conclusion

The present work shows the capability of fluorescence spectroscopy combined with PLS and SVM multivariate regression analysis in the determination of the quality of olive oils according to their storage life.

The advantage of this technique is that it doesn't use environmentally harmful reagents and it is characterized by the rapidity of its use.

Last but not leastthe development of new environment-friendly analytical techniques and the modification of pre-existing methods are key points in advancing the concept of green analytical chemistry.

Reference

1. K. Bouarroudj, A. Tamendjari, R. Larbat, Ind Crops Prod, **8** (2016).
2. E.G. Shendi,D.S. Ozay, M.T. Ozkaya, N.F.Ustunel, OCL - Oilseeds and fats, Crops and Lipids, **6** (2020).
3. M. Farrés-Cebrián, R.Seró, J. Saurina, O.Núñez, J Separations, **13** (2016).
4. S. Vichi, A.I. Castellote,L. Pizzale, L.S. Conte, S. Buxaderas, E. López-Tamames, J Chromatogr A, **15**(2003).
5. N. Sinelli, M.S. Cosio, C. Gigliotti,E. Casiraghi, Anal ChimActa, **7**(2007).
6. A. Gredilla, S.Fdez-Ortiz de Vallejuelo, N. A. Elejoste, de Diego, J.M. Madariaga,TrAC Trends Anal Chem, **39** (2016)
7. A. El Orche, M. Bouatia, S. Yanisse, H. Labjar, M. Mouhsin, M.Bouha,M. Oubenali, M. Mbarki, J Spectrosc,**9** (2020)
8. A. El Orche, M. Bouatia, M. Mbarki, JAMC, **9** (2020)
9. W. Svante, S. Michael, E. Lennart, J Chemometrics, **22** (2001).
10. A. El Orche, M. Bouatia, H. Labjar, M. Maaouni, M. Mbarki, IEEE 6th International Conference on Optimization and Applications, BeniMellal, Morocco, **4** (2020).

Chemical characterization of almond meal as a co-product of the mechanical extraction of almond oil.

Nadia Houmy[1,5*], *Reda* Melhaoui[1], *Kamal* Belhaj[1], *Aurore* Richel[2], *Marianne* Sindic[3], *Christophe* Hano[4], *Souhayla* Kodad[1], *Aatika* Mihamou[1],*Mohamed* Addi[1], *Malika* Abid[1] and *Ahmed* Elamrani[1].

[1] Laboratory for Agricultural Productions Improvement, Biotechnology and Environment, Faculty of Sciences, University of Mohammed Premier, Oujda, Morocco
[2] Laboratory of Biological and Industrial Chemistry, University of Liège, Gembloux, Belgium
[3] Laboratory QSPA, Gembloux Agro Bio Tech, University of Liège, Gembloux, Belgium
[4]Laboratoire de Biologie des Ligneux et des Grandes Cultures, OrleansUniversity, France
[5] Laboratory of Food Technology and Quality, CRRA, INRA, Oujda, Morocco.

Abstract. Almonds harvesting, along with precleaning, hulling, shelling, and sorting, generate recoverable co-products such as double, malformed, or damaged almonds. The common way that generates more benefit is the extraction of almond oil for food and cosmetic uses. The aim of this study is to compare composition and nutritional value of almond meals of the main varieties from eastern Morocco, *Marcona* (M), *Fournat de Breznaud* (FNB), *Ferragnes*and*Ferraduel* (FF) and *Beldi* a local ecotype (B), during three consecutive crop years (2016-2018). Significant differences were observed, which mainly concern the residual oil in almond cakes and especially their fibres, sugars, and proteins. Differences depend on the quality of the raw material and the effects of the crop year on the variety. Besides, from a nutritional point of view, analysed almond meals still contain significant amounts of residual oil and the major constituents of almond seeds (sugars, fibres, and proteins). The amino acid profile presents eight essential amino acids (Cysteine, Methionine, Isoleucine, Leucine, Phenylalanine, Tyrosine, and Lysine Valine) and the Protein Digestibility Corrected Amino Acid Score (PDCAAS) ranges between 24.78% for (M) and 62.17% for (FF). We conclude the analysed almond meals present interesting nutritional values as a specific ingredient for gluten-free culinary preparations and light foodstuffs.

1 Introduction

Almonds are the edible seeds of the *Prunus amygdalus Dulcis* tree, more commonly known as the almond tree, which is native from its wild ancestors of the central plateau of Asia, currently domesticated, and widely cultivated in Mediterranean climate areas [1]. Although

*Corresponding author:houmy.nadia@gmail.com

almonds are commonly referred to as nuts, they are actually the seeds found at the centre of the almond fruit. Almond fruit consists of four parts: kernel or Seed or "meat", seed coat (a thin leathery layer known as brown skin of the seed), almond shell, and almond hull which corresponds to the outer green shell cover [2].

The kernel or seed is the edible part of the nut, with a high nutritional value. Almond kernels can be consumed raw or cooked, blanched, ground into flour and even made into non-dairy milk. It can be also incorporated, as an ingredient, into culinary recipes. They are very rich in fat, making them a perfect source of edible oil and oil for cosmetic [3].

In eastern Morocco, thanks to the Moroccan green plan and the PROFAO (Solidarity farming Project of Almond Sector in Eastern Morocco, 2011-2018); 6000 hectares of new almond orchards were created. PROFAO, as an integrated project, allows the organisation of almond's farmer cooperatives and facilitates the creation of economic interest grouping for almond post-harvest facilities, processing, production and marketing of almonds and derivatives [4].

Almond nuts crushing, shelling and sorting operations generate co-products like doubles malformed almonds or mechanically damaged (split and broken kernels) and defects seeds. The best way for exploiting this co-product and deriving more benefits and add value is the mechanical extraction of almond oil for food and cosmetic uses. This process of almond oil extraction also generates another by-product (Almond residue or almond meal) which represents at least 50% by gross mass of the raw material of damaged almonds. Thus, the aim of this study is to assess the nutritional value of these almond residues by analysing the chemical composition of almond meals from fourintroduced varieties, *Marcona* (M), *Fournat de Breznaud*(FNB), *Ferragnes*and*Ferraduel*(FF) and *Beldi* a local ecotype (B), during three consecutive crop years (CY1: 2016; CY2: 2017; CY3: 2018).

2 Material and methods

2.1 Plant material and sampling

In eastern Morocco, producers and processors of almonds in the Sidi Bouhria region are organised into an economic interest grouping (GIE- Sidi Bouhria). In order to derive maximum benefit from their almond production with the perspective of zero waste and using everything the almonds orchard grows. Thus, the almond kernels grading and sorting co-product, corresponding to split & broken malformed and double almonds, is intended for the extraction of almond oil for food and cosmetics. In addition to almond oil, the mechanical process produces a residue "the almond cake /"meal". This study concerns the chemical characterisation of the almond meals of the five most cultivated varieties in the region of eastern Morocco. The almonds come from five-year-old almond trees, installed on a limestone soil located in Sidi Bouhria, in the east of Morocco (GPS coordinates: 34° 44′13.6″N, 002° 20′15.0″ W). The orchards are rain-fed with a minimum of additional irrigation under semi-arid climatic conditions (Table 1). The GIE-Sidi-Bouhria provided the almond meals used in this study.

Table 1. The average temperature and annual cumulated precipitation registered at the closest meteorological station to Sidi Bouhria region (GPS coordinates: 34° 44′13.6″N, 002° 20′15.0″ W) during the three crop years (CY) 2016-2018

Sidi Bouhria crop year	Average temperature (°C)		Annual cumulated precipitation(mm)
	Minimum	Maximum	
CY1 (2016)	11.5	24.5	289.6
CY2 (2017)	12.5	25.6	231.5
CY3 (2018)	12	26.0	205.4

2.2 Mechanical extraction of almond oil

The mechanical extraction of almond oil was carried out by a mechanical press system (IBG MonfortsOekotec GmbH & Co.KG, made in Germany). The screw speed was 60 RPM, and the fine particles were removed from the almond oil by filtration followed by centrifugation at 3000 RPM for 15min. The GIE-Sidi Bouhria provided almond meal samples for the Physico-chemical analyses. Samples were crushed and sieved through a 1 mm mesh sieve, then stored at 4°C until use. Analyses were made in triplicate for the almond meal of each variety.

2.3 Physico-chemical analyses

2.3.1 Dry matter

Dry matterwas carried out in accordance with the thermogravimetric method [5]. It consists to dry 5 to 20 g of the homogenized sample at 103° ± 2° C until obtaining a constant weight.

2.3.2 Residual oil

Residual oil in almond meal of deferent varieties was performed by chemical extraction using a Soxhlet apparatus, where 200 ml of hexane was used for a total oil extraction of 5g of powdered almond meal then the solvent was evaporated under vacuum using a rotary evaporator and the oil was collected [6].

2.3.3 Total sugars

Total sugars contentsis determined according to the Bertrand method with little modifications in four main steps:The first step performed by acid hydrolysis and neutralization of the entire hydrolysate to transform all the polysaccharides into simple reducing sugars. The second step is a sample deproteinisation. The third step consistinto precipitation and filtrationof sugars extract. Finally, titration with potassium permanganate was carried out for determination of total sugar contents in analysed samples of almond meal. The result is expressed as a percentage of mass of glucose in 100 g of sample [7].

2.3.4 Dietary fibres

The determination of dietary fibreswas carried out according to the AOAC 991 method [8]. Three successive enzymatic attacks were carried out (α-amylase, protease and α - amyloglucosidase) to eliminate starch and proteins. The fibresare precipitated by technical ethanol. The obtained value was corrected by determining the residual ash (Muffle furnace), proteins (Kjeldahl) and by the blank.

2.3.5 Ash

Ash is determined according to the AACC 2001method [9]. The aim is to determine the mineral content of the sample by heating and so destroying the organic material. Thus, 5 g of the already dried sample was introduced in a pre-tared crucible. Then the sample was introduced into a Muffle furnace at 550 °C overnight, then allowed to cool in a desiccator and finally the weight difference was measured.

2.3.6 Proteins

Total nitrogen content was estimated by the Kjeldahl method [10], then protein content was calculated by multiplying the total nitrogen content by a conversion factor of 5.18 [11].

2.3.7 Amino acid profiles

The determination of amino acids by HPLC is carried out according to the method of Spackman *et al.,* [12] by acid hydrolysis of almond meal proteins. This analysis depends on the protein content in sample (500 mg for a sample containing 2% of nitrogen). In our case, 130 mg of each almond meal sample was taken and dissolved in 10 ml of 6 N-HCl containing 0.1% phenol. The samples were mixed and sealed in the presence of nitrogen, and then were heated in oven at 110° C for 24 hours. After cooling in ice, a solution of 30 ml of a citrate buffer at pH 2.2 was added, then a pH adjustment was carried out with 7.5 N and 1 N of NaOH respectively in order to obtain a pH of 2.2. The solutions thus obtained were diluted with citrate buffer (pH = 2.2), after adding 1 ml of Norleucine at 50 µM / ml as internal standard. The solutions were partially filtered by 0.2 µm filter. A 20-µl of aliquot was injected into the HPLC using a sodium oxidase column, cation exchange resin, a column post for the derivation of amino acids into ninhydrin and a UV detector at 570 nm except for Proline at one length 440 nm wave.

2.3.8 Protein quality calculations

The protein quality evaluation was studied by calculation of sum essential amino acid (ΣEAA), sum of non-essential amino acid (ΣNEAA) (indicate the importance of the nutritional value of proteins), Chemical Index (CI) and Proteins Digestibility of Corrected Amino Acid Score (PDCAAS). This later has been adopted as the chosen method for the measurement of the protein value in human nutrition [13].
The Chemical Index (CI) was calculated on the basis of the procedure described previously by Sujak et al (2006),based on following calculation[$(\%\text{ limiting amino acid})_{sample}/(\%\text{limiting amino acid})_{protein\ of\ reference})]x100$ [14]. In our case, limiting amino acidswere Met + Cys and the protein referenceused of amino acids forhuman food protein composition appropriate for an adult human was FAO/WHO/UNU requirements(2007) (Table 2) [13].

Table 2. Protein of reference from FAO/WHO/UNU

Essential amino acids	% AA
Cysteine + Methionine	1.7
Histidine	1.6
Isoleucine	1.3
Leucine	1.9
Lysine	1.6
Phenylalanine + tyrosine	1.9
Threonine	0.9
Tryptophan	0.5
Valine	1.3
Sum of EAA	12.7

AA: amino acid (g of 100 g of protein)

Protein digestibility Corrected Amino Acid Score (PDCAAS) was determined according to Schaafsma [15], where it is CI × true faecal digestibility of almond meal. The true faecal digestibility in our study was used from Ahrens et al (2005) results [16].

2.4 Statistical analysis

The chemical characterization of almond meal obtained from four varieties and one local ecotype, during three consecutive crop years, was carried out on triplicate for each sample and each parameter. Statistical analyses were conducted using Statistical Package for the Social Sciences (SPSS for Windows, version 21, SPSS Inc., Chicago, IL, USA). The normal distribution was verified according to Shapiro Wilk test. Two-way ANOVA statistical analysis and Tuckey's post-hoc test was used for means comparison, the difference was considered significant (at $P<0.05$).

3 Results and discussion

Almond processing coproducts might represent a precious source of raw materials for food, cosmetic, and pharmaceutical industries, however, their valorisation needs a fine physicochemical characterization for the development of novel procedures to be implemented to the sectorial industries. Indeed, providing innovative products from Almond pressed-oil residues (communally named almond oilcake or almond meal) could provide benefit to the overall almond production. Table 3 shows chemical composition of almond meals, corresponding to almond pressed-oil residue, of five most cultivated varieties in the regions of eastern Morocco [*Marcona* (M), *Fournat de Breznaud* (FNB), *Ferragnes and Ferraduel* (FF) and the *Beldi* (B)]. These results (table 3) concern three consecutive crop years and six analysed parameters of Almond meals: dry matter, residual oil, sugars, fibres,ash andproteins contents.In general, the chemical composition of almond oil meal change between varieties and crop years mainly for dry matter, sugars and ash.

3.1 Dry matter

Dry matter data registered for almond meals (Table 3), range between a minimum of 89.29% for FF-CY1 and a maximum of 94.56 % for M-CY2. Indeed, dry matter values for analysed almond meals of each varieties are slightly different; these differences seem to be mainly related to the harvest year, the variety, and probably to the technique of post-harvest drying. The ANOVA analysis (Table 3) confirms this observation and shows a significant difference at $P<0.05$. In general, the results are similar to the literature, for instance, Arbouche R (2013) found 94.6% dry matter in the almond meal [17], which is very similar to our findings for M-CY2 (94.56 %). These results are comparable to those of several other types of meals (oil cakes of soybean, groundnut, coconut, cottonseed, sunflower, sesame), whose dry matter contents vary between 85% to 94% [18].

3.2 Residual oil

Oil cakes/meals are solid residues, obtained after the extraction of oil from the plant part such as seeds or nuts. In general, residual oil in meals depends on quality of the raw material and on the processing of oil extraction. In spite of the fact that the process used for extraction of almond oils is the same for all varieties, the quantities of residual oil extracted from different almond meals analysed (Table 3) vary widely between a minimum of 7.45% (M -CY1) to a maximum of 23.58% (M-CY2). F-F and FNB present the highest residual oil content comparing to (M) and (B). The ANOVA analysis (Table 4) confirms this observation and shows a significant difference at $P<0.001$. Since the press system is the same, the difference should be related to the raw material, particularly to the physical state of the kernels, moisture and kernels defects but also to the almonds variety [19]. In this regard, Rabadan et al (2017) reported comparable results that range from 15.24% and 25.5 % of residual oil in almond meals[20] and his concluded that residual oil of meals depends on efficiency of mechanical oil extraction. Besides, Savoire et al. (2013) reported that residual oil yield was influenced by many parameters such as kernels variety; press system parameters and pre-treatments preceded this process [21].

3.3 Sugars

Whatever the crop year,almond oil cake of the variety *Marcona* (M) had the lowest sugar content (Table 3)respectively for the three studied years (12.60%; 12.57%; 15.02%), while the variety *Fournat*(FNB) seems to present the highest sugar content (13.84 /21.65 % / 19.21%). The sugar contents of almond oil meals from the three other varieties are between these limit values. It is certainly here genotype-related differences, which are confirmed by ANOVA analysis (Table 4) that shows significant differences between genotypes at $P<0.05$.

The almond oil cake, although considered simply as a by-product of almond oil extraction, is an excellent source of nutrients such as proteins, vitamins, minerals, fibres and bioactive components that may offer a number of health benefits. In this regard recently, in a research collaboration, but also from the literature [22-25], we show that high phenolic content remains in this coproduct making it usable for improvement of nutritional and functional properties of certain food products. In addition, this coproduct contains significant amounts of other phytochemicals compounds including tocopherols, flavonoids, tannins, and can be used for cosmetics and pharmaceutical purposes.
In the present study, analyses focus on proteins content, amino acids profile ash and dietary fibres.

3.4 Dietary fibers and Ash

Dietary fibres can act by changing the nature of the contents of the gastrointestinal tract and by changing how other nutrients and chemicals are absorbed. Fibres and minerals bring nutritional benefits to consumers. The results of the analysed almond cakes for the dietary fibres and ash contents are presented in Table 3. They vary between a minimum of 31.52% /DM registered for *Marcona* (M-CY2) and a maximum of 42.3%/ DM for *Beldi* ecotype (B-CY2). The calculated average value of dietary fibres content in analysed almond meals is 35.5%/ DM for the three crop years of this study, this value is in accordance with the literature.

Analysis of ANOVA (Table 4) show no significant difference for the factor Year, whereas the factors Variety and Year * Variety interaction present a significant difference between the samples at $P<0.05$. Finally, the fibres' content of analysed almond meals is clearly higher than those given, by the literature, such as those of sunflower seeds (28.41%) and soybean (17.8%) [19].

Table 3.Summary of chemical compositions of almond mealfrom five most cultivated varieties in eastern Morocco. Results of analyses concern almond meal contents (%) for :dry matter, residual oil, sugars and fibres contents, ash, proteins content and amino acid profile for three consecutive crop years (CY: Crop year; M: *Marcona*: FNB: *Fournat de Breznaud*; FF: *Ferragnes – Ferraduel* ; B: *Beldi*)

Sample	Dry matter %	Residual oil %	Sugars %	Dietary fibres %	Ash %	Proteins %
M-CY1	90.44 ± 0.40ab	7.45 ± 0.20a	12.60 ± 0.64a	33.04 ± 2.43a	7.28 ± 0.00e	43.24 ± 5.38e
FNB-CY1	91.51 ± 0.19b	7.66 ±0.30a	13.84 ± 0.20a	32.26 ± 2.93a	7.14 ± 0.01e	39.58 ± 0.34de
F-F-CY1	89.29 ± 1.70a	8.52 ± 0.55ab	13.74 ± 0.28a	37.71 ± 3.36ab	8.03 ± 0.08fg	37.59 ± 0.34bcd
B-CY1	91.16 ± 0.25b	8.07 ±0.35ab	14.20 ± 0.99a	34.86 ± 2.56ab	7.89 ± 0.10f	38.79 ± 0.21cd
M-CY2	94.56 ± 1.19d	23.58 ± 1.22f	12.57 ± 0.71a	31.52 ± 5.04a	6.80 ± 0.06d	34.79 ± 2.80abc
FNB-CY2	93.37 ± 0.63d	12.17 ± 0.56cd	21.65 ± 0.32d	38.42 ± 3.97ab	8.15 ± 0.11g	31.15 ± 0.13a
F-F-CY2	94.06 ± 0.88d	22.48 ± 1.03e	14.48 ± 0.85ab	31.94 ± 2.93a	6.84 ± 0.03d	33.39 ± 0.01ab
B-CY2	93.23 ± 0.98cd	8.71 ± 0.65ab	18.85 ± 2.90cd	42.30 ± 1.58b	7.98 ± 0.03fg	38.78 ± 0.24cd
M-CY3	93.55 ± 6.44d	8.95 ± 0.84b	15.02 ± 1.30ab	35.45 ± 2.13ab	4.75 ± 0.01a	36.61 ± 1.05bcd
FNB-CY3	91.80 ± 8.22bc	12.81 ± 0.38e	19.21 ± 0.55cd	39.75 ± 0.51ab	7.25 ± 0.03e	36.54 ± 0.12bcd
F-F-CY3	90.74 ± 9.26ab	11.9 ± 0.11cd	15.29 ± 0.44ab	34.50 ± 0.56ab	5.88 ± 0.05b	40.18 ± 1.13de
B-CY3	91.85 ± 8.15bc	10.41 ± 0.09c	17.48 ± 0.97bc	35.05 ± 2.39ab	6.14 ± 0.12c	36.50 ± 0.26bcd

Ash contents of the analysed almond meals from the three crop years (Table 3) varied between a minimum of 4.75 % DM for (M-CY3) and 8.15 % DM for (FNB-CY2). Analysis of ANOVA (Table 4) shows that there is a significant difference between the analysed samples at $P<0.05$.However, the observed differences can not be related to almond genotype or crop year effect. Almond meal is rich in macro elements, namely, potassium, phosphorus and magnesium (data not shown) which make it an important source of these essential elementsRabadan et al. (2017) [20]. In this regard, Karaman et al (2015), recommend, whenever possible, to incorporate a low dose (1 to 2%) of almond flour in food products (for instance baked goods) to improve their nutritional quality and enrich their mineral content[24].

Table 4. Variance Analyses "ANOVA two way" of constituents of almond cake/ meal, from five most cultivated varieties in eastern Morocco (NS=Non-significant, +, ++ and +++ stand for significate difference at 0.05, 0.01 and 0.001 respectively

	Dry matter	Residual oil	Sugars	Dietary fibres	Ash	Proteins
Year	+++	+++	+++	NS	+++	+++
Variety	+++	++	+++	+	+++	++
Year*Variety	+++	+++	+++	++	+++	+++

3.5 Protein content and amino acid profiles of almond oil cake

The growing interest in almonds and derivatives is due to their nutritious ingredients such as lipids, proteins, carbohydrates, vitamins and minerals, as well as secondary metabolites. Most studies focus on almond oil, for food uses and especially in cosmetics, that is more profitable, however, few studies concern the protein fraction, which after almond oil extraction constitutes the major component of the almond cake / meal. According to US department of agriculture (USDA) database, the protein content ranges from 16.0 to 22.0% of kernel weight amongst different varieties. Therefore, theoretically after extraction of the oil, which constitutes about 50% of the weight of the kernel, protein contents must double in almond cake / meal. Indeed, protein content of the analysed almond meals from the three cropyears (Table 3) ranges between a minimum of 31.15% for (FNB-CY2) and 43.24% for (M-CY1). Similar values have been reported by Rabadan et al. (2017) [20] for almond meals of Ferragnes (40.50%) and Ferraduel (44.85%) varieties, but a high protein content (55.93%) for almond meal of *Marcona*, which is known for its low oil richness.

ANOVA test (Table 4) shows a significant difference between the analysed almond meal samples. The significant differences observed seem to be due to both the genetic factor (almond variety) and the environment (crop year) with significant effects respectively at $P<0.01$ and at $P<0.001$, whereas the factor "Crop Year * Variety" interaction present a highly significant effect at $P<0.001$.

Amino acid (AA)profiles ofanalysed almond cake / mealarepresented in Table 5. Quantitative and qualitative analysis of AA profiles allows the identification of 17AA. The observed AA profiles of almond mealsare dominated by glutamic acid (Glu), aspartic acid (Asp), and arginine (Arg). Respectively,Glu, Asp and Arg contents (% DM) range between a minimum –maximum values of 8.46 % for (FNB-CY1) - 13.20 % for (M-CY1), 4.25 % for (FF- CY1) - 5.70 % for (M- CY2) and 3.82 % for (FNB- CY2) -5.31% for (M-CY2) (Table4). However, cysteine (Cys: 0.01 - 0.22 % DM) and methionine (Met: 0.23-0.35% DM) have the lowest contents. Regarding the essential amino acids (EAA), leucine (Leu)is the most abundant, followed by valine (Val). Finally, from a qualitative point of view, registered compositions for AA profiles of analysed almond cakes are in agreement with the finding reported by Sociasi Company et al (2008) [3]. Moreover, from these results, we conclude that the genotype (almond variety) strongly influences the quality (AA profiles) and the protein richness of almond seeds and consequently those of almond's oil cakes.

Table 5. Amino acid profiles (Composition % of crude protein) of almond oil cake of five most cultivated varieties in eastern Morocco (M: *Marcona*: FNB: *Fournat de Breznaud*; FF: *Ferragnes-Ferraduel*, B: *Beldi*).Analyses for Almonds of three consecutive crop years (CY)

	CY1				CY2				CY3			
	M	FNB	FF	B	M	FNB	FF	B	M	FNB	FF	B
Asp	4.50	4.32	4.25	4.35	5.70	4.36	4.79	5.22	4.78	4.70	5.09	5.02
Thr	1.40	1.35	1.34	1.34	1.62	1.43	1.38	1.45	1.32	1.39	1.50	1.41
Ser	1.80	1.73	1.69	1.71	2.42	1.98	2.08	2.20	1.92	1.94	2.14	1.98
Glu	8.86	8.46	8.49	8.63	13.20	9.74	11.06	11.81	10.58	10.44	11.21	10.10
Pro	1.63	1.56	1.57	1.51	2.10	1.66	1.74	1.98	1.82	1.79	2.01	1.93
Gly	2.53	2.40	2.43	2.42	3.30	2.75	2.76	2.88	2.66	2.77	2.87	2.78
Ala	1.91	1.86	1.81	1.83	2.43	1.96	1.98	2.18	1.90	1.92	2.13	1.99
Cys	0.10	0.12	0.07	0.07	0.03	-	-	0.01	0.07	0.06	0.22	0.02
Val	2.48	2.38	2.35	2.39	2.07	1.73	1.75	1.93	2.10	2.18	2.42	2.30
Met	0.23	0.30	0.32	0.24	0.23	0.26	0.24	0.23	0.27	0.32	0.35	0.26
Ile	1.94	1.89	1.84	1.83	1.66	1.34	1.39	1.48	1.65	1.68	1.84	1.73
Leu	3.19	3.07	3.02	3.03	3.47	2.80	2.91	3.18	3.08	3.09	3.45	3.20
Tyr	1.47	1.39	1.37	1.38	1.40	1.14	1.16	1.29	1.30	1.26	1.46	1.34
Phe	2.45	2.37	2.30	2.31	2.56	2.04	2.14	2.27	2.29	2.30	2.47	2.28
His	1.42	1.38	1.35	1.35	1.42	1.11	1.19	1.24	1.40	1.40	1.50	1.37
Lys	1.42	1.41	1.43	1.40	1.55	1.52	1.38	1.40	1.36	1.57	1.58	1.48
Arg	4.55	4.36	4.24	4.27	5.31	3.82	4.48	4.76	4.79	4.56	5.03	4.57

From a nutritional point of view, the analysed almond meal samples show interesting protein values and a medium to pretty good amino acid profile, which is deficient in certain essential amino acids. This finding is in agreement with the results reported by Ahrens et al (2005) [16]. The mean values of chemical indices (CI) and PDCAAS (Proteins Digestibility of Corrected Amino Acid Score) are shown in Table 6. Like all vegetable proteins, which often have one or more limiting amino acids, the analysed samples of almond cakes have relatively low chemical indices (CI),which vary between a minimum of 28.22% for (M-CY2) and a maximum value of 70.81% (FF-CY3). These "CI" values are comparable to those of proteins of vegetable origin and which are generally lower than the CI of proteins of animal origin. The gap between CI values may be due to the genetic and environmental factors, which strongly influence the amino acid profiles in plants [11].All the calculated PDCAAS values for the analysed samples of almond meals are less than 100 but close to reported values for proteins of maize, rice and wheat [21]. "PDCAAS values" for almond meals range between 24.78% for (M-CY2) and 62.17% for (FF-CY3). Besides, for whole almonds,Boye et al (2012) [21], published PDCAAS values for three almond varieties (Carmel, Mission, and Nonpareil) range between 22 and 24 % and are lower than PDCCAS values for almond oil cakes of the five studied varieties (M: *Marcona*; FNB: *Fournat de Breznaud*; FF: *Ferragnes-Ferraduel*; B: *Beldi*). These differences could be explained by genetic and environmental factors.

Finally, we concluded that almond meal coproducts are interesting sources of essential amino acids (EAA, Table 6). They are therefore new good sustainable nutritional alternatives to animal proteins.

Table 6. Total essential Amino acids (ΣEAA g%); non-essential Amino acids (ΣNEAA g%); chemical index (CI), and protein digestibility corrected amino acid score (PDCAAS) for almond meal coproducts of five studied almond varieties (M: *Marcona*; FNB: *Fournat de Breznaud*; FF: *Ferragnes-Ferraduel*, B: *Beldi*)

	ΣNEAA	ΣEAA	CI (Met + Cys)	PDCAAS (%) (Met + Cys)
M- CY1	25.76 ± 0.56	16.10 ± 0.12	47.58 ± 7.95	41.78 ± 6.98
FNB- CY1	24.69 ± 0.16	15.66 ± 0.10	60.06 ± 0.78	52.74 ± 0.68
FF- CY1	24.47 ± 0.71	15.40 ± 0.38	58.50 ± 0.25	51.37 ± 0.22
B- CY1	24.72 ± 0.13	15.34 ± 0.08	44.65 ± 1.56	39.20 ± 1.37
M- CY2	34.47 ± 0.45	16.00 ± 0.24	28.22 ± 3.59	24.78 ± 3.15
FNB- CY2	26.27 ±0.08	13.37 ± 0.08	38.42 ± 4.75	33.74 ± 4.17
FF- CY2	28.90 ± 0.19	13.53 ± 0.07	32.86 ± 1.71	28.85 ± 1.50
B- CY2	31.03 ± 0.10	14.47 ± 0.11	30.49 ± 1.12	26.77 ± 0.99
M- CY3	28.45 ± 0.17	14.82 ± 0.09	45.37 ± 1.43	39.84 ± 1.26
FN- CY3	28.13 ± 0.19	15.24 ± 0.04	51.40 ± 2.94	45.13 ± 2.58
FF- CY3	30.48 ± 0.87	16.79 ± 0.56	70.81 ± 10.12	62.17 ± 8.89
B- CY3	28.37 ± 0.24	15.39 ± 0.10	37.24 ± 0.29	32.70 ± 0.26
True faecal digestibility (%)		87.81		

4 Conclusion

This study focused on the characteristics of oil cake as an extraction residue from almond oils and showed, in view of their major constituents (sugars, proteins, fibres, and minerals), that they can be used as an ingredient or food additive, therefore, with an economic advantage for producers, but also from a zero-waste perspective of great environmental interest.

Although this research was more oriented to almonds harvesting and post-harvesting processing stages, the impact of the genotype and the crop year effect, seem very visible, since the quality of the harvest almonds and their coproducts differ from year to year. This depends on the impact of agricultural conditions and interactions between almond varietiesand pedo-climatic factors. Thus, for the studied oilcake samples of different almond varieties and except their fibre contents, significant differences were observed for all the other analysed parameters (Residual oil, Dry matter, Sugars, Proteins, Ash).

The chemical composition, of the residues of mechanically pressed almond oil (almond meal), depends mainly on the variety of almond that, in turn, is affected by soil and climatic conditions, which determines the quality of the raw material used for the almond oil extraction.

Moreover, the almond meal coproducts contain beneficial nutrients and bioactive molecules; they constitute an interesting source of protein (EAA), dietary fibres and minerals; they have also the advantage to be a gluten-free product. Therefore, these co-products of almond oil extraction would be an interesting ingredient for specific gluten-free culinary preparations and in the formulation of light foodstuffs. Indeed, world protein demand is growing strongly and it seems necessary to offer new sustainable and good nutritional alternatives to animal protein. Legume cakes, such as soybean and rapeseed meals, because of their relatively balanced essential amino acid profile (EAA), are widely used in the formulation of animal feed and vegan food. Almond oil cakes, despite their relatively unbalanced essential amino acid (EAA) profile, have the advantage of being gluten-free and could be a new ingredient for the formulation of new light foodstuffs and vegan foods.

Acknowledgements -We are grateful for the Moroccan Higher Education Ministry and Wallonia Brussels International (WBI Project 2-4) for supporting this research.The assistance of PRODIGIA Company for the mechanical extraction of almond oils and the almond growers in eastern morocco for providing the almond samples for this study.

5 References

1. C. Grasselly. CIHEAM, Options Méditerranéennes., 28- 43, **32** (1976).
2. A. J. Esfahlana, R. Jameia, R. J. Esfahlan. Food. Chem.,349–360,**120** (2010)
3. R. Sociasi Company, O. Kodad, J.M. Alonso and T.M. Gradziel. Almond Quality: A Breeding Perspective, in J. Janick (Eds.), Horticultural Reviews, New York, (2008).
4. A.Elamrani, R.Melhaoui, N.Houmy, M. Addi, M.Abid, A.Mihamou, M.L.Fauconnier, M.Sindic& H.Serghini-Caid. *Valorisation Challenges to Almonds and their Co-products: Characterization of Oils Extracted from Broken Almonds Springing from Shelling and Sorting Operations.*In proceeding of the fourth American Moroccan Agricultural Sciences Conference, 9-11May 2018, Meknes, Morocco (2018).
5. ISO 662. International Organisation of Standardisation. Corps gras d'origines animale et végétale — Détermination de la teneur en eau et en matières volatiles (1998)
6. C.S. James. *Analytical Chemistry of Foods: Determination of fat by the Soxhlet methods.* Blackie Academic and Professional, ed., C.S. James, London, (1995).
7. State standard of Russian Federation GOST5672-68. Method for determining the mass fraction of sugar. Bread and bakery products. (1968)
8. A.O.A.C. 991. 43. Association of Official Analytical Chemists: Total, Soluble, and Insoluble Dietary Fibre in Foods: Enzymatic-Gravimetric Method, Mes-Tris Buffer (1996).
9. A.A.C.C. 08-01. American Association of Cereal Chemists. Ash: Basic method (2001).
10. A.F.NOR. NF.V04-407. L'Association Française de Normalisation. Détermination de la teneur en azote total et calcul de la teneur en protéines – Méthode Kjeldahl (2002).
11. A. Rabadán, M. Álvarez-Ortí, J.E. Pardo. Sci. Hortic. Amsterdam.,218-224, **244** (2019).
12. D. H. Spackman, W. H. Stein, and Stanford. Anal. Chem., 1190–1206,**30**, 7 (1958).
13. FAO/WHO/ONU. Food and Agriculture Organisation/ World Health Organization/ Organization of United Nations. Protein and amino acid requirements in human nutrition. Technical Report Series 935. Geneva, Switzerland. (2007).
14. A. Sujak, A. Kotlarz, W. Strobel. Food Chem., 711-719,**98** (2006)
15. G. Schaafsma*The Protein Digestibility–Corrected Amino Acid Score*, in Proceedings of the symposium Criteria and Significance of Dietary Protein Sources in Humans, 4 October 1999, San Francisco, USA (1999).
16. S. Ahrens, M. Venkatachalam, A. M. Mistry, K. Lapsley& S. K. Sathe. Plant. Food. Hum. Nutr., 123–128,**60** (2005).
17. R. Arbouche,*Incorporation du tourteau d'amande d'abricot en substitution au tourteau de soja dans l'alimentation des animaux domestiques: Ruminants (cas des ovins à l'engrais) et des monogastriques (cas du poulet de chair).*Doctoral Thesis,ElHadj Lakhdar University, Batna, Algeria (2013).
18. S. Ramachandran, S. K. Singh, C. Larroche,C. R. Soccol, A. Pande. Bioresource. Technol.,2000–2009, **98** (2007)
19. S. Sivaramakrishnan& D. Gangadharan. Chapter 13: Edible Oil Cakes. Biotechnology Division, National Institute for Interdisciplinary Science and Technology (NIIST), CSIR, Trivandrum 695 019, Kerala, India (2009).

20. A. Rabadán, M. Álvarez-Ortí, R. Gómez, A. Pardo-Giménez, J. E. Pardo. Sci. Hortic. Amsterdam., 539–546,**225** (2017).
21. R. Savoire, J.L. Lanoisellé, E. Vorobiev. Food. Bioprocess. Tech., 1–16,**6** (2013)
22. D.Tungmunnithum, A. Elamrani, M. Abid, S.Drouet R.Kiani, L.Garros, A.Kabra, M.Addi and C. Hano.Quick, Green and Simple Ultrasound-Assisted Extraction for the Valorization of Antioxidant Phenolic Acids from Moroccan Almond Cold-Pressed Oil Residues, Appl. Sci. **10** (2020).
23. D.Tungmunnithum, M. Abid, A.Elamrani, S.Drouet, M.Addi and C.Hano. Life.**10** (2020).
24. S. Karaman, S. Karasu,F.Tornuk, O.Toker, U.Geçgel, O. Sagdic. J.Agr. Food. Chem., 2305–2313,**63**(2015).
25. S. Čolić, G. Zec, M.Natić, and M.Fotirić-Akšić. Almond (*Prunus dulcis*) oil, in e-book "Fruit Oils: Chemistry and Functionality". M. F. Ramadan Editor (Springer Nature Switzerland, AG 2019).
26. J. Boye, R. Wijesinha-Bettoni and B. Burlingame. Brit. J. Nutr., S183–S211, **108** (2012).

Evaluation of pomological and biochemicalquality of Moroccan almond native genetic resources for conservation of biodiversity

Souhayla Kodad*, *Reda* Melhaoui, *Nadia* Houmy, *Mohamed* Addi, *Hana* Serghini-Caid, *Ahmed* Elamrani, *Malika* Abid, and *Aatika* Mihamou

Laboratory of Agricultural Production Improvement, Biotechnology and Environment LAPABE, Faculty of Sciences, Mohammed Premier University, Oujda, Morocco

Abstract. The edible and tasty almond {*Prunus amygdalus* [syn. P. dulcis (Mill.) D.A.Webb]} kernel is not anymore the main reason for its production and uses, other properties related to its by-products, especially the sweet edible almond oil, which is mainly used as a salad dressing or as an ingredient in a highly appreciated Moroccan almond-based recipes. The studied almond populations are derived from seedling, the trees of these plantations are very heterogeneous in terms of vigor and flowering time and are facing a genetic bottleneck due to an attack by an insect 'bark beetle', as well as the introduction of new foreigner cultivars. Our preliminary study concerns the evaluation of fruits quality of locally commercializedalmond ecotypes, with a shelling rate ranging from 22.37% to 34.21%and it comparison with the introducedFrench cultivars couple Ferragnes/Ferraduel (Fg/Fd) grown in the same region, that presents higher commercial values but yet lowest nutritional traits compared to local populations.Since Beldi almonds showed higher total tochopherols content with avalue equal to 511.64 mg/kg, while for Fg/Fd cultivars it presents only 433.1 mg/kg. Hence the interest of a detailed study to prove both the genetic variability of local Beldi almond populations for the conservation of biodiversity, and also the biochemical propertiesof almond oils and cakes lead to post-harvest recovery of these products by increasing it commercial value in favour of local farmers and cooperatives.

1 Introduction

Almond (*Prunus dulcis*)), is one of the oldest domesticated fruit trees and it has been adapted to various microclimate in many countries over the globe[1]. It is also a major nut tree grown in the Mediterranean climate [2]. The traditional cultivation of almonds used open-pollinated seedlings [3] which, with self-incompatibility, produced very strong heterozygosity within this species [4].This great variability provided a useful genetic pool for almond germoplasm evolution allowing in each growing region a selection of well adapted almond cultivars [5].World almond production was 2.2 million tons in 2017, and only the United States provided 46% of the total production (FAOSTAT.2018). The production also involves many Mediterranean climate areas. The most important major growth areas in modern times are Spain, Italy, Morocco, Tunisia, Greece and Portugal[6]. Spain, Iran and Morocco are considered to account for 22% of the world's total production.

Morocco has seen through 'Plan Maroc Vert' a massive introduction of new varieties of almond trees with late blooming cultivars allowing them to escape late frosts and ensure better productionwith fruits of a high physical quality. The introduction of the new varieties has made local "Beldi" almond orchards less abandoned or replaced, causing genetic erosion and regression of the indigenous almond populations while they have demonstrated through decades their tolerance to drought and adaptation to poor soils in these fragile agro-ecological zones. Moreover, some physicochemical analysis of the local almonds fruits quality collected from different regions of Morocco has already revealed a high quality of these almonds[7], [8] suggesting other potential post-harvest uses such as marzipan production, almond meal and oil. This could increase the market value of these fruits and improve the income of local farmers. Thus, finding and characterizing the old genotypes for the conservation of biodiversity is of definite interest in the diversity of cultures within these almond groves. The results could also be useful for the management of almond germplasm and for the selection of the best parental genotypes for breeding programs.

*Corresponding author: souhaila.kodad@gmail.com

More than half of the almond trees grown in Morocco originate from seedlings, mostly located in the north and in the south of the country. Great diversity in almond trees resulting from the variability of the environmental conditions and the climate changes has been shown in each production region. Numerous studies had been undertaken to evaluate genetic variability and quality of nuts within the Moroccan almond population using morphological and phonological characteristics [9], but there are minimal studies on the chemical composition of almonds, mainly concerning kernels, the edible part which have a very significant nutritional value. The knowledge of its chemical composition will allow the assessment of not only quality criteria but also consumption criteria due to the incidence of certain compositional parameters on the shelled almond's nutritional and health values [6]. It will be essential to characterize the local Beldi almond chosen for their performance and their productivity, according to their pomological characterization, in order to identify the genotypic richness, the relationship between the local populations and the introduced cultivars, and also to prove the genetic variability which can be exploited for the conservation of the biodiversity. Concurrently, the analysis of the organoleptic and nutritional properties of almond oils and cakes will lead to the postharvest development of these products in favor of pilot cooperatives. The search for distinctive characteristics, whatever the nutritional characteristics of Beldi almonds, according to production zones, will also make it possible to highlight the characteristics linked to the territory. Since, nowadays determining the authenticity and origin of food is a key issue in food quality control and management[10]. The main objective of this work is to assess pomologicaly characterized almonds of the Beldi ecotypes in Northeastern Morocc oby biochemical characterization of its oils and cakes in order to highlight the different components of the almond kernel and to emphasize its importance in considering almonds quality and to prospect its evaluation in breeding programs.

2 Materiel and methods

2.1 Pomological characterization of almond nuts and kernels

The different Beldi ecotypes batches (L1, L2, L3, L4, L5, and L6) considered in the present study have been bought in a local niche market in eastern Morocco from local fruit growers. The pomological characterization was carried out on fruits (almonds and kernels), according to the descriptor of the International Board for Plant Genetic Resources [11].The sphericity (Ψ) of the nuts and grains was determined using equation (1) The geometric mean diameter (Dg) is calculated according to equation (2)[12], and the volume of the nuts and grains (V) can be calculated using the following equation (3)[13]:

$$\varphi = \sqrt{(3\ \&\ LWT)} \times 100 / L \quad (1)$$

$$Dg = \sqrt[3]{LWT} \quad (2)$$

$$V = (\pi [Dg]^3) / 6 \quad (3)$$

2.2 Biochemical analyses of almond oils

The extraction of the oils was carried out at a cooperative of SIDI BOUHRIA in eastern morocco by using an oil screw press extractor KOMET model DD85G. The analyses were carried out one week after the extraction.

2.2.1 *Oil content*

Oil content was calculated using the weight of the bottle containing the oil minus the weight of the same empty bottle divided by the weight of the dry kernels of four studied Beldi ecotypes according totheir shelling rates, which are L1,L2,L4 and L5. All of the oil extractions were performed in triplicate.

2.2.2 *Acidity index*

Acidity (g/100 g oleic acid) was defined as the percentage of free fatty acids that are present in the oil resulting from the hydrolysis of triglycerides. To define the acidity index of almond oil the following protocol was applied. 5g of almond oil (P) are dissolved in 50ml of absolute ethanol, to which 8 drops of phenolphthalein (1%) are added. The free fatty acids are titrated with stirring with an alcoholic solution of potassium hydroxide at 0.1N (the pink color of the phenolphthalein persists for at least 10 seconds).
The free acidity is calculated according to the following formula:
% of free acid = (V (NaOH) × N (NaOH) x molar mass x 100) / P x 1000
V: volume of titration NaOH in l
N (NaOH): normality of NaOH
P: test sample

2.2.3 *Peroxide index*

The peroxide index (PI) was used to estimate the content of hydroperoxides present in the oil (primary oil oxidation products). This index was determined by the so-called acetic acid and chloroform method corresponding to the AOCS Cd 8-53 standard. According to the following protocol:5g (P) of almond oil were dissolved in 30 ml of a chloroform acetic acid mixture (3:2 v / v) then 0.5 ml of a saturated solution of potassium iodide were added to the mixed. Finally 30ml of distilled water are added after exactly one minute of stirring. This solution was then titrated with 0.01N sodium thiosulfate using starch as a color indicator.
The peroxide value is expressed in milliequivalents peroxide per 1000g of oil according to the following formula:
IP (meq O2 / Kg) = [(VE x M× 1000] / P
VE: volume of sodium thiosulfate for sample titration
P: test sample

2.2.4 *Colorimetric determination of phenolic contents*

Total phenolic contents were determined according to the Folin-Ciocalteu method by using caffeic acid as a standard, and by spectrophotometric absorbance measurement at 760 nm. Folin Ciocalteu's reagent is a yellow acid consisting of a mixture of phosphotungstic acid (H3PW12O40) and phosphomolybdic acid (H3PMo12O40). It is reduced, during the oxidation of phenols, to a mixture of blue oxides of tungsten and molybdenum [5]. The coloring produced, whose maximum absorption at 760 nm, is proportional to the quantity of polyphenols present in the almond oil extracts [5].

2.2.5 *Determination of total proteins*

The total protein determination consists in measuring the total nitrogen by the Kjeldahl method and in multiplying this nitrogen content by a conventional factor 6.25 [5]. The principle of this method consists of mineralization of a sample with sulfuric acid in the presence of a catalyst, alkalization of the reaction products, distillation of the released ammonia collected in a solution of boric acid, and titrated by sulfuric acid solution.

2.3 Biochemical analyses of Beldi almond oil cakes

2.3.1 *Dry matter content*

The container were placed in the oven at 103 ° C for 1 hour 30 minutes, then cooled in a desiccator away from ambient humidity before being tarred (P1) in a precision balance to 3 decimal places. Then 1g (P2) of almond oiscakes wasweighed in each container. Then they were placed in the oven at 103 ° C ± 2 ° C until the weight is stable. They were then cooled in a desiccator away from ambient humidity and then weighed (P3).
Expression of the result:
The dry matter and the humidity are calculated according to the following formulas:
% Dry matter (MS) = [100 × (P3 - P1)] / P2
%Humidity rate= 100 -% (DM)

2.3.2 *Total ash content*

The silica crucible previously weighed was placed in the muffle furnace at 550 ° C overnight to ensure that the impurities are burned. Then, they have been cooled in a desiccator for 30 minutes before being tarred (P1) in a precision balance to 3 decimal places. About 1g (P2) of the almond cakewas weighed into each crucible. These were placed in the muffle furnace at 550 ° C for 8 hours. The crucibles were then cooled in the desiccator away from ambient humidity and then weighed (P3). The expression of the total ash content results relative to the dry matter according to the following formula:
% Ash = [100 × (P3 - P1)] / P2

2.3.3 *Composition of major mineral elements*

The major elements analyzed in this study were potassium (K), magnesium (Mg), calcium (Ca), and phosphorus (P). They were measured by different methods,UV-visible spectrophotometry at 700 nm for phosphorus, titration for magnesium (Mg) and calcium (Ca), while for potassium (K), it was measured by a SHERWOOD model 360 flame photometer. The mineralization of samples was by measuring 2g of almond cake in a Teflon container and

https://doi.org/10.1051/e3sconf/202018304005

placed in a muffle oven for 5 hours at 550 ° C. Then 6 ml of 65% nitric acid and 1 ml of 35% hydrogen peroxide were added. The samples were subjected to mineralization in a high performance microwave digestion unit (MLS 1200 meg) using the following program: 250 W: 2 min; 0 W: 2 min; 250 W: 6 min; 400 W: 5 min; 600 W: 5 min; ventilation: 10 min. The program used is based on the manufacturer's recommendations for use (microwave, January, 1995). The rotor was then cooled for one hour.

2.3.4 *Crude cellulose determination*

The crude fiber analysis was performed using the FIWE fibre analyser (VELP SCIENTIFICA). With the 1g of almond oil cake samples remaining inside the crucibles to avoid the loss of fibre due to the transfer according to AOAC,1990[5].

2.3.5 *Determination of total sugars*

Total sugar was measured according to the Bertrand method, the principle of which is that the glucose partially reduces Fehling's liquor in excess the cuprous oxide formed (red precipitate) is dosed by manganimetry. A table gives the correspondence between the mass of copper and the mass of glucose. The reaction must take place hot and for three minutes from boiling to respect the correspondence of the tables. A quantity of glucose reacts with the excess copper (II) ions to form a brick-red precipitate. Excess copper (II) ions were eliminated. The precipitate reacts with an excess of iron (III) ions to dissolve it. Iron (II) ions were obtained, and dosed with a solution of potassium permanganate.

2.4 Comparison between the Beldi ecotypes and Fg/Fd cultivars according to the major fatty acids and the tocopherol isomers grown in eastern Morocco

2.4.1 *Tocopherols profile analysis*

According to the official method, AOCS 8-89 (AOCS 1989), tocopherols analyses were performed by using HPLC-FLD, (Agilent Technologies series 1200 system, Agilent Technologies). It was equipped by an automatic injector, on a Uptisphere 120A° NH2 column (150 mm *3 mm, 3 µm) Interchim (Montluçon, France). 10 µL was injected. The temperature was maintained at 30 °C. The mobile phase was hexane/2-propanol (99:1, v/v) with a flow rate of 1 mL min-1. The different tocopherol isoforms (α-, β-, γ-and δ-tocopherols) were evaluated using oil/hexane solution of both the studied Beldi ecotypes oils, and Fg/Fd cultivars oils extracted by using the same screw press extractor. The obtained tocopherols from Sigma-Aldrich (Steinheim, Germany) were used as external standardization for the identification of tocopherols at 292 nm. Tocopherol homologs were quantified by comparing each sample peak response to that of the corresponding standard.

2.4.2 *Fatty acid analysis*

The present analysis was also conducted for both Beldi almond ecotypes and Fg/Fd cultivars. The fatty acids were converted to fatty acid methyl esters before analysis by mixing a solution of 10 mg of oil in 0.2 mL of hexane with 0.5 mL of another solution composed by 55 mL of dry methanol added to 20 mL of pentane and 25 mL of BF3 at 14% weight in methanol. The mixture was placed in a water bath for 90 min at 75 °C. 0.6 mL of saturated NaCl and 0.2 mL of 10% H2SO4 (V: V) was added. HP 6890 series gas chromatography system was used for analyzing the fatty acid methyl esters. The system was equipped with an FID detector and a capillary column (Supelcowax: 30.0 m 9 250 mm 9 0.25 lm). The carrier gas was nitrogen. The temperatures were maintained at 150 and 250 °C, for the injector and detector, respectively. While for the oven temperature, it was set at 210 °C. A comparison with the retention time of the respective ester standards was utilized for the identification of the individual fatty acid methyl esters.

2.5 Statistical analysis

The experimental design used in this study was completely randomized design. Each determination was performed in triplicate. All data were analyzed by Analysis of Variance (ANOVA), and Duncan's multiple range with SPSS software for Windows (SPSS.21, USA), the values of the various parameters were expressed as an average ($\bar{X}$). Only variables with a confidence level greater than 95% ($P < 0.05$) were considered significant.

3 Results and discussion

3.1 Pomological characteristics of Beldi almond ecotypes

The descriptive statistics of the Beldi ecotype samples taken from the local marketwere presented in Tables 1 and 2. For the majority of the almond nuts parameters such as weight, width, length, thickness, Dg, and volume, the statistical analyzes showed that the genotype L5 represents the highest values with the exception of sphericity, where the highest value was observed in the L6 genotype. Several forms have been observed for the different studies ecotypes round, ovate, oblong, and cordate (Table 1). The sizes ranged from medium to large, and all the hulls were hard during the crushing process. The weight of almond nuts varied between 3.2g and 6.12g. The width, length, and thickness of the almond fruits ranged from 17.39 mm to 25.02 mm, from 27.01 mm to 30.47 mm and from 14.71 mm to 19.25 mm, respectively. The parameters calculated from the mathematical formulas, sphericity, and volume varied from 69.61% to 81.67%, 4086.07 mm^3 to 7838.92 mm^3 for almonds. While the yield and percentage of doubles varied from 22.37% to 34.21%, from 6 to 100%, and for the shape of the almond, four categories were found, intermediate, narrow, broad, and extremely broad (Table 2). Our results are comparable to those reported by Melhaoui et al. (2018) [7] on almond cultivars Marcona, Ferragnes, Ferraduel, and Fournat introduced in the same region. The studied almond genotypes show a great variability in morphological and pomological characters. Flowering was from early January to late March. This three-month interval justifies the great variability of the plant material. The yield for most almond ecotypes was low ranging from 22.37% to34.21%. It must be taken into account that these trees grow abundantly, without pruning, fertilizing or pesticide applications.

Table 1. Beldi almond nuts pomological characteristics.

Lot	Weight (g)	Width (mm)	Lenght (mm)	Thickness (mm)	Dg	Sphericity	Volume	Nut shape	Nut size	Nut color	Marking of outershell
L1	3.20a	19.74a	29.84 bc	15.00a	20.64ab	69.61a	4086.07 a	Oblong	Large	Intermediate	Sparsely pored
L2	3.63ab	17.39ab	28.48 b	15.21a	19.78a	69.77a	4741.25 a	Oblong	Medium	Light	Intermediate
L3	3.82ab	20.55ab	27.01a	16.14b	20.74ab	77.74bc	5172.55 a	Cordate	Large	Dark	Densely pored
L4	4.003 b	21.68 b	28.49 b	14.71ab	20.85ab	73.22ab	5318.31 a	Ovate	Large	Dark	Densely pored
L5	6.12c	25.02c	30.47c	19.25c	24.46c	80.39c	7838.92 b	Round	Large	Light	Sparsely pored
L6	3.94b	22.24 bc	27.37a	18.35c	22.35 bc	81.67c	5869.71 ab	Round	Medium	Dark	Sparsely pored

Table 2. Beldi almond kernels pomological characteristics.

Lot	Weight (g)	Width (mm)	Lenght (mm)	Thickness (mm)	Dg	Sphericity	Volume	Yield of shelling %	Double kernels %	Kernel shape	Kernel colour intensity
L1	1.006ab	12.55a	23.26 b	7.94a	13.12 a	57.46a	1190.22 ab	34.21b	71	Intermediate	Dark
L2	0.86ab	10.79ab	21.34 b	8.14ab	12.01 a	56.32a	959.29a	24.82a	22	Narrow	Dark
L3	0.75a	12.64ab	17.52 a	7.79ab	11.98 a	68.62b	942.14a	24.33a	10	Extremely Broad	Dark
L4	0.86ab	12.80b	21.02 b	6.77b	12.15 a	58.32a	990.52a	23.14a	6	Broad	Dark
L5	1.16b	15.14c	21.54	8.31b	13.90 a	64.56ab	1429.12 b	22.37a	44	Extremely Broad	Dark

L6	-	-	-	-	-	-	-	-	100	-	Dark

3.2 Oil analyses of the four studied almond genotypes

3.2.1 *Beldi almonds oils content*

This study shows significant differences in the analyzed almond oils content, ranging from 38.14 to 47.35%, with L2 presenting the highest quantity of oils (Table 3). The obtained results show richness in oils of Beldi almond using screw press extraction comparable to the results obtained by using chemical extraction for other almond cultivars, were the oil content ranges from 35.27% to 40% [14], 40 to 65% [15], 48% to 67% [16], from 36.7% to 79% [17]. The obtained result shows, kernel rich in lipids, generally considered to be desirable for a healthy diet, or even preferable in the confectionery industry, because higher oil contents result in less water absorption by the almond paste, which makes local almond very appreciated in bakery and in the making of almond-based recipes [18].

3.2.2 *Acidity index*

The acidity indexis one of the most important parameters that define the quality of oils. The analysis of the studied almond oils showed a low content of free fatty acids which did not exceed 0.54% (Table 3). With an order grow according to the genotypes, L5 with 0.17%, L4 and L1with 0.19%, and L2 with 0.54%. The results of the acidity index obtained are relatively low compared to those reported in the literature by M.M. Özcan et al. (2011)[19] for the Guara, Cristomorto, Tuono, Guara, Ferragnes, Nonpareil varieties, which are around 2,279%, 3.55%, 1.38% 2.39% 1.89% respectively, and similar to those reported by Sakar et al., JMES, (2017)[20]. This index reflects the degree of degradation of the lipid reserves in the kernels in particular the triacyglycéroles. This degradation causes an increase in free fatty acids, the major fatty acid is oleicacid for the studied almond oils. During storage or extraction, the triacylglycerols can be degraded by endogenous or exogenous lipases brought about by microbial contamination. The low value of the acidity index obtained for the studied oils therefore shows a good quality of the oils produced in eastern Morocco.

3.2.3 *Peroxide index*

The peroxide index (PI) is a crude indicator of primary lipids amount of oxidation, it is a tool for measuring the amount of hydroperoxides formed by oxidation of fatty acids, especially polyunsaturated fatty acids. However, obtaining a low PI does not necessarily mean that the oil is not oxidized. Indeed in some cases, the kinetics of the oxidation of oils is very advanced and the hydroperoxides (dosed during the determination of the IP), are themselves transformed into secondary products some of which are volatile (the oil is then rancid). The peroxide index of the samples of almond oils studied varied between 6.82 meq O2 /kg and 13.25 meq O2 /kg (Table 3). The peroxide index of the L5, L4, and L1 genotypes is less than 8.89 meq O2 / kg, while the IP value of the oil extracted from the seeds of the L2 genotype was the highest (13.25 meq O2 / kg) compared to the other genotypes. The results of the peroxide index are relatively low compared to those reported in the literature by M.M. Özcan et al. (2011) [19]for other cultivars which varied between 7.58 meq O2 / kg and 15.59 meq O2 / kg. The results of the IP obtained could signify a better quality of the oils extracted by pressure by KOMET Model DD85G compared to the Soxhlet process used by M.M. Özcan et al. (2011)[19] which has the disadvantage of being a hot chemical extraction process which promotes lipid peroxidation.

3.2.4 *Total polyphenol content*

The results of the total polyphenols analysis for the four studied almond oils were shown in Table3. The L5 genotype has the highest polyphenol content (437.85 $mg.kg^{-1}$), followed by L4 (423.8 $mg.kg^{-1}$), then with L1 and L2 366.26 $mg.kg^{-1}$and 353.58 $mg.kg^{-1}$, respectively. These values obtained were similar to those reported by Sakar et al. , JMES, (2017)[20]for other almond cultivars, while for other species phenols present lower values mainly for the most important fruit culture in eastern Morocco,which according to Mansouri et al., 2013 [21]virgin olive oil of autochthonous cultivars present only 316.59 $mg.kg^{-1}$.Polyphenols make a significant contribution to the quality of almonds. Almond polyphenols are primarily composed of tannins and flavonoids, which contribute to both nutrition value and health-promoting behavior according to Bolling, et al 2017 [22].Generally, greater phenolic content and relatively high antioxidant activity of cold-pressed nut oils eventually make them superior to those achieved by the solvent/soxhlet method[23].

Tables 3. Analysis of oil content, total phenolic content, acid index and peroxide index of Moroccan Beldi almonds.

Lot	Oil content%	Acid value (% oleic acid)	Peroxide value (meq O2/kg)	Total phenolic contentmg/ Kg
L1	40.27± 0.2	0.19a ± 0.02	8.89a ± 0.02	366.26ab ± 0.3
L2	47.35± 0.01	0.54b ± 0.02	13.25b ± 0.04	353.58a ± 0.2
L4	38.14± 0.03	0.19a ± 0.02	8.58a ± 0.02	413.58ab ± 0.1
L5	39.49± 0.09	0.17a ± 0.002	6.82a ± 0.01	437.85b ± 0.1

3.3 Characterization of the Bledi almond cakes

3.3.1 Dry matter and moisture content of almond cakes.

The dry matter of a sample corresponds to its mass after the complete evaporation of the free water. The results obtained from the oil cake samples (Table 4) have shown that there are significant differences between the varieties ($p < 0.05$), with a content that varies between 92.03% for L5and 92.42% for L1.The variation of the dry matter implies the variation of the humidity rate. In fact, for the four genotypes, the humidity levels were on average of 7.58% for L1, 7.74% for L5, 7.96% for L2, and 8.59% for L4 (Data not shown). The values obtained are slightly lower than those for almond cake reported by M.M. Özcan et al. (2011)[19]. These results show that the oil cakes obtained from the kernels of the four studied genotypes were favorable for long-term conservation. In fact, on a microbiological level, low water contents limit the development of microorganisms, with the exception of molds [24]. The content of the dry matter is strongly influenced by the maturity stage of the seeds, the process of harvesting, and the storage conditions. Water content has a significant effect on the weight of the cake without affecting its nutritional value.[25].

3.3.2 Ash content

Ash represents the mineral part of almond cake and gives an idea of its contribution in mineral salts.Statistical analysis (ANOVA) of the total ash content revealed significant differences ($p < 0.05$) between the cakes of the different genotypes studied. The total ash contents were between 4.76%for L2 and 6.36% for L4. The values were higher than those reported by other studies concerning other varieties of almonds which have ash contents varying between 2.74% and 3.05% according to M.M. Özcan et al. (2011)[19].

3.3.3 Total protein content

Proteinsare the main constituent of muscle tissue, it's very important for growing or reproducing animals which they need a diet rich in protein for mass gain. The level of protein in a food is described as a percentage of total weight and often expressed as a percentage of crude protein. Statistical analysis of the results of the total protein content of the studied genotypes almond cakes ($p < 0.05$), showed significant differences between the genotypes represented in table 4 by different letters (a-b).The almond cakeof L4 genotype has the highest content (44.61 %), followed respectively by that of L1 (36.84%), L2 (36.67 %), and L5(34.09%)(Table 4). According to the results obtained, the significant differences could be explained by the genetic variability.Our results are comparable to those reported by Z. Ahmad (2010)[26].

3.3.4 Crude cellulose content

In this part of the study, we only analyzed one type of fibers, which is thecrude cellulose (Table 4).The results were presented in table 4. The crude cellulose content varied between 9.67% (L5) and 21.75% (L2). However, too much fiber in the diet can reduce its nutritional value, resulting in stunted growth and reproductive performance and even death under conditions such as pregnancy toxemia [27].

 https://doi.org/10.1051/e3sconf/202018304005

3.3.5 Total sugar composition

Energy is obtained from carbohydrates, such as simple sugars, starch and fiber. These food components provide energy for animals to maintain body heat and to perform normal bodily functions and activities [27]. The total sugar content (Table4), expressed as a percentage of dry matter, varied between 16.73% for L5 and 17.42% for L1. These results are comparable to those reported by Houmy (2014)[28]. Statistical analysis of the results of the total sugars of almond cakes, showed that there are significant differences between the genotypes (<0.05).

3.3.6 Content of major mineral elements

Statistical analysis (ANOVA) of the major mineral content of the oil cakes of the four almond genotypes studied shows significant differences (p <0.05). The results of the mineral macro-elements analysis in Beldi almond cakes have shown that potassium is the major element in this fraction in all the analyzed samples, followed by phosphorus then Magnesium, and Calcium (Figure 1). High potassium content was observed in L1 (9.1 $mg.g^{-1}$) while the lowest value was obtained withL4 (6.08 $mg.g^{-1}$). Potassium partially controls the metabolism of carbohydrates, promoting the synthesis of adenosinetriphosphoric acid and phosphocreatine as well as the release of pyruvic acid. It influences growth and participates in the activity of fundamental enzymatic systems such as that of carboxylases, peptidases [29].Calciumand magnesium contents were between 3.05 $mg.g^{-1}$ for L2 and 2.25 $mg.g^{-1}$for L4 and 3.23mg.g-1 for L5 and 7.37mg .g-1 for L2, respectively. The magnesium values were comparable to those reported by M.M. Özcan et al. (2011)[19]. Magnesium, like calcium, is involved in the process of neuromuscular excitability. In this sense, the deficiency in magnesium leads to hyperexcitability. While for the phosphorus contents it varied between 5.11 $mg.g^{-1}$ (L5) and 7.07 $mg.g^{-1}$ (L1). These values are lower than those reported by other studies on almonds [19]. The L1 genotype is the richest in major mineral elements compared to the other studied genotypes and the L5 genotype showed the lowest values (Figure 1).Phosphorus is a component of deoxyribonucleic acid, which is basic for cell development and division. Similar to phospholipids, it contributes to the fluidity of the cell membrane; in the form of phosphate. It also helps preserve osmotic balance and plays a key role in several metabolic functions.

Table 4. Analysis ofdry matter, ash content, crude cellulose, and total sugars and protein of Beldi almonds cakes.

Lot	Dry Matter %	Ash%	Crude Cellulose%	Total Sugar%	Total Protein% (N x 6,25)
L1	92.42d ± 0.03	5.52b ± 0.01	10.03a ± 0.3	17.42b ± 0.3	36.84a ± 0.1
L2	92.26c ± 0.008	4.76a ± 0.04	21.75b ± 0.3	16.85a ± 0.2	36.67a ± 0.2
L4	92.40a ± 0.03	6.36c ± 0.01	11.76ab ± 0.4	16.74a ± 0.4	44.61b ± 0.4
L5	92.03b ± 0.04	5.31b ± 0.04	9.67a ± 0.5	16.73a ± 0.3	34.09a ± 0.2

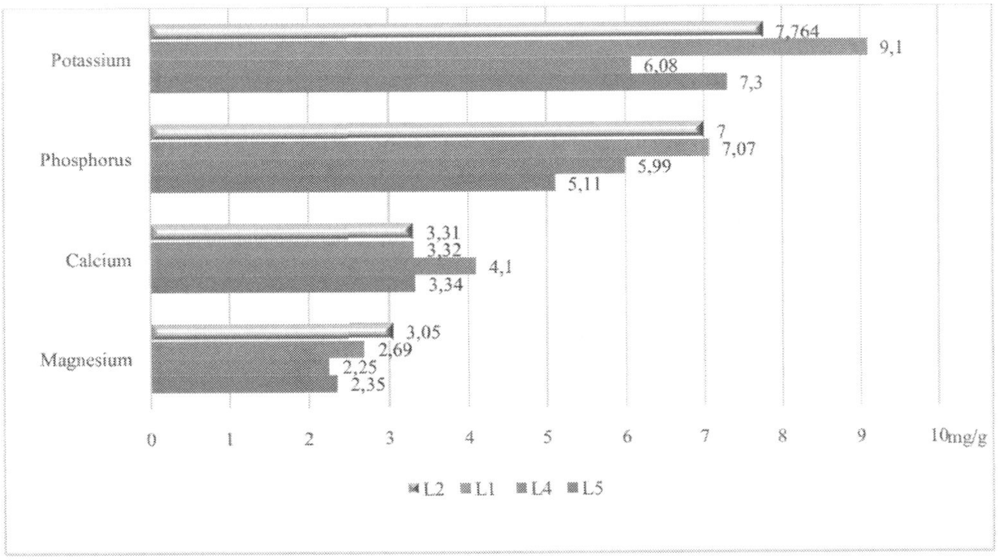

Fig. 1. Major mineral elements composition of Beldi almond cakes.

3.4 Comparison between the Beldi ecotypes and Fg/Fdcultivars according to the major fatty acids and the tocopherol isomers grown in eastern Morocco

Fatty acids and tocopherols are key nutrients that donate almonds different helpful capacities in numerous wellbeing viewpoints, in this manner, almond fatty acids and tocopherols are exceptionally critical for industry and researchers alike.

3.4.1 Major fatty acids of Beldi almond and Fg/Fd produced in eastern Morocco

Table 5 shows the relative proportions of the various fatty acid methyl esters (FAMEs) in the oils. Palmitic, oleic, linoleic, and the Stearic acid were the main FAMEs identified for both Beldi almond and Fg/Fd cultivars, with a dominance of Oleic acid. The highest value of saturated fatty acids was obtained with Fg/Fd cultivarsand represents 11.22%, while for Beldi almond it represents 10.39% of SFA, with the majority being palmitic acid 8.24 and 8.03% respectively. The content of monounsaturated (MUFA) fatty acids in the oils was higher for beldi Almond with 66.18%, while Fg/Fd it represents only 60.29%. The unsaturated fatty acids were principally represented by oleic acid 65.27–58.96%, followed bylinoleic acid 23.42-28.46% for Beldi and Fg/Fd respectively. Similar results were reported by Mattha and Icp-aes (2020) [17].

3.4.2 Composition of different tocopherol isomers of Beldi almond ecotypes and Fg/Fd cultivars grown in eastern Morocco

Tocopherols presentedin table 6 show that the Moroccan Beldi almonds hold the highest total tocopherol contents with 499.37 mg.kg-1 of oil, compared to the introduced cultivars Fg/Fd with 425.03mg.kg-1 of oil. Similar results were obtained byMelhaoui, R (2018)[26] comparing to other cultivars grown in the same regions. Tocopherol analysis and the separation by HPLC revealed the presence of three tocopherol forms: α, B and Ɣ with a dominance of α-tocopherol for both Beldi almonds and Fg/Fd cultivars, representing 97.6 and 98.13% of the total tocopherols content, respectively. The variability in the amounts of tocopherols obtained could beexplained by the environmental conditions and genotype effect[17].

Table 5. Major fatty acids of Beldi almond and Fg/Fdcultivars produced in eastern Morocco.

Fatty acids	Fg/Fd	Beldi
Palmitic acid C16:0	8.24a	8.03b
Palmitoleic acid C16:1	1.24b	0.84c
C17:0	0.03a	0.04a

C17:1	0.1a	0.08a
Stearic acid C18:0	2.93c	2.32b
Oleic acid C18:1	58.95c	65.26b
Linoleic acid C18:2n6	28.46b	23.42a
SFA	11.22	10.39
MUFA	60.29	66.18
PUFA	28.46	23.42

Table 6. Oil content and composition of different tocopherol isomers of beldi almond ecotypes and Fg/Fd cultivars produced in eastern Morocco.

Tocopherols mg/kg	Fg/Fd	Beldi
α-tocopherol	425.03a	499.37a
β-tocopherol	2.28c	2.82c
Ɣ-tocopherol	5,80b	9.45b
δ-tocopherols	ND	ND
Total tocopherols	433.11	511,64

4 Conclusion

Studies carried out on the quality of the fruit in these populations have revealed that the low commercial value it's explained by the small size of kernels, it wrinkled, its dark color and also by high percentages of double kernels. These negative traits reduce the marketable value of this ecotype because they do not meet the physical quality standards required by the market[16]. However, the chemical composition of these kernels has reached a very high quality suggesting other post-harvest potentials. These results are consistent with those obtained by[8].The multiple analyses showed that the studied almond cakes are potentially recoverable in the food industry. Indeed, it is an excellent source of fiber, protein, sugars, and minerals. For the studied genotypes, a considerable genetic diversity was observed in terms of phenological, pomological characters, and the physicochemical composition indicating that there is rich and precious plant material that makes these cakes potentially valuable in agro-foodand for the conservation of biodiversity.

References

1. A. Fernandez, I. Martı, C.F. i Forcada, K. Kamali, M. J. Rubio-Cabetas, M. Wirthensohn, R. S. i Company, GRCE, 205–219 (2015)
2. R. Kester, D.E. & Asay, Almonds, Advances in Fruit Breeding, 387–419 (1975)
3. N. Richter-Dyn and N. S. Goel, Theor. Popul. Biol,**3**, 4, 406–433 (1972)
4. R. S. Company and A. J. Felipe, J. Hortic. Sci.,**67**, 3, 313–317 (2016)
5. [AOAC], Assn. of Of Analytical Chemists. Official methods of analysis (1990)
6. T. M. Gradziel, R. S.Company, Almonds Botany, Production and Uses (2017)
7. R. Melhaoui, M-L. Fauconnier, M. Sindic, M.Addi, M Abid, A Mihamou, H. Serghini-Caid, A. Elamrani , Int. J. Fruit Sci.,**00**, 00, 1–10 (2018)
8. O. Kodad, G. Estopañán, T. Juan, and R. Socias I Company, JAOCS, J. Am. Oil Chem. Soc, **90**, 2, 243–252(2013)
9. A. Lansari, A. F. Iezzoni, and D. E. Kester, Euph.,**78**, 27-41, (1994)
10. D. Barile, J. D. Coı, M. Arlorio, M. Rinaldi, F.C.,**17**, 197–206 (2006)
11. R. Gülcan, Almond descriptors (revised),(1985)
12. E. E. Perez, G. H. Crapiste, and A. A. Carelli, Biosyst. Eng., **96**, 1, 41–45 (2007)
13. A. H. Mirzabe, J. Khazaei, G. R. Chegini, and O. Gholami, Agric. Eng. Int. CIGR J., **15**, 2, 256–265 (2013)
14. C. Aydin, J. F. Eng., 60 **60**, 315–320 (2003)
15. O. Kodad, R. S. Company, M. S. Prats, and M. C. L. Ortiz, HSB.,**0316**, (2016)

16. S. I. COMPANY, J. Agric. Food Chem.,**56**, 4096–4101 (2008)
17. B. Mattha and G. Á. H. Á. Icp-aes, J F Sc Tech,(2020)
18. R. Socias i Company, O. Kodad, J. M. Alonso, and C. Font-Forcada, Séries A. Mediterr. Semin., **243**, 94, 235–243 (2010)
19. D. A. Mehmet Musa Özcana, A. Ünvera, E. Erkanb, Sci. Hortic., **127**, 330–333(2011)
20. E. H. Sakar, M. El Yamani, and Y. Rharrabti,JMES, **8**, 8, 2679–2686 (2017)
21. F. Mansouri, A. Benmoumen, G. Richard, M.L. Fauconnier, M. Sindic3, H. Serghini-Caid, A. Elamrani, Riv. Ital. delle Sostanze Grasse, **93**, 1, 21–30(2016)
22. B. W. Bolling, IFT., **00**, 1–23 (2017)
23. K. Ghafoor, F. Al Juhaimi, and S. Hussain, JFST, (2018)
24. F. Chapeland-Leclerc, N. Papon, T. Noël, and J. Villard, Rev. Francoph. des Lab., **2005**, 373, 61–66(2005)
25. A.B. Siddique and D. Wright, AJPS., 2, 983–992 (2003)
26. Z. Ahmad, Ther. Clin. Pract., **16**,1, 10–12(2010)
27. Hinton,Landlinks Press (2007)
28. N. Houmy, S. ElMoudden, M. Boujnah. orbi.uliege.be (2014)
29. L. Hennaux, L'alimentation minérale du bétail au Congo belge. l'Institut National pour l'Etude Agronomique du Congo Belge(1956)

E3S Web of Conferences **183**, 05001 (2020) https://doi.org/10.1051/e3sconf/202018305001
I2CNP 2020

Effect of physicochemical parameters on magnetic treatment of water

Redouane Mghaiouini[1,2*], *Mohammed* Salah[3], *Mohamed* Monkad[2] and *Abdeslam* El Bouari[1]

[1]Department of Chemistry, Physical Chemistry Laboratory Applied Materials, Faculty of Sciences-Ben M'sik, Hassan II University, Casablanca Morocco.
[2]Department of Physics, Condensed Matter Laboratory, Faculty of Sciences, Chouaib Doukkali University, Eljadida Morocco.
[3]Department of Chemistry, Molecular Modeling and Spectroscopy Research Team, Faculty of Sciences, Chouaïb Doukkali University, El Jadida, Morocco.

Abstract: The main objective of this work is to offer a new knowledge of the water magnetism phenomenon, indeed the present study focused on the search of the most decisive parameters, which influence the process of magnetization of water. The main parameters studied in the present paper are the temperature, the salinity, the pH and the linear flow speed of the water. The increases of temperature, water velocity or salinity extend the magnetization time, however the variation of pH does not have any influence. The outcomes of this work will be very helpful for engineers in agriculture and other industrial fields to a better use of magnetized water (MW).

1 Introduction

The properties of water and their changes under the action of a magnetic field were approved by many studies in literature [1-5]. Indeed, water treated by a magnetic field, undergo a lot of changes in its physical properties, optics, electromagnetism, thermodynamics and mechanics, compared with those of pure water [6-12]. Xiao Feng Pang et al. [13] have studied the influences of magnetic field on physical properties of water, including soaking effect, electric conductivity, specific heat, surface tension force, and refraction index. They found that the magnetic fields reduce the specific heat of water, increase the soaking degree and hydrophobicity of water to materials, and depress its surface tension force and increase refractive index and electric conductivity of water relative to those of pure water.

According to them, these changes are caused by the variation of microscopic structures and distribution of water molecules under the action of a magnetic field. The physical water treatment in the magnetic field has emerged as an alternative solution in place of the chemical water treatment and other conventional process. Thus, water treated by a magnetic field has extensive applications in many fields as biology [14], electrochemistry [15], building construction [16-18], agriculture [19-20] and other fields [21-27]. In the same spirit, the present work, focuses mostly on the effect of electromagnetic field on the water, and offer more knowledge of magnetism phenomenon of this liquid. Indeed, the present paper focuses on the search of the most decisive parameters, which influence the process of magnetization of water using a special magnetic device, which generates a low electromagnetic field of 40 nT. The outcomes of this work will be very helpful for engineers in agriculture and in other industrial fields.

2 Experimental background

2.1 Materials description

The aqua-4D Magnetic water treatment is a physical process based on quantum mechanics and electromagnetism [28]. To detect the action of the electromagnetic flux density on water, a specific equipment was realized in the laboratory. It consists of an apparatus that generates the electromagnetic field, a pump, sensors connected to a computer to follow the magnetization of the water (Figure 1). The experiment consists of pumping the water from one tank to the other by passing it through the electromagnetic apparatus.

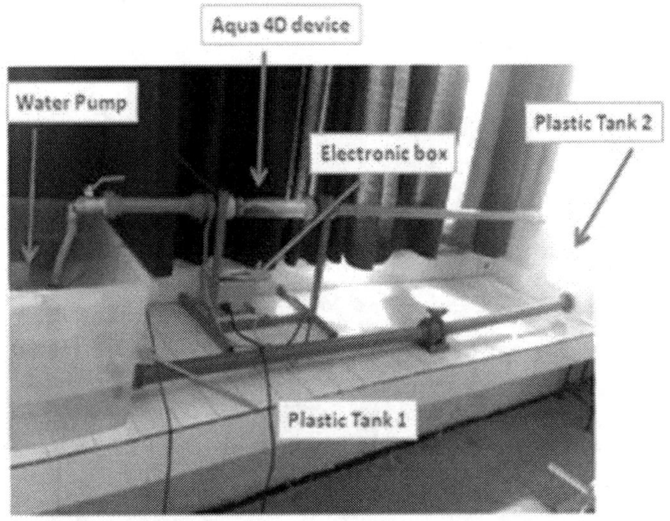

Fig. 1. Equipment of aqua-4D and Electromagnetic field generator.

The figure 2shows the method of heating the magnetized water and tap water to do a comparative study by visualizing the electromagnetic field at equilibrium, during heating and cooling of both types of water. The monitoring of the electromagnetic field was done by the Kobra software. The data were automatically recorded.

 https://doi.org/10.1051/e3sconf/202018305001

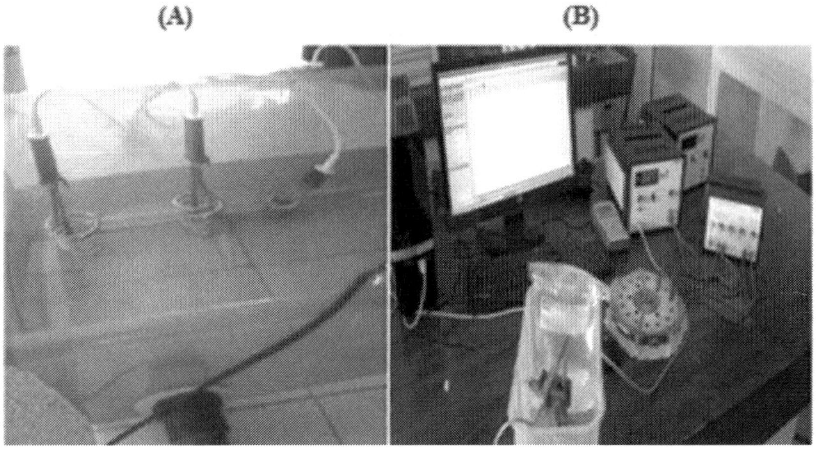

Fig. 2. (A)Method of heating of MW and measurement of electromagnetic. (B) Field intensity as a function of time.

2.2 The rearrangement of moleculesH_2O

The figure 3 depicts the main elements of water, which are hydrogen bond (Figure 3a) and dipole moment (Figure 3b), ensuring cohesion between water clusters without electromagnetic field and in the existence of the magnetic field flux. They respond to the influence of the energies coming from outside.

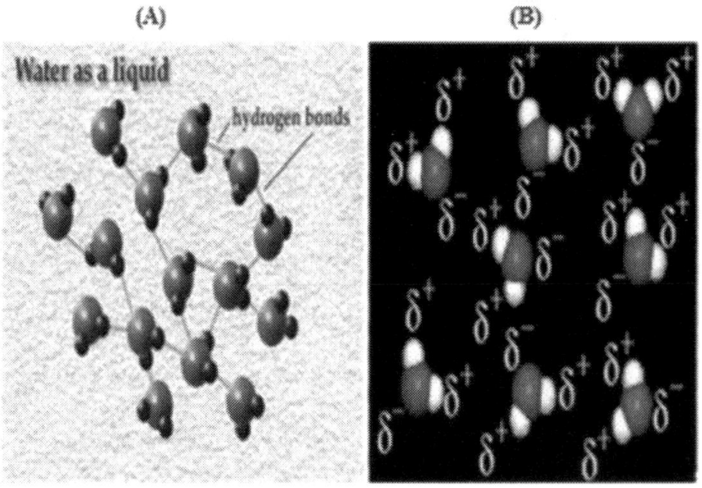

Fig.3. Water as a liquid without (Right) and with electromagnetic field (Left).

3 Results and discussion

3.1 magnetization phenomenon of water

In order to check the capacity of storing water for the electromagnetic energy, we have investigated the evolution of the magnetic field received, during treatment by the electromagnetic field, generated by the Aqua-4D device (Figure4).

The linear regression of the fluctuations of the magnetic field received by the tap water shows a decrease in a non-zero value (0.0075 mT) after eight hours (28800 s), which shows the magnetization phenomenon of water. Consequently, the water is a substance having the ability to store electromagnetic energy.

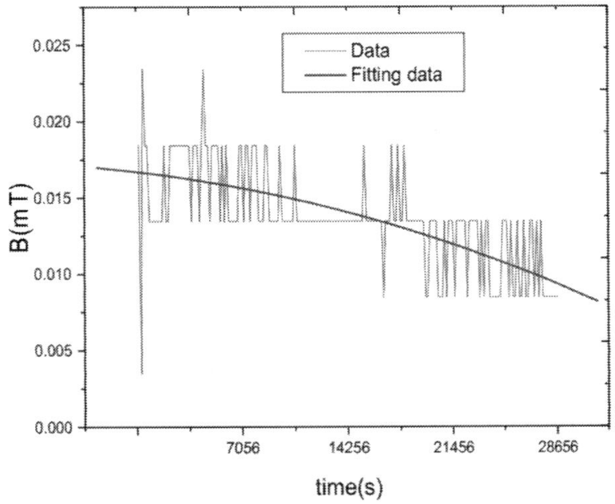

Fig. 4. Monitoring of fluctuations in the static magnetic field of magnetized water

During the 15h of electromagnetic treatment, the recorded magnetic field fluctuations, are relatively low (Bmax=0.0230mT and Bmin=-0.025mT) with an average value of 0.0015mT. This variation is slightly greater than the value measured before the application of the electromagnetic field. These assessments of the flux density of the magnetic field allowed us to monitor possible fluctuations in this field after the exposure of both types of water (magnetized water (MW) and tap water (TW)) to the electromagnetic field. The experiments indicate that tap water has indeed become magnetized after the treatment by the flux density of the magnetic field generated by the aqua-4D device. The evolution of the static magnetic field after the electromagnetic treatment generated by the aqua-4D device was followed. The experimental results show that during the 15h of electromagnetic treatment, the recorded magnetic field fluctuations are relatively low (Bmax=0. 023mT and Bmin=0. 008mT) worth an average of 0.0155mT for magnetic water, whereas for tap water the obtained values are Bmax=0. 0185mT and Bmin= 0.005mT with a Bavgof 0.01175mT. The measurements of magnetic field allow to monitor the possible fluctuations of this field after the exposure of the samples to the electromagnetic field.

3.2 Effect of temperature on EMF intensity and magnetization time

According to literature [29], the thermal agitation of water causes a rotational motion of hydrogens (positively charged) around the oxygen nucleus (negatively charged), which forms an electric current with a short duration generating a Lorentz force in a uniform static magnetic field which influence on chemical and physical properties of water. Indeed, for a better concrete of the effecting of temperature on the magnetization water phenomenon, we have proceeded to a variation of the temperature and observe the influence of this variation on the time of magnetization of water.

The figures 5 illustrate the influence of temperature on the effect of magnetization by the electromagnetic field generated by the Aqua 4D apparatus. According to the previous figure, the water passes through an electromagnetic field, becomes magnetized water

(MW). The time of magnetization at 26.5 °C, 40°C are 15 h, 12h respectively, so there is a difference of 3h as TM between the two conditions of temperature. Consequently, the increase of temperature decrease the time of magnetization of water. Concerning the relation between the intensity of magnetic field and temperature we have observed that the Bavg at 26.5°C is equal to 0.020 mT, but at 40°C the value of Bavg does not exceed a value of 0.002 mT. So, the increase of temperature decreases dramatically the intensity of magnetic field. This experimental observation is in conformity with the fundamental laws of electromagnetisms magnetic moments. The thermal agitation of water molecules causes a disorder by Brownian motion and the disappearance of the magnetic order.

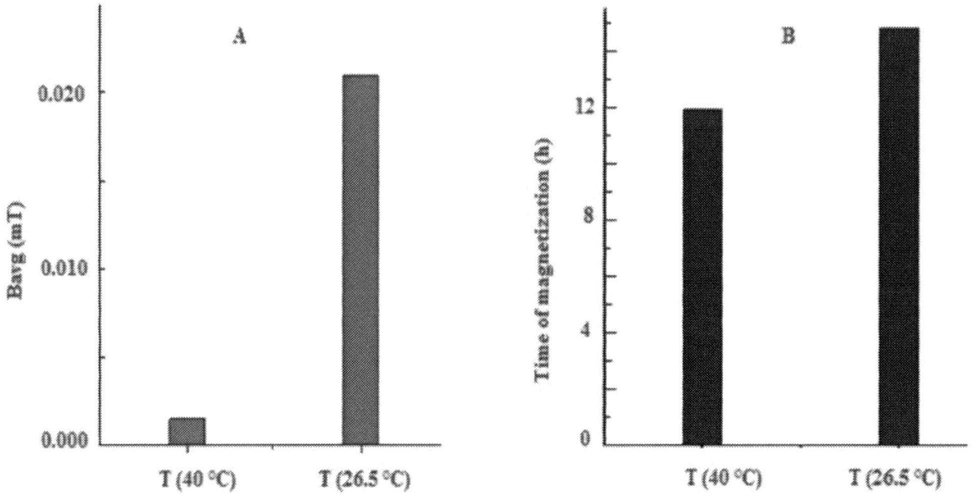

Fig.5. (A) Intensity of the average magnetic field of the water according to temperature. (B) Magnetization time as a function of water temperature.

3.3 Effect of Salinity on EMF intensity and magnetization time

The water is diamagnetic, in the presence of a magnet; the water's diamagnetism will repel the magnet. When salt is added, it increases the intensity of the water's magnetic field. So, we have tried to study this phenomenon. Indeed, we proceed to a variation of the salinity at a fixed temperature of 26.5 °C leaving the conductivity of the water increases during the action of electromagnetic field. The profile of variation of the electromagnetic field as a function of time in the case of salty water in comparison with that of tap water indicates that the electromagnetic field is affected by the salinity. The impact of salinity on the intensity of EMF is very weak (Figure 6), indeed the intensity of EMF at salinity (0.048g/l) is B_{avg} = 0.020 mT, the increase of salinity by 35.952 g/l caused a slight increase of intensity of magnetic field (IMF) by 0.0025 mT so, the impact of salinity on IMF is very weak. Concerning the time of magnetization (TM) of water, we observe the decrease of TM with salinity, therefore at salinity of 0.048 g/l, the time of magnetization is equal to 11.5 h, but at salinity of 36 g/l we obtained 15 h as TM, so the increase of salinity extended the TM (Figure 6). Thus, the factor of salinity promotes the magnetization of water by extending the time of magnetization of water with a weak impact on intensity of magnetic field.

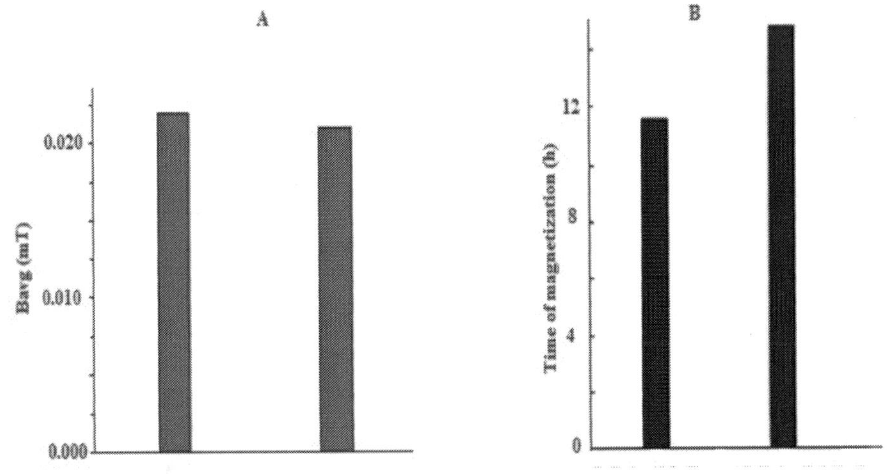

Fig.6. (A) Intensity of the average magnetic field of water as a function of salinity. (B) Variation of magnetization time according to the salinity of water.

3.4Effect of pH on EMF intensity and magnetization time

In literature [30], some works have studied the effect of magnetic field on water characteristics, specially the variation in the pH on treated and no treated water. In our case we have focused on the effect of pH on EMF intensity and magnetization time. The results are depicted at figure 7.

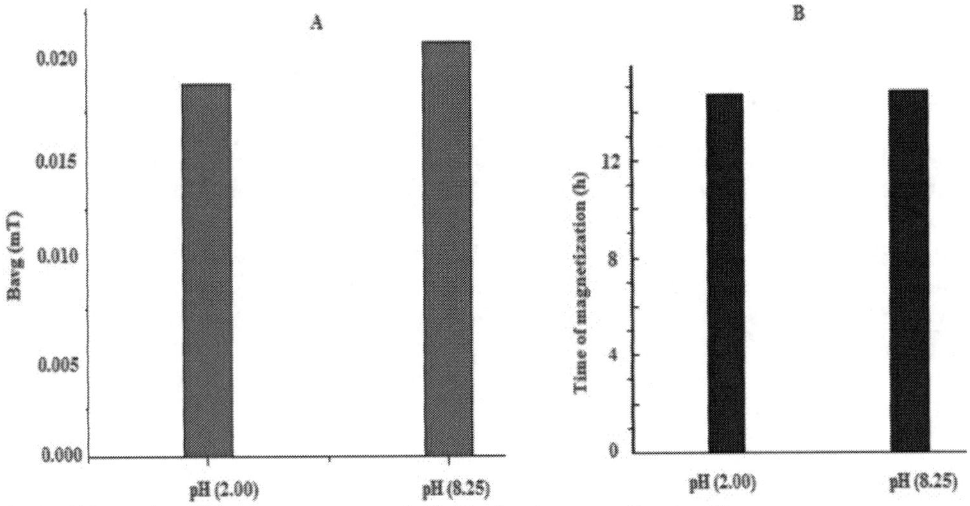

Fig.7. (A)Intensity of the average magnetic field of water according to pH.
(B)Magnetization time and water pH.

According to figure 7, the intensity of magnetic field at pH=2 (very acidic environment) the intensity of magnetic field is equal to 0.018 mT. The increase of pH from acidic to basic pH=8.25, increases the intensity of magnetic field by 0.002 mT, so the impact of variation of pH on IMF is weak. in other side, concerning the impact of pH on the time of magnetization of water we observe the following. The time of magnetization at pH=2.00 is15 h, but at pH=8.25 we obtained a value of 15.30 h as time of magnetization, so the influence of the pH at time of magnetization is very slight.

3.5 Effect of water velocity on EMF and magnetization time

Among the most parameters cited in the studies of water magnetization, we found the velocity parameter. For example, F. Alimi et al. [31] have tested the variation of the efficiency of the magnetic treatment when the flow velocity was changed. In our case we have studied the effect of water velocity on EMF and magnetization time. The results are depicted in figure8.

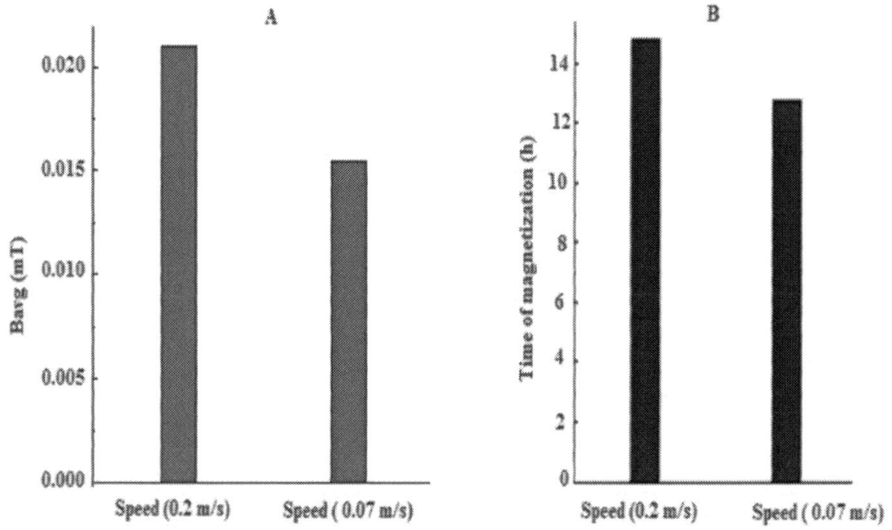

Fig. 8. (A) Intensity of the average magnetic field of water as a function of rate flow.(B) Magnetization according of both flow rates of tap water

According to our results we have observed the following. At 0.2 m/s the intensity of EMF is 0.0025 mT but at 0.07 m/s we obtained 0.0016 mT, consequently the decrease of velocity of water by 0.193 m/s decreases the intensity of magnetic field from 0.025 mT at V=0.2 m/s to 0.0152 mT at 0.07 m/s, consequently, the water velocity influenced the intensity of magnetic field of water. Concerning the impact of water velocity on magnetization time we observe that the TM is equal to 15 h at 0.2 m/s but at 0.07 m/s we found 12 h only as time of magnetization, so the velocity of water affects its magnetic field and its time of magnetization. The variation of water velocity by 0.13 m/s caused an increase of 0.004 mT of IMF and extended the TM by 2 hours.

4 Conclusion

According to the results of the present work, we can conclude the following: The water owns the ability to store the electromagnetic energy. The important factors which increase the performance of electromagnetic devices are the salinity, the speed of water and temperature. The increase of salinity causes a variation of the intensity of magnetic field and extends the magnetization time by 3 hours. The thermal agitation of water declines the intensity of the magnetic field and extends the magnetization time by 3 hours. The variation of pH has a slight effect on the intensity of magnetic field but not on the magnetism time. The variation in the water velocity affects the magnetic field of water and the time of magnetization. The variation of water velocity by 0.13 m/s caused an increase of 0.004 mT and extended the time of magnetization by 2 hours.

Acknowledgements

We would like to thank company planet horizons technologies SA (Switzerland), to provide assistance in Aqua-4D electromagnetic processing equipment.

References

1. M.C. Amiri, A.A. Dadkhah, Colloids Surf. Physicochem. Eng. Asp. **278**, 252 (2006)
2.J.S. Baker, S.J. Judd,Water Res**30**,(1996)
https://doi.org/10.1016/0043-1354(95)00184-0
3. A. Bubnov, V. Novotná, V. Hamplová, M. Kašpar, , M. Glogarová, J. Mol. Struct. **892**, 151 (2008)
4. R. Cai, H. Yang, J. He, W. Zhu, J. Mol. Struct. **938**, 15 (2009)
5. A.C.Cefalas, E. Sarantopoulou, Z. Kollia, C. Riziotis, G. Dražic, S. Kobe, J. Stražišar,A.Meden,J.Comput.Theor.Nanosci(2010).https://doi.org/info:doi/10.1166/jctn.2010.1544
6. K.T. Chang, C.I. Weng, J. Appl. Phys. 100, (2006) 043917.https://doi.org/10.1063/1.233597
7. M.C. Chang, C.Y. Tai, Chem. Eng. J. **164**, 1 (2010)
8. E. Chibowski, L. Hołysz, , A. Szcześ, Water Res. **37**, 4685 (2003)
9. E. Chibowski,A. Szcześ, Chemosphere **203**, 54(2018)
10. P. X. Feng, D. Bo, Sci China Ser G-PhysMechAstron, **51**, 1621(2008)
11. L. Holysz, A. Szczes, E. Chibowski J. Colloid Interface Sci,**316**,996 (2007)
12. K.X. Zhou, G.W. Lu, Q.C. Zhou, J.H. Song, S.T. Jiang, and H. R. Xia J Appl Phys, **88**, 1802 (2000); https://doi.org/10.1063/1.1305324
13. X.F. Pang, B. Deng and B. Tang,ModPhys Lett B, 26, (2012) DOI:10.1142/S0217984912500698.
14. M. Molouk, K. Alkhazan and A.A. Nasser Saddiq, J Evolution Biol
2, 7(2010)
15. G.B.chyus, A. Ruchinskene, and V. Deninis Protection of Metals, **39**, 443(2003)
16. N. Su, C.F. Wu.Cem. Concr. Compos,**25**,681(2003)
17. H.Wei,Y.Wang, Jianjun Luo,Constr Build Mater 91-100.
(2017)
18. I.J. lin and J. Yotvat, J. Magn. Magn. Mater,**83**, (1990)
19. J. Bogatine, Environ. Sci. Technol., **33**, 1280(1999)
20. R.Noran,U.Shani and I.Lin, Magn Electr Separ vol.7, pp.109-122(1996)
21.R.Mghaiouini,Ecol. Eng,**21**, 245 (2020) doi:10.12911/22998993/119816.
22. R. Mghaiouini, A.Graich, A.Elaoud, T. Garmim, et al. Indian J Sci Technol, **13**, 1172 (2020)
23. R. Mghaiouini, A. Elmelouky, and A. Elaoud , Int. j. adv. appl. sci, **7**,91(2020).
24. R.Mghaiouini,A.Elmelouky,A.El Moznine, M. Monkade, A. El Bouari MedJChem **10**, 1 (2020)
25. R. Mghaiouini, N.Benzbiria, M.E. Belghiti, H.E. Belghiti, M. Monkade, A.El bouari. (2020),matpr.https://doi.org/.matpr.2020.04.518
26.R. Mghaiouini, A. Elaouad, H Taimoury, I. Sabir et al. Int. J. Curr. Eng. Technol**10**,1(2020)
27.R.Mghaiouini,A. Elaoud,T. Garmim , M.E. Belghiti et alInt. j. adv. appl. sci(2020),DOI: https://doi.org/10.14741/ijcet/v.10.1.4

28. O. Toshiaki ,T. Hozumi, J. Appl. Phys., **106**,1(2009)
29. H.B. Amor, A. Elaoud, M. Hozayn, Asian Res. J. Agric, **8**,1 (2018)
30. F. Alimi, M. Tlili, C.Gabrielli, M. Georges, M.B. Amor,Water Res, **40**, 1 (2006)

Synthesis of N-CNT/TiO_2 composites thin films: surface analysis and optoelectronic properties

Hamza Belkhanchi[1], *Younes* Ziat[1*], *Maryama* Hammi[2**], *Charaf* Laghlimi[3], A*bdelaziz* Moutcine[3], *Anas* Benyounes[4], and *Fouzia* Kzaiber[1]

[1]Laboratory of Engineering and Applied Technologies, Higher School of Technology, Sultan Moulay Slimane University, Beni Mellal, Morocco
[2]Laboratory of Materials, Nanotechnologies and Environment, Department of Chemistry, Faculty of Sciences, University of Mohammed V, Rabat, Morocco
[3] Molecular Electrochemistry and Inorganic Materials Team, Beni Mellal, Faculty of Science and Technology, Sultan Moulay Slimane University, Morocco
[4]Department of Chemistry, Faculty of Sciences, University of Mohammed V, Rabat, Morocco

Abstract. In this study, we have investigated the surface analysis and optoelectronic properties on the synthesis of N-CNT/TiO_2 composites thin films, using sol gel method for a dye synthetized solar cell (DSSC) which is found to be simple and economical route. The titanium dioxide based solar cells are an exciting photovoltaic candidate; they are promising for the realization of large area devices. That can be synthetized by room temperature solution processing, with high photoactive performance. In the present work, we stated comparable efficiencies by directing our investigation on obtaining Sol Gel thin films based on N-CNT/TiO_2, by dispersing nitrogen (N) doped carbon nanotubes (N-CNTs) powders in titanium tetraisopropoxyde (TTIP). The samples were assessed in terms of optical properties, using UV—visible absorption spectroscopic techniques. After careful analysis of the results, we have concluded that the mentioned route is good and more efficient in terms of optoelectronic properties. The gap of "the neat" 0.00w% N-CNT/TiO_2 is of 3eV, which is in a good agreement with similar gap of semiconductors. The incorporated "w%N-CNTs" led to diminishing the *Eg* with increasing N-CNTs amount. These consequences are very encouraging for optoelectronic field.

1 Introduction

Carbon nanotubes are the subject of many investigations in various areas such as composite materials [1-3], in the (Li ion) batteries to increase their performance [4], the storage of the energy [5], optic [6] and photovoltaic [7]. And for the photovoltaic field, there are several important characteristics that need to be considered for a photocatalyst as TiO_2 in dye sensitized solar cells (DSSC), it is crucial to be able to transfer holes from the sensitizing dye after the dye has injected electrons in the titanium dioxide (TiO_2). On the other hand, DSSCs

Corresponding author:*ziat.estbm@gmail.com; **mar.hammi@yahoo.com based on nanostructure materials offer very low cost and relatively efficient photovoltaics' energy conversion [8]. The used routes don't embody any inconvenient to the dye layer on the titanium dioxide nano crystallites [9]. Due to its ecofriendly production, the DSSCs are extensively studied for the reliable and safe energy source [10].

In this study, we have shown that the TiO_2 and N-CNT composed an active layer, and the electrons recombination procedure may affect the best conversion of DSSCs' efficiency. Given that the nitrogen doped CNT has a special structure and can conduct electrons at room temperature (RT). Where the electrons are generated from TiO_2 under light irradiation, the N-CNT compound can also easily conduct photoelectrons. The analysis of the present results concluded that the sol gel route is good and more efficient in terms of optoelectronic properties.

This work provides new insights into the development extrinsic semiconductors, which attracts great interest in the current solar nanotechnology. Indeed, nitrogen doped carbon nanotubes (N-CNTs) are known by their large surface area, high electrical conductivity and chemical stability. These properties have provided to N-CNTs potential applications in N-CNT-based composites which are used as candidates for nanoelectronic devices. In fact, the synergistic effect of N-CNTs and TiO_2 is caused by the improved photocatalytic activity through reducing recombination efficiency of electron-hole pairs.

2 Synthesis

The "hybrid CNTs" were synthesized within catalytic chemical vapor deposition in a fluidized room reactor by ethylene C_2H_4 as carbon source and acetonitrile/N_2 as nitrogen/carbon source. The experiment carried out initially with C_2H_4 (650 ml/min) for ½ hour aimed at producing CNT. And the N-CNTs were produced from acetonitrile/N_2 for 35 min. The carbon nanotubes were purified via aqueous solution of Sulfuric acid 'H_2SO_4' – 50 vol % under refluxes for 3h 10min.

For titanium isopropoxyde ($C_{12}H_{28}O_4Ti$), the coated solution is obtained by dissolving 3ml of $C_{12}H_{28}O_4Ti$ from Aldrich 98% 6ml of isopropanol: C_3H_8O, then different 'weight' percent (%) of N-CNTs powders (0.09; 0.25 and 0.4wt %) were dispersed in the obtained solution. This mixture is stirred at 65°C for 12min. Then, 6ml of acetic acid: CH_3COOH was added. After that, other mixture was stirred in 20min. Within 2hours, the final mixture is being stirred after adding 12ml of methanol: CH_3OH to have the Sol Gel solution. This latter was spincoated on indiumtinoxide (ITO) glass substrates. The ITO should be etched with detergent, followed by ultrasonic cleaning for few minutes, then rinsed with deionized water and finally steaming at 110 °C for 10 minutes. Then, the annealing temperature has been elevated from 100 – 500°C.

3 Specific area and pores diameter within BET and BJH schemes

The mesoporous structure of TiO_2 was examined within the pores diameter calculations and surface area. The multipoint N_2gas sorption is experimented at 77K to examine the specific area and pores diameter of the synthesized TiO_2 via the "micromeritics ASAP2010". The specific surface calculations are based on Brunauer, Emmett et Teller (BET), which is expressed by [11]:

$$S_{BET} = n_m^a \times N \times a_m \quad (1)$$

The n_m^a is the monolayer capacity, where a_m presents the molecular crosssectional area. N stands for *Avogadro constant*. The pore size distribution was obtained by t the Barrett, Joyner, and Halenda (BJH) scheme [12]. For BJH, the sorption data in the relative pressure range from 0.05-0.25, the calculating pore size distributions is from empirical isotherms via the Kelvin model of pore filling. It is applicable only to the small macro pore size and mesopore range. Hence, the Table 1 shows the evolution of "surface area" and "pores diameter" versus annealing temperatures.

Table 1. Pore diameter and specific surface area of TiO_2 associated to annealing temperature.

Annealing temperature (°C)	Pore diameter PD(nm)	Specific area (SBETm^2.g^{-1})
100	1.95	205
300	2.09	160
500	4.86	125

As noticeable through Table 1, The increase of annealing temperature augments give rise to pores diameter (P_D) which reaches the maximum up to 500 °C. Thermal treatment of the powders at 500°C leads to a systematic decrease of the SBET values and a considerable enlarging of the average pore size. So, to keep high SBET, the annealing temperature is limited to 300 °C at which crystal growth was slightly affected, these data suggest the difference in porosity can affects their sorption and photocatalytic behaviors of TiO_2.

4 Optical characterization and Swanepol method

The absorption spectroscopic technique, enable us to record the optical transmission of the layers as a function of wavelength, and determining the value of the optical gap (*Eg*) of the layer as well as the refractive indices of the films. For this purpose, the following formulas are given by the method of Swanepol and Manifacier [13-14].

$$d = \left(\frac{\lambda_1 \lambda_2}{2(\lambda_1 n_2 - \lambda_2 n_1)}\right) \quad (2)$$

where d is the thickness of the film. n_1 and n_2 present, respectively, the refractive indices of the films for λ_1 and λ_2 "adjacent wavelengths". Titanium oxide is a transparent material, whose refractive index in bulk form is equal to 2.5 (anatase phase) [15]. Whereas, in the form of a thin layer, its refractive index varies according to the conditions of preparation. According to the authors [16-17], the refractive index varies between 1.9 and 2.5.

In this work, we assume that the deposited thin films are homogeneous. The porosity of thin films is calculated from the following equation [18].

$$Porosity = \left(1 - \frac{n^2 - 1}{n^2_D - 1}\right).100(\%) \quad (3)$$

Where n_D is the refractive index of pore-free TiO_2[19], and n is the refractive index of porous thin films. From this equation, we can assume the relationship between the refractive index and the porosity. Low porosity lead to the increase of n. Thus, in order to have remarkably higher refractive indexes, our films must be highly densified.

The refractive indices in the spectral region of transparency is expressed by:

$$n_1 = \left[N_1 + (N_1^2 - S^2)^{1/2}\right]^{1/2} n_2 = \left[N_2 + (N_2^2 - S^2)^{1/2}\right]^{1/2} \quad (4)$$

Where N is the *Swanepoel* coefficient in the transparent spectral region. And S is the refractive index of the glass.

$$N_1 = 2S\left(\frac{T_{max1} - T_{min}}{T_{max1} \times T_{min}}\right) + \left(\frac{S^2+1}{2}\right) N_2 = 2S\left(\frac{T_{max2} - T_{min}}{T_{max1} \times T_{min}}\right) + \left(\frac{S^2+1}{2}\right) \quad (5)$$

Where T_{max} and T_{min} are the maxima and minima of the transmittance curve. To predict the absorption (α) in terms of transmittance (T), we have the following formula

$$\alpha = \frac{1}{d} \ln \left(\frac{100}{T}\right) \qquad (6)$$

4.1 Transmittance and the optical gap E_g

For the fields of spectroscopy, we can distinguish the ultraviolet (UV), infrared (IR) and microwave. In the present study, we used a double beam recording spectrophotometer for UV—Visible spectral "300—800nm". The transmittance (T) presents the ratio of transmitted light intensity to the incident light intensity. To plot the spectra, a blank substrate was used in the reference beam of the spectrophotometer, a computer device is connected to this unit enable us to visualize imitates the spectra which are representing the transmittance (T) of the film.

The obtained spectra of transmittance (T) connected to N-CNTs/TiO_2 composites were plotted (Fig. 1). This figure shows typical spectra of N-CNTs/TiO_2 in the range of 300—800nm with different "N-CNTs wt %". At this point, it is concluded that the T decreases with increasing the incorporated N-CNTs in wt%. The obtained film shows a high transmittance for TiO_2thin film, it is of 77.5%. And $T$is of 61% for the "0.40 wt% N-CNT/TiO_2"composite.

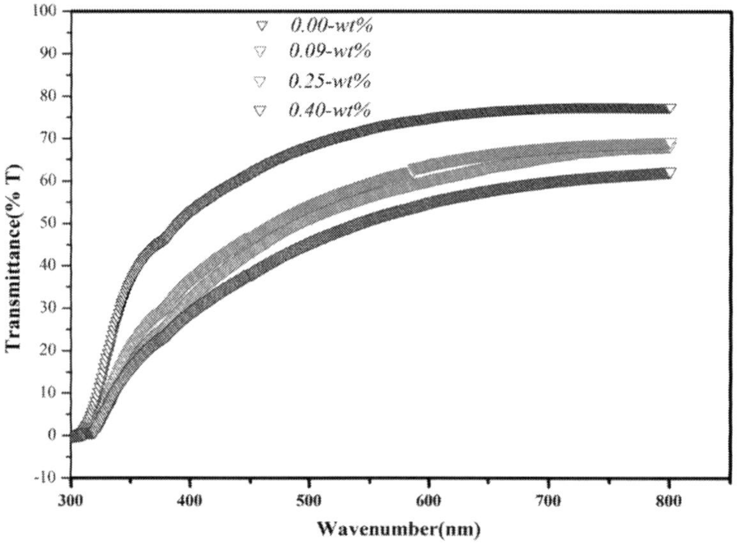

Fig. 1. The transmittance (T) versus wavenumber for 0.00; 0.09; 0.25 and 0.40 wt% N-CNTs/TiO_2 composite

In the UV range, the abrupt decrease in film transmittance is due to the fundamental absorption of light. It corresponds to transitions between valence band towards the conduction band when there is excitation charge carriers. Beyond the fundamental absorption threshold, the absorption coefficient follows a variation with the energy ΔEg, which is expressed by the following relation [20]:

$$\Delta E_g = \frac{h^2}{8m^*}\left(\frac{3n}{\pi}\right)^{2/3} \qquad (7)$$

Where h stands for the Planck constant, m^* refers to the effective mass of the carriers and $n$is the concentration of free electrons.

From the transmittance spectra, we can calculate the optical gap (*Eg*) value of semiconductors from Tauc method [21]. Where the wavelength composing white light are scanned via the spectrophotometer and electromagnetic radiation interacts with matter. The IR causes vibrational transitions in molecules while the UV—Vis cause electronic transitions. Then the absorptions start to consider the band gap of the semiconductors. Graphically, it is possible to define the optical gap within the followed formula [22-24].

$$(\alpha h\nu) = B(h\nu - Eg)^n \quad (8)$$

h, ν, α, n and B present the "Planck constant, vibration frequency, absorption, the power factor and B is a proportional constant", respectively. The curves of $(\alpha h\nu)^2$ as function of energy $(h\nu)$ characterize a linear portion whose intersection with the axis of hν, gives the optical gap. From the *T* spectra, we have the *Eg* of the N-CNTs/TiO_2 composites, see (Fig. 2).

N-CNTs fillers contents were indicated on (Fig. 2. a, b, c and d). The gap of "the neat" 0.00w% N-CNT/TiO_2 is of 3eV, which is in a good agreement with similar gap of semiconductors [25]. Due to the augmentation of the fillers content, the optical gap values are reduced from (Fig. 2.a to Fig. 2.d). Clearly in the same figure, we showed the determination of optical gap by the extrapolation from variation of $(\alpha h\nu)^2$ versus hν. Hence, the incorporated "N-CNTs wt%" led to diminishing the *Eg* with increasing N-CNTs amount.

Similar behavior of band gap energy is expressed by TiO_2/MWCNT nanocomposites [26]. However, our studied composites revealed larger absorption and shift towards shorter wavelengths.

These consequences are very encouraging, therefore we have focused our attention on the front of a solar cell component, by performing the electrical measurements of the obtained thin films.

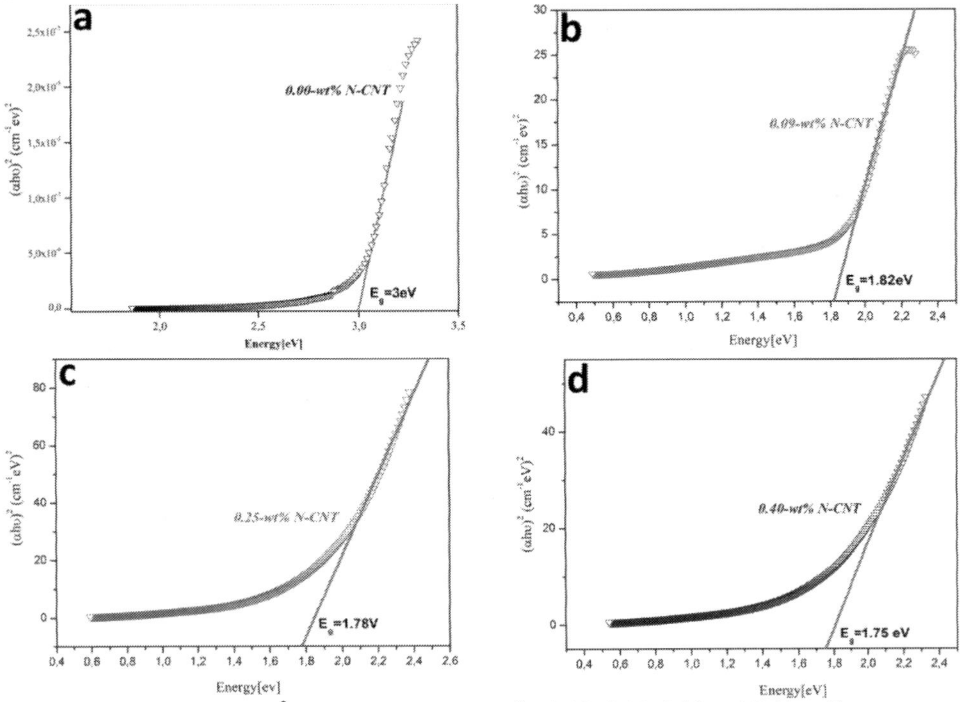

Fig. 2. The behavior of $(\alpha h\nu)^2$ versus photon energy for 0.00; 0.09; 0.25 and 0.40 wt%

4.2 Sheet conductivity behaviour in "wt% N-CNTs/TiO_2" composite gap

The obtained N-CNTs/TiO_2 composite were investigated in the surface conductivity point of view, which is extracted from surface resistivity recognized as sheet resistance defined by the followed formula [17]:

$$R_s = \rho \times d^{-1} \quad (9)$$

Where ρ is the resistivity. The film prepared by solgel route using different amount of % N-CNT exhibited an early threshold conduction at 9%, which gives a good promising thin film in the terms of the conductivity electric, see (Fig. 3).

It is noted that the reduction of resistivity involves an augmentation in the carrier concentration. Increasing the quantity of carriers leads to an augmentation in the visible absorption, then increasing the figure of merit (FOM) of the present thin films based on N-CNTs/TiO_2. The transparent conductive oxides (TCO) are extensively used in solar cell fields, to get high quality TCO films, it is important to define the (FOM) for qualifying those layers of films and predicting the influence of the used TCO films on solar cells process. The obtained results are validated with the optical properties of the elaborated "N-CNTs/TiO_2" composite. These findings supposed that those thin films are good candidates for the optoelectronics field.

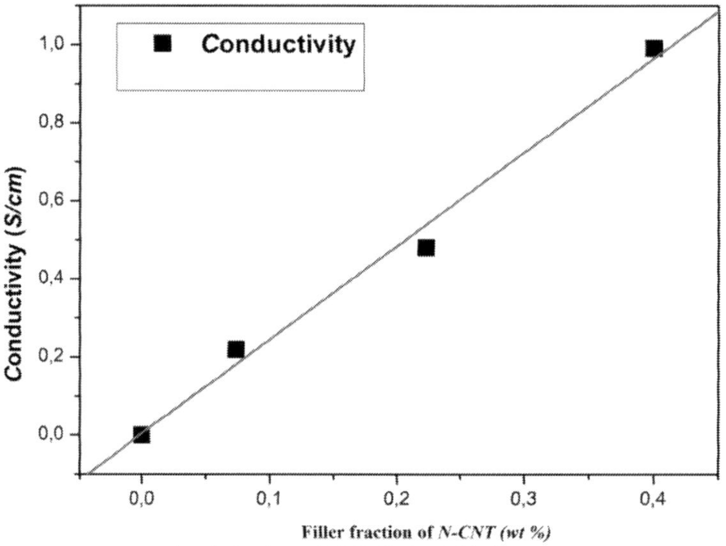

Fig. 3. Variation of average conductivity in wt% "N-CNTs/TiO_2"

According to (Fig. 3), it is noteworthy that electrical conductivity of unloaded thin film is of the order of 10-9S/cm. It is linearly enhanced when the concentration of incorporated N-CNTs are elevated. Typical behavior is expressed by TiO2 based thin films as shown by [27], they observed decrease in sheet resistance with increasing number of deposited bilayers indicates that the MWCNT/TiO2 films become more conductive. They attribute this increase in conductivity to an increase in MWCNTs in the films and, more importantly, to the increased percolation of MWCNTs within the nanocomposite film as more bilayers are deposited. This evolution is more probably due to number carriers increase within TiO2 host matrix. Therefore, the gap must drop significantly, in good agreement with Burstein - Moss theory [28, 29].

5 Conclusion

For the N-CNT/TiO_2 composites, the synthesis, specific area and pores diameter within BET and BJH schemes were described. The current study was based on the solgel elaboration, since it is efficient route. The optical characterization and Swanepol method were discussed. The obtained spectra of *T* connected to N-CNTs/TiO_2 composites shows typical spectra in the range of 300—800nm with different "N-CNTs wt %". The obtained film shows a high transmittance for TiO_2 thin film, it is of 77.5%. And T of 61% for the "0.40 wt% N-CNT/TiO_2"composite. The gap of "the neat" 0.00w% N-CNT/TiO_2 is of 3eV, which is in a good agreement with similar gap of semiconductors. The incorporated "w%N-CNTs" led to diminishing the *Eg* with increasing N-CNTs amount. It is noted that the reduction of resistivity involves an augmentation in the carrier concentration. Increasing the quantity of carriers leads to an augmentation in the visible absorption. These consequences are very encouraging for optoelectronic application.

Acknowledgements: The authors would like to thank Ms. ***Maryem*** **Errachid** from "Economy Management, Mohammed V University of Rabat, Faculty of Law, Economics and Social Sciences-Souissi, Rabat, Morocco" and Ms. ***Saloua*** **Rzaoudi**, from "Economic sciences and of management, Polydisciplinary Faculty, Sultan Moulay Sliman University, Beni Mellal, Morocco"to accept the language revision.
Our thanks are extended to "the mechatronic department, Higher School of Technology, Sultan Moulay Slimane University, Beni Mellal, Morocco", especially, the professors ***Fatima Zahra*** **Baghli** and ***Yassine*** **Lakhal** for the fruitful discussion.

References

1. D. Xu, G. Luo, J. Yu, W. Chen, C. Zhang, D. Ouyang, Y. Fang, X. Yu, J. Alloys Compd**702**, 499(2017)
2. S. Aror, N. Kumari, C. Srivastava, J. Alloys Compd**801**, 449(2019)
3. T. Zhao, X. Peng, X. Zhao, J. Hu,T. Jiang, X. Lu, H. Zhang, T. Li, I. Ahmadc, J. Alloys Compd**817**, 153057 (2020)
4. Y. Ziat, A. Benyounes, O. El Rhazouani, C. Laghlimi, M. Hammi, Turk. J. Mater **3**, 61 (2018)
5. C. T. Hsieh, H. Teng, W. Y. Chen, Y. S. Cheng, Carbon **48**, 4219(2010)
6. A. Shabaneh, S. Girei, P. Arasu, M. Mahdi, S. Rashid, S. Paiman, M, Yaacob, Sensors **15**, 10452 (2015)
7. H .Derbal-Habak , C. Bergeret, J. Cousseau, J. M. Nunzi, Sol. Energy Mater. Sol. Cells **95**,S53(2011)
8. B. Kiliç, E. Gür, S. Tüzemen, J. Nanomater**7**, ID 474656(2012)
9. M. N. Amalina, N. A. Rasheid, and M. Rusop, J. Nanomater**6**, ID 637637 (2012)
10. M.S. Ahmad, A. K. Pandey, N.A. Rahim, Renew. Sust. Energ. Rev **77**,89(2017)
11. J. Rouquerol, D. Avnir, C. W. Fairbridge, D. H. Everett, et al, Pure. Appl. Chem **66**, 1739 (1994)
12. R. Bardestani, G. S. Patience, S. Kaliaguine, Can. J. Chem. Eng**97**, 2781(2019)
13. R. Swanepoel, J. Phys. E: Sci. Instrum**16**, 1214(1983)
14. J. C. Manifacier, D. M. Murcia, J.P. Fillard, E. Vicario, Thin solid films **41**, 127(1977)
15. M. D. Hernandez-Alonso M. D. Hernandez-Alonso Isabel T. Tejedor, J. M. Coronado, J. Soria, M. A. Anderson, Thin Solid Films **502**, 125 (2006)
16. A. Turković, M. Ivanda, J. Tudorić-Ghemo, N.Godinović, I. Sorić, In Proceedings of Symposium A 3 of the International Conference on Advanced Materials (ICAM 91) Non-Stoichiometry in Semiconductors(1992)
17. S. Balaji, Y. Djaoued, J. Robichaud, J. Raman Spectrosc**37**, 1416 (2006)

18. T. Ohsaka, S. Yanaoka, O. Shimomura, Solid State Commun**30**, 345 (1979)
19. M. H. Liao, J. Solid. State. Chem **179**, 2020 (2006)
20. J. Mass, P. Bhattacharya, R.S. Katiyar, Mat. Sci. Eng. B-Adv **103**, 9 (2003)
21. J. Tauc, A. Menth, J. Non-Cryst. Solids **8**, 569 (1972)
22. Y. Ziat, A. Abbassi, M. Hammi, A. Ait Raiss, O. El Rhazouani, Opt. Quantum Electron**48**, 511 (2016)
23. A. Benyounes, M. Hammi, Y. Ziat, A. Slassi, N. Zahra, Appl. Phys. A **90**, 124 (2018)
24. Z. Zarhri, M. ÁngelAvilésCardos, Y. Ziat, M. Hammi, O. El Rhazouani, J. Cesar Cruz Argüello, D. Avellaneda Avellaneda, J. Alloys Compd**819**, 153010 (2019)
25. M. Hammi , A. Abbassi, Mater. Lett**166**, 206(2016)
26. M. B. Askari, Z.T. Banizi, M. Seifi, S. B. Dehaghi, P. Veisi, Optik**149**, 447 (2017)
27. K. E. Tettey, M. Q. Yee, D. Lee, ACS Appl. Mater. Interfaces **2**, 2646 (2010)
28. B. Šantić, A. Moguš-Milanković, D.E. Day, J. Non-Cryst. Solids**296**, 65 (2001)
29. D. M.Tawati, M. J. B. Adlan, M. J. Abdullah , J.Non-Cryst. Solids **357**, 2152 (2011)

AUTHOR INDEX